It's a jungle out there – the strange animals of economic organization in agri-food value chains

It's a jungle out there – the strange animals of economic organization in agri-food value chains

edited by:
Gaetano Martino, Konstantinos Karantininis,
Stefano Pascucci, Liesbeth Dries and
Jean Marie Codron

Wageningen Academic
Publishers

EAN: 9789086863013
e-EAN: 9789086868445
ISBN: 978-90-8686-301-3
eISBN: 978-90-8686-844-5
DOI: 110.3920/978-90-8686-844-5

First published, 2017

Table of contents

17 Sampling strategies in the agrifood context

Foreword

This book originates from an intensive debate carried out under the Jean Monnet project 'Governing and organizing the agri-food supply systems in the European Union countries: institutional environments, governance modes and policy patterns' (542857-LLP-1-2013-1-IT-AJM-IC), at the 140th EAAE Seminar 'Theories and empirical applications on policy and governance of agri-food value chains' and during the workshop 'It's a jungle out there! The strange animals of economic organization in agri-food value chains' organized in the context of the European Association of Agricultural Economists Congress 2014.

In the new institutional economics, the economics of governance is a growing, fruitful field of scientific inquiry. In recent decades, an increasing number of contributions have provided extensive new outcomes and evidence. Scholars have built on robust theoretical perspectives that provide detailed knowledge of the modes of organizing the production and the exchange of goods and services. They have also illustrated how these modes are designed and adopted through the choices of economic actors and how they relate to the economic context. Research has elucidated neglected solutions to the problem of coordination among economic agents. These results progressively delineate a conceptual architecture linking the general achievements of different theories to operational concepts and constructs. Furthermore, the evidence points to a central role for organizational choice and institutions. Hence, the set of theoretical instruments convincingly showed the importance of accounting for the 'institutional structure of production' (Coase, 1992) and at the same time delineated challenging suggestions.

The key task of this volume is to explore the variety of organizational forms that exists in the European agri-food sector and to identify an appropriate theoretical framework that allows readers to obtain a comprehensive understanding and a systematized set of conceptual instruments to analyse this variety and to explore the relationship between organizations and the extent of the regulatory domain.

Organizational issues have been extensively addressed in agribusiness economics in recent decades. While this leads to an enhanced knowledge of the characteristics of the sector, a systematic analytical basis is still lacking, leaving many areas unexplored. The volume aims to contribute to these directions with the objective of presenting and contrasting different theoretical and conceptual frameworks to establish a comprehensive basis for the analysis of coordination, organization and governance in European agri-food supply systems.

The focus of the volume is primarily on governance modes from two perspectives: First, the book wants to illustrate and explain the variety of organizational forms and their internal complexity. Second, the volume analyses the characteristics of these forms, seeking to provide elements for an operational taxonomy. More precisely, the

G. Martino *et al.* (eds.) **It's a jungle out there – the strange animals of economic organization in agri-food value chains**
DOI 10.3920/978-90-8686-844-5_0, © Wageningen Academic Publishers 2017

objective of the volume is to identify concepts that allow one to analyse and to explain the variety and complexity of the organizational modes in agri-food supply systems.

The economics of governance is basically grounded in transaction cost economics (TCE). However, the volume intends to broaden this perspective by also capturing insights coming from contributions at the frontier of other fields. The reason for this is that the TCE framework has to date focused on relatively narrow issues and needs to be enriched to explain more subtle differences in economic organizations. Similarly, the works collected also recognize that there is a need for a more encompassing framework based on different theories of economic organization to address the variety of economic organizations observed in the agri-food sector.

Moving from the contributions of international leading scholars (Part I: Theories and conceptual approaches), the debates concentrated on the main conceptual challenges, the emergence of a variety of organizational modes and the identification of patterns of analysis. The theoretical introduction defines the general framework based on the concepts of transaction cost economics and economic organization theory, including vertical coordination, uncertainty, hybrids, relational contracts and rights sharing. Part II (Case studies on organizational diversity) presents an analysis of cases in which the introduced conceptual apparatus highlights the characteristics, risks and performance of the organizational solutions observable in agri-food value chains. The cases cover just a small part of the great organizational variety at stake, but they both provide insights into the organizational forms studied and highlight the possibility of progressively identifying operational models of applied research in the field. Part III: (Case studies on organizational change) collects studies on the dynamics in agri-food chains and gathers empirical information, which can be organized under the conceptual tools elaborated in the first two parts. Therefore, these cases will try to identify some of the forces potentially stimulating organizational change and map further potential study areas.

Ménard sets basic analytical tools needed to connect the apparent 'jungle' of organizational arrangements to their underlying logic, emphasizing how this diversity structures our economies (Chapter 1; Ménard, 2017). He does so in the continuity of Coase, who pointed out that coordination and transactions are tightly interwoven. The chapter develops four interdependent themes: contracts as coordinating tools; a transaction model to capture the various ways through which contracts operate; the institutional embeddedness of the organizational arrangements of which these contracts are key components; and, the different institutional layers that frame this institutional environment. Making clear that contracts create commons grounds delineating 'acceptance zones' of action within which coordination develops, Ménard contrasts alternative theoretical understanding of contracts. He submits that variations in the characteristics and use of contracts arise from the interaction between organizational arrangements and their institutional environment.

Two key points of this chapter deserve particular attention in our view. First, Ménard suggests that contracts need being 'dimensionalized' according to the way rights are allocated and to the nature and modalities of transactions at stake. A central theoretical proposition of the chapter is that the distinction between property rights and decision rights provides means to deal with this issue. Referring to relational contracts, he submits that the combination of these two dimensions, the allocation of these rights and how they allow dealing with different transactions, frame the analytical space of the organizational variety.

Second, the chapter proposes a differentiated approach to the 'institutional environment' through an identification of its layers. Ménard particularly emphasizes the need to conceptualize the intermediate institutional layer, linking the macro-institutional layer at which general rules are established and the layer of organizational arrangements. He identifies this intermediate layer as the 'meso-institutional layer', through which the general rules are interpreted and implemented, thus framing the actual domain within which organizations operate. In doing so, he shed new light on the institutional context of Agrifood chain governance. An impressive set of institutional bodies (from certification agencies to inter-branch agreements) can then be better understood, highlighting their role in governance choices.

Research pathways thus stem from this inquiry about the variety of organizational arrangements and of the institutional layers within which they are embedded, providing the reader with a basic conceptual apparatus to develop a promising research agenda.

Grandori starts by identifying the necessity of addressing three basic analytical problems: the description, the classification and the evaluation of organizational forms (Chapter 2; Grandori, 2017). It is recognized that the variety of the agri-food value chains provides an articulated and rich set of empirical cases that may promote analytical developments. 'Non-hierarchical firms, hierarchical contracts, hybrids between markets and democracies, networks that are firms': these 'monsters' in the agri-food organizational jungle trigger reflection and allow one to set up a three dimensional taxonomy. The allocation of property rights, organizational coordination mechanisms, and the legal and contractual foundations are conceptualized as the dimensions of the classification. Multiple *continua* among the organizational forms can be identified. The resulting taxonomy calls for a complementarity between knowledge generation and adaptation. The study opens new perspectives by discussing the concept of panarchy in agri-food studies and emphasizes the varieties of the contract as well as the opportunities provided by configurational analysis approaches.

The article of Hobbs concentrates on hybrids as the most diffused governance mode in agri-food chains (Chapter 3; Hobbs, 2017). This study also seeks to elaborate on existing theories and to identify models of analysis suitable for the complexity of the sector at stake. A three-tiered approach to the analysis of hybrid modes of organization is discussed: analysis that examines the influence of the external (institutional) environment; analysis that examines the inter-firm environment, drawing upon insights from transaction cost economics, agency theory, and the property rights approach; and analysis that considers the intra-firm (managerial) environment, including insights from the resource-based view of the firm and elements of strategic management theory. A case study concerning biotechnology industries provides an illustrative application of the approach. Complexity, uncertainty and asset specificity are recognized as the main drivers of organizational forms, while a specific role for entrepreneurship appears to be necessary. The study concludes by identifying three challenges for future research that require one to address the variety of hybrids, to connect theory and empirics in a stronger way and to account for the dynamics of organizational patterns.

Chlebicka, Fałkowski and Łopaciuk-Gonczaryk (Chapter 4; Chlebicka *et al.*, 2017) address the field of motivation to cooperate. The chapter develops an analysis of why cooperative behaviour between farmers may or may not occur. The study examines this problem from a double institutional perspective. At one level, it is recognized that the possibility of cooperation is embedded in the institutional environment and that cooperative behaviour can vary across regions. The second level of analysis offers a comparison between informal and formal governance mechanisms. The chapter also contributes to the debate about the low willingness to cooperate in ex-Communist countries, which is commonly linked to the legacy of communist dictatorship. The authors start by pointing out that formal and informal enforcement mechanisms encouraging cooperation may depend on different types of trust. According to this view, interpersonal relationships and mutual trust between individuals should facilitate the establishment of groups of a relatively small size whereas trust in formal institutions should support collective action in larger organizations. Different types of enforcement mechanisms are also determined by group specialization patterns, as more complex products are more dependent on formal regulations. The chapter presents the results of a survey conducted among a sample of Polish farmers. The authors find that lack of trust may indeed be perceived as an obstacle to establishing producer organizations in Poland. They also provide evidence in line with the theoretical arguments linking different types of trust with the level of cooperation and groups' specialization patterns. In addition, the picture emerging from this investigation documents differences in collaborative equilibria across Polish regions.

D'Arienzo, Ventura and Martino (Chapter 5; D'Arienzo *et al.*, 2017) focus on the 'network contract', a contract typology that was recently introduced in Italian law. The network contract basically allows farmers to pool resources and information in order to cope with competition challenges. Through the network contract, a group of farmers can seek to increase their own innovative and competitive capabilities by elaborating a common plan for using resources, sharing information and undertaking joint activities. The article aims at elaborating a conceptual interpretation of the network contract in light of transaction cost economics. On the basis of this conceptual framework, the study discusses the main characteristics of the contract and underlines a potential approach for its design.

Stranieri and Banterle concentrate on European traceability standards and their effects on vertical coordination in the agri-food chain. In their contribution, traceability standards are considered from a supply chain perspective with a focus on information management (Chapter 6; Stranieri and Banterle, 2017). More precisely, the analysis indicates that existing standards can be classified in terms of their effects on the dyadic and supply chain relationships. Within this approach, the authors introduce six criteria to differentiate voluntary traceability standards and to provide a classification of these standards. By identifying *ex-ante* and *ex-post* transaction costs and the incentives (market, chain and regulatory) to adopt traceability standards, the study states that the complexity of the standards depends on the costs (*ex-ante* and *ex-post*) and the incentives the firm could face. Because of the increase in both physical and human asset specificity, traceability standards increase transaction costs in dyadic relationships, while the increase in transparency reduces transaction uncertainty and the related transaction costs. Based on transaction cost variation and on the existing governance of the supply chain, the adoption of voluntary standards also determines the reorganization of the food supply chain

De Souza and Bánkuti focus on the function of measurement in agri-food transactions and adopt TCE and Measurement Cost Economics (MCE) in a complementary way (Chapter 7; De Souza and Bánkuti, 2017). In the agri-food sector, measurement difficulties are widespread in many areas of activity. These difficulties bring about transaction costs *ex-ante* and *ex-post*, to guarantee the economic rights to assets and to maintain the ownership of the asset within the firm. The ease of application of the measurement – the authors argue – would enable contracts as a coordination arrangement. On this basis, the chapter proposes a conceptual framework concerning the institutional arrangements under both TCE and MCE. This framework is then applied to the dairy and coffee sectors in Brazil. The analysis shows the role of asset specificity and property rights in shaping the variety of governance structures adopted by agents in both the sectors investigated. In the case of the milk sector in Paraná state, the informal agreement dominates, mainly as a result of the use of informal enforcement for settling disputes. The limited possibility of small producers to expand the volume of production is recognized as a motivation for the role of informal agreements. Governance structures seem to provide sufficient, even though not fully efficient, mechanisms for coordination.

Russo, Perito and Di Fonzo (Chapter 8; Russo *et al.*, 2017) consider private food safety standards (PFSS) and identify a pooling paradox: large retail chains largely invest in food safety, but they do not put a correlated effort into informing consumers about this strategy. The study carries out an economic analysis of this paradox in light of contract theory. Two strategies are considered: pure private standards and collective standards. The principal-agent model proposed is aimed at explaining why a profit-maximizing supermarket may decide to implement PFSS without extensive advertising to its customers. The main result is that, regardless of the impact of the PFSS on consumers, standards are a supply management tool ensuring that suppliers comply with the buyers' requirements in the presence of incomplete information and incomplete contracts. The potential loss from not achieving a critical food safety level drives retailer strategies: contract purchasing prevails if the expected loss is large or, alternatively, a mix of contract and spot market purchasing is convenient. Moreover, product differentiation may bring about complex outcomes.

Watanabe, Lourenzani and Bànkuti (Chapter 9; Watanabe *et al.*, 2017) examine contract farming in light of transaction cost economics. The authors start from the observation that the lack of a specific legal framework may prevent agents from adopting this type of agreement. The study assumes that in contract farming, the parties are interdependent but legally independent. Accordingly, the authors explore the use of the hybrid governance structure concept to analyse contract farming. Comparing different types of contracting drawn from the literature, the authors first conclude that contract farming cannot be held as a marketing contract or a work contract. Second, the authors recognize that the lack of a clear legal framework may favour a role for informal institutions – for example, connected to local traditions, also concerning technology. Two case studies, milk and biodiesel in Brazil, are proposed. The study underlines the complexity of the reciprocal obligations of the parties, the role of power asymmetries and the parties' reputations.

Paoloni and Gioia focus on the organizational arrangements that have received increased attention in the Common Agricultural Policy reform for the period 2014-2020 (Chapter 10; Paoloni and Gioia, 2017). Among others, they analyse producer organizations (POs), interprofessional organizations (IOs) and protection consortia, of which the latter are not directly addressed in the European Union's recent legislation. Considering the Italian case – in which the protection consortia' s main role is to safeguard product certification in protected designations of origin (PDO) and protected geographical indications (PGI) – the chapter first recognizes that they may act both as PO and IO, especially in the wine sector. Representativeness and *erga omnes* principles are both directed to preserve and protect PDO and PGI product quality and to support coordination. The chapter extensively discusses a central issue affecting the possibility of spreading the model of protection consortia through the entire agri-food sector: while the consortia's nature requires it to transfer decision rights from the participants to the consortia and to act *erga omnes*, the competition law tends to limit the consortia's working capability and scope.

Furesi, Madau and Pulina (Chapter 11; Furesi *et al.*, 2017) analyse the territorial brand – a form of location-related brand that refers to food characteristics derived from being produced in a given place – conceptualizing it as an 'umbrella' brand contributing to the governance of the chain and to the coordination upon which value creation is based. The study focuses on the 'Porto Conte Regional Park' brand, a case regarding products of a territory in the northwest of Sardinia (Italy). The brand was recently promoted by the park management and requires the producers to comply with a set of rules concerning product quality and environmental values. Three wine sector firms are considered as case studies. The study examines the motivations of the three firms to adopt the brand and the effects of the adoption on the marketing outcomes and governance. The complexity of transactions managed under the brand umbrella seems to be able to influence the advantages in terms of value creation. The study concludes that the territorial brand may raise specific changes in the transaction governance in different firms.

The chapter of Bojnec examines the dairy value chain in Slovenia, recognizing the importance of the sector and of quality and purchasing conditions as factors affecting the farm-gate price (Chapter 12; Bojnec, 2017). Two further developments in the sector are the entry of retailers and the intra-industry trade that emerged since the beginning of the 1990s. HACCP standards and compliance with competition policy represent two important drivers of change influencing the sector, while the changes that occurred in the last two decades in the value chain (reduction of the number of small farms and farm concentration) tend to trigger the competitive process. A variety of organizational arrangements have been chosen to cope with the uncertainty raised by the market dynamics. This variety seems to provide opportunities to cope with market shocks due to the abolition of the EU quota system and supports the long-term stability of the sector.

Toccaceli considers the theme of corporate social responsibility in agricultural cooperatives (Chapter 13; Toccaceli, 2017). The case study – the cooperative Fermes de Figeac in France – provides information on how to combine environmental and economic objectives with social initiatives by setting relationships with other enterprises and regional public bodies. The chapter first illustrates the historical evolution of the cooperative and investigates its connection to the territory through both an institutional and an organizational perspective. It then provides an interpreting scheme based on three analytical elements: the firm can reach financial benefits by implementing adequate corporate social responsibility strategies; it is worthwhile to the cooperative firm to make a strategy fitting the regulatory environment; and the process of social value creation can be framed in a strategy aimed at multifunctionality. The interpretative scheme is then contrasted with both territorial studies and the new institutional economics, emphasizing the role of the institutional environment and of the meso-institutions on the cooperative strategy design.

Martino, Toth and Frascarelli (Chapter 14; Martino *et al.*, 2017) consider stakeholder beliefs about recent changes in the Common Agricultural Policy (CAP). The chapter examines this theme as a specific issue in the field of organizational impacts of the new CAP. Private orderings are seen as new tools for achieving private and social CAP objectives. This represents a novelty in the landscape of CAP interventions in recent decades. From an institutional perspective, the article investigates stakeholder beliefs about such a change, assuming that these beliefs are an important driver of effective changes. The empirical analysis is based on the Delphi approach and indicates that the respondents recognize the potential impact of the policy change on organizational changes.

Ciliberti and Frascarelli investigate the organizational changes in the tobacco chain in Italy triggered by the Common Agricultural Policy reform for the period 2014-2020 (Chapter 15; Ciliberti and Frascarelli, 2017). The authors assume that asset specificity and uncertainty play an important role in determining the organizational arrangements chosen by the agents in this sector. Accordingly, it is recognized that hybrid structures are usually chosen and framed within the new institutional environment. The chapter extensively describes the complex arrangements set up by tobacco chain actors and argue that these arrangements support chain viability as well as the actors' connection with the institutional framework. The variety of contracts observed in the case study and the contractual framework are recognized as effective tools for coping with uncertainty.

Adopting a political economy perspective, the chapter of Corsinovi and Gaeta considers market failures and regulatory interventions in the European wine sector (Chapter 16; Corsinovi and Gaeta, 2017). The focus is on the effects of the constraints the European Union imposes on the competitive capabilities of European firms and on price volatility. The study first analyses the European policy in the wine sector. Then, it considers the effect of European regulation on agricultural land allocated to wine production and the intensity of price dynamics. The study suggests that the strategic behaviour of wine producers can cause market instability in the short term but may eventually lead to market stabilization in the medium term. New policy interventions focusing on competition and innovation are expected to allow farms to face global market challenges.

Riganelli and Marchini investigate firm strategy changes and consumer reactions triggered by the information provided through voluntary certification (Chapter 17; Riganelli and Marchini, 2017). The focus of the study is on the mandatory introduction of palm oil in the list of food ingredients as set by the European Union Reg. 1169/2011. The chapter focuses on the Roundtable on Sustainable Palm Oil (RSPO). The demand for environmentally self-regulated products is increasing, and correspondingly increasing is the importance of information disclosure, i.e. the effort to measure and report verifiable product quality information. Environmental labels are usually thought of as a tool available to the firm in this specific context. Using data extracted from the AIDA database, the study shows how information disclosure causes a positive reaction at the consumer level.

Nazzaro, Marotta and Stanco (Chapter 18; Nazzaro *et al.*, 2017) address the theme of agri-food short chains – especially direct selling – as a way to internalize the production of public goods in the agricultural sector. The farms engaged in short chains are thought of as multifunctional farms tackling the changing demand triggered by post-modern consumers. Environmental, cultural and traditional values, seen as positive externalities, are recognized as the main drivers of the multifunctional characteristics sustained by short food chains. Virtuous behaviours of the famers and of the consumers are the forces establishing the process of shared value creation. The empirical analysis investigates the variables influencing the shared value creation in a sample of wine producers in the Telese Valley (South of Italy). Human and social capital, technology and local public goods provisioning are considered. Human capital, in particular, appears to be necessary both to directly create value and to foster social capital. These forms of capital sustain the farm's capability to provide public goods, which, in turn, fosters real competitive levers for value creation.

Marangon and Troiano (Chapter 19; Marangon and Troiano, 2017) focus on the farm as a unit of production of private and public goods and analyse an incentive-based instrument (payment for ecosystem services, PES) to support the provision of ecosystem services and the sustainability of agri-food chains. The chapter first discusses a number of examples of PES to reflect both a Coasean and a Pigouvian approach. Second, it examines the role of PES in sustaining biodiversity conservation in rural areas. Innovative approaches combining a PES scheme with eco-certification are then illustrated, and the importance of cooperation in supporting these instruments is emphasized. A specific role for public decision-makers is finally noted, in particular as regards the coordination between the suppliers and users of ecosystem services.

Bitzios, Jack, Krzianial and Xu address the issues raised by the potential gaps in traceability legislation (Chapter 20; Bitzios *et al.*, 2017). Drawing from an economic sociology theoretical background, the chapter adopts the concept of creative dissonance and identifies possibilities for enhancing food company strategies in global chains. The chapter starts by considering three main challenges faced by a company in the United Kingdom, as an example, to gaining access to the global food market: meeting consumer requirements, as well as the responsibilities of food business operators; coping with the differences between enforcement regimes; and addressing different requirements for product traceability. The study concentrates on the potential dissonance between the regulatory system and the related risk of fraud. The article first provides examples of how the dissonance between regulatory regimes could challenge the European company. Then, food fraud is included within the drivers to adopt a traceability system. Two case studies on genetically modified crops and trade with China are discussed.

References

Bitzios, M., Jack, L., Krzianial, S.M. and Xu, M., 2017. Dissonance in the food traceability regulatory environment and food fraud. Chapter 20. In: Martino, G., Karantininis, K., Pascucci, S., Dries, L. and Codron, J.-M. (eds.) It's a jungle out there – the strange animals of economic organization in agri-food value chains. Wageningen Academic Publishers, Wageningen, the Netherlands, pp 375-393.

Bojnec, Š., 2017. Functioning of the dairy value chain in Slovenia. Chapter 12. In: Martino, G., Karantininis, K., Pascucci, S., Dries, L. and Codron, J.-M. (eds.) It's a jungle out there – the strange animals of economic organization in agri-food value chains. Wageningen Academic Publishers, Wageningen, the Netherlands, pp 229-238.

Chlebicka, A., Fałkowski, J. and Łopaciuk-Gonczaryk, B., 2017 Horizontal integration between farmers – governing cooperation through different enforcement mechanisms. Chapter 4. In: Martino, G., Karantininis, K., Pascucci, S., Dries, L. and Codron, J.-M. (eds.) It's a jungle out there – the strange animals of economic organization in agri-food value chains. Wageningen Academic Publishers, Wageningen, the Netherlands, pp 85-104.

Ciliberti, S. and Frascarelli, A., 2017. Collective arrangements in the agro-food supply chain: the case of the interbranch organization Tabacco Italia. Chapter 15. In: Martino, G., Karantininis, K., Pascucci, S., Dries, L. and Codron, J.-M. (eds.) It's a jungle out there – the strange animals of economic organization in agri-food value chains. Wageningen Academic Publishers, Wageningen, the Netherlands, pp 283-302.

Coase, R., 1992. The institutional structure of production. American Economic Review 82: 713-719.

Corsinovi, P. and Gaeta, D., 2017. Interventionist policies and market failures of European wine regulations. Chapter 16. In: Martino, G., Karantininis, K., Pascucci, S., Dries, L. and Codron, J.-M. (eds.) It's a jungle out there – the strange animals of economic organization in agri-food value chains. Wageningen Academic Publishers, Wageningen, the Netherlands, pp 303-321.

De Souza, J.P. and Bánkuti, S.M.S., 2017. Transactions in coffee and milk agri-food value chains: why measure? Chapter 7. In: Martino, G., Karantininis, K., Pascucci, S., Dries, L. and Codron, J.-M. (eds.) It's a jungle out there – the strange animals of economic organization in agri-food value chains. Wageningen Academic Publishers, Wageningen, the Netherlands, pp 139-160.

Furesi, R., Madau, F.A. and Pulina, P., 2017. Value in territorial branding, agro-food production and governance patterns: a case study. Chapter 11. In: Martino, G., Karantininis, K., Pascucci, S., Dries, L. and Codron, J.-M. (eds.) It's a jungle out there – the strange animals of economic organization in agri-food value chains. Wageningen Academic Publishers, Wageningen, the Netherlands, pp 215-228.

Grandori, A., 2017. Linnaeus in the jungle: configurational lenses for discerning forms of economic organization in agri-business. Chapter 2. In: Martino, G., Karantininis, K., Pascucci, S., Dries, L. and Codron, J.-M. (eds.) It's a jungle out there – the strange animals of economic organization in agri-food value chains. Wageningen Academic Publishers, Wageningen, the Netherlands, pp 51-66.

Hobbs, J.E., 2017. A three-tiered approach to the economics of hybrids: an application to the agricultural innovation sector. Chapter 3. In: Martino, G., Karantininis, K., Pascucci, S., Dries, L. and Codron, J.-M. (eds.) It's a jungle out there – the strange animals of economic organization in agri-food value chains. Wageningen Academic Publishers, Wageningen, the Netherlands, pp 67-82

Marangon, F. and Troiano, S., 2017. Policy and governance of payments for ecosystem services in agri-food value chains. Chapter 19. In: Martino, G., Karantininis, K., Pascucci, S., Dries, L. and Codron, J.-M. (eds.) It's a jungle out there – the strange animals of economic organization in agri-food value chains. Wageningen Academic Publishers, Wageningen, the Netherlands, pp 357-373.

Martino, G., Tóth, J. and Frascarelli, A., 2017. Concepts in European policy intervention and food exchange: a Delphi approach. Chapter 14. In: Martino, G., Karantininis, K., Pascucci, S., Dries, L. and Codron, J.-M. (eds.) It's a jungle out there – the strange animals of economic organization in agri-food value chains. Wageningen Academic Publishers, Wageningen, the Netherlands, pp 263-281.

Ménard, C., 2017. Finding our way in the jungle: insights from organization theory. Chapter 1. In: Martino, G., Karantininis, K., Pascucci, S., Dries, L. and Codron, J.-M. (eds.) It's a jungle out there – the strange animals of economic organization in agri-food value chains. Wageningen Academic Publishers, Wageningen, the Netherlands, pp 27-50.

Nazzaro, C., Marotta, G. and Stanco, M., 2017. Short food supply chains and shared value on the multifunctional farm: an analysis of determinants. Chapter 18. In: Martino, G., Karantininis, K., Pascucci, S., Dries, L. and Codron, J.-M. (eds.) It's a jungle out there – the strange animals of economic organization in agri-food value chains. Wageningen Academic Publishers, Wageningen, the Netherlands, pp 337-355.

Paoloni, L. and Marianita, G., 2017. Contractual scope in the agri-food market: the case of protection consortia in the wine sector. Chapter 10. In: Martino, G., Karantininis, K., Pascucci, S., Dries, L. and Codron, J.-M. (eds.) It's a jungle out there – the strange animals of economic organization in agri-food value chains. Wageningen Academic Publishers, Wageningen, the Netherlands, pp 197-211.

Riganelli, C. and Marchini, A., 2017. Labelling strategies in the agri-food context: the case study of RSPO certification. Chapter 17. In: Martino, G., Karantininis, K., Pascucci, S., Dries, L. and Codron, J.-M. (eds.) It's a jungle out there – the strange animals of economic organization in agri-food value chains. Wageningen Academic Publishers, Wageningen, the Netherlands, pp 323-336.

Russo, C., Perito, M.A. and Di Fonzo, A., 2017. The apparent paradox of unadvertised private food safety standards. Chapter 8. In: Martino, G., Karantininis, K., Pascucci, S., Dries, L. and Codron, J.-M. (eds.) It's a jungle out there – the strange animals of economic organization in agri-food value chains. Wageningen Academic Publishers, Wageningen, the Netherlands, pp 161-176.

Stranieri, S. and Banterle, A., 2017. Traceability standards and vertical coordination. Chapter 6. In: Martino, G., Karantininis, K., Pascucci, S., Dries, L. and Codron, J.-M. (eds.) It's a jungle out there – the strange animals of economic organization in agri-food value chains. Wageningen Academic Publishers, Wageningen, the Netherlands, pp 119-138.

Toccaceli, D., 2017. Agricultural cooperation and corporate social responsibility: alternative approaches in a literature review. Chapter 13. In: Martino, G., Karantininis, K., Pascucci, S., Dries, L. and Codron, J.-M. (eds.) It's a jungle out there – the strange animals of economic organization in agri-food value chains. Wageningen Academic Publishers, Wageningen, the Netherlands, pp 239-261.

Ventura, F., D'Arienzo, M.C. and Martino, G., 2017. Italian network contracts viewed from the perspective of transaction cost economics. Chapter 5. In: Martino, G., Karantininis, K., Pascucci, S., Dries, L. and Codron, J.-M. (eds.) It's a jungle out there – the strange animals of economic organization in agri-food value chains. Wageningen Academic Publishers, Wageningen, the Netherlands, pp 105-117.

Watanabe, K., Lourenzani, A.E. and Bankuti, S.M.S., 2017. Contract farming, a strange type of contract: more than obligation 'to give'. Chapter 9. In: Martino, G., Karantininis, K., Pascucci, S., Dries, L. and Codron, J.-M. (eds.) It's a jungle out there – the strange animals of economic organization in agri-food value chains. Wageningen Academic Publishers, Wageningen, the Netherlands, pp 177-196.

Part I – **Theories and conceptual approaches**

1. Finding our way in the jungle: insights from organization theory[1]

C. Ménard

Centre d'Economie de la Sorbonne, Université de Paris, Panthéon-Sorbonne, 75005 Paris, France; claude.menard@univ-paris1.fr

Abstract

This contribution is about the variety of modalities of coordination of economic activities, with application to the agri-food sector. Building on recent developments in organization theory and institutional analysis, it explores two dimensions along which coordination operates: it proposes an extended transactional model to explain the variety of coordination devices, with a special emphasis on contracts and their role in hybrid arrangements; and it characterizes the different layers composing the institutional setting in which these arrangements are embedded, with an emphasis on the long ignored or neglected role of the intermediate, 'meso-institutional' layer. Insights are also provided on the technological background to coordination and on some policy issues.

Keywords: transaction, contracts, property rights, hybrids organizations, meso-institutions

[1] This paper originated, like most papers in this book, from a pre-congress symposium of the EAAE organized in Ljubljana in 2014. Further development owes much to keynote speeches delivered at the University of Ghent, the Belgium Association of Agricultural Economists, the Perugia conference on Contracts in Agrifood Systems, and the Conference of the International Co-operative Alliance held in Stirling. I am grateful to the different organizers for the opportunity to develop these ideas and for the extremely useful comments and suggestions provided by numerous participants. Special thanks go to Jean-Marie Codron, Liesbeth Dries, Guido van Huylenbroek, Kostas Karantininis, Gaetano Martino, Stefano Pascucci, and Richard Simmons. The usual disclaimer fully applies.

1.1 Introduction

A central issue in organization theory and more generally in economics, going back to Adam Smith (1776) and beyond, is about the nature and modalities of coordination and how these modalities make transactions possible in a world characterized by division of labour and specialization of tasks. Indeed, as rightly pointed out by Coase in his comment on Smith (Coase, 1994), coordination and transactions are tightly interwoven.

An important difficulty in that respect comes from the variety of modalities that can provide coordination. Although Smith was much aware of this variety, as shown in his discussion of the organization of production in a manufacture compared to a network of independent artisans, the issue of organizational diversity has long been put under the bushel, largely because of the neoclassical focus on the price mechanism. This issue has been reinitiated by Coase and substantiated by Williamson, among others, and well summarized in their respective Nobel lectures from 1991 and 2009. By giving content to what Coase identified as 'the institutional structure of production', modern organization theory thereafter acknowledged the richness of the arrangements that structure our economies (for a synthesis, see Gibbons and Roberts, 2013). It also faces a key puzzle: is there some logic at work in that jungle? With one subsidiary question: what forces determine the choice of a specific arrangement in that diversity?

What makes things even more complex is that these modalities of coordination change over time. In that respect, there have been major evolutions within the last decades in our economies, and particularly in the agri-food sector. These changes can be well illustrated by the passage from segmented activities largely coordinated through independent and relatively isolated local or regional markets to the worldwide organization of activities through tightly controlled value chains paired with vertical coordination. 'Globalization' is the catch word for this evolution although it says close to nothing about its content.

Indeed, these changes and their technological support have introduced profound modifications in the way economic transactions are structured, the modalities of their coordination, and the devices and mechanisms through which basic functions can be regulated. A vivid illustration is that of securing the quality of perishable agricultural products in worldwide supply chains. As those familiar with this example know well, there are many institutional layers involved in facing problems such as food safety, beside the layer at which organizations operate. First, there are rules and norms framing the activities of the various organizational arrangements, defined at what can be called the macro-institutional level. For a long time these rules and norms were shaped mainly at the national level; they are increasingly dependent on international agreements of different types. Second, these rules and norms are embedded in beliefs and behaviours of individual as well as collective actors on the one hand, and they need to be implemented and enforced on the other hand. This complex interaction raises the issue of the transmission mechanisms linking the macro-institutional level at which rules and norms are defined and the organizational level at which rules and norms are actually transformed into actions. I have suggested identifying this intermediate layer as the 'meso-institutional' level (Ménard, in press-a). The resulting institutional setting within which the agri-food sector is structured, and this is true of course for all other sectors of the economy, is much more complex than the one initially considered by institutionalists of different obedience.

What I develop hereafter is far from pretending to review extensively what we have learned about organizational changes and the different institutional layers in which they are embedded. I do not even deliver a survey on specific aspects. I rather submit a very selective approach to what I consider as important lessons from recent researches with a clear focus on the organizational level and insights on the nature and status of meso-institutions. In doing so, my main goal is to point out a set of concepts that can provide tools to make progress in our knowledge of the jungle that so many chapters in this book illustrate.

These considerations are organized along four themes. Section 1.2 takes contracting practices as a starting point since contracts provide important tools to coordinate, although their role and significance vary widely across sectors and regions. Section 1.3 proposes an extended transactional model to explain these variations in contracts as coordination mechanisms, with a special emphasis on their role in hybrid arrangements. Section 1.4 switches attention to the institutional embeddedness of organizations and their contractual component and briefly explores characteristics of the different layers composing this institutional setting, with an emphasis on the long ignored or neglected role of the intermediate, 'meso-institutional' layer, hypothesizing its key role in understanding the diversity of contractual practices and their performance. Section 1.5 concludes.

1.2 Why contracts matter ... but don't do it all!

The rapid development of impersonal transactions (North, 1990a) has led to an increasing role of contracts as a tool to organize and secure these transactions. Hereafter, 'contracts are understood as formal agreements that explicitly determine binding rules for transferring rights to use goods or services among well-defined economic units (the 'parties' to a contract)'. In what follows I first provide insights on their significance in the agri-food sector of some major economies. I then turn to data suggesting their limited role as coordinating tools, thus raising the issue: what else? Unfortunately, in doing so, I rely on a very limited set of data, concentrated on a very small number of countries. An initiative that originated at the OECD in 2008 to collect extensive data on contracts in agriculture was never carried to its final phase. However, preliminary results based on nine OECD members showed the scarcity of data available and the difficulty to collect new data (see preliminary indications in Ménard and Vavra, 2008; Vavra, 2009).

1.2.1 The (slow) diffusion of contracts...

Notwithstanding these limits, available data suggest that the use of contracts to coordinate activities has been a striking change in the agri-food sector over recent decades, with particular impact at the farm level. Let me illustrate with two leading production zones: USA[2] and the European Community.

In the USA, the value of production under contract has more than tripled since the 1970s: it represented 11% of the total value in 1969 and close to 40% in 2008 (MacDonald and Korb, 2011). However, it is noticeable that this evolution concerned mostly large, often corporate farms. Indeed, only 11% of the total number of farms was concerned (same source). The practice of contracting also varies widely with respect to the type of contract and the type of the activity, as Table 1.1 suggests.

In Europe, there is also a very uneven diffusion of contractual agreements. In Germany, according to Drescher (2000), the majority of contracts in the agri-food sector are 'short term contracts'. In the UK, contracts tend also to be short term (e.g. yearly in the milk industry) and mostly developed in highly concentrated sectors such as the poultry industry, entirely dominated by 4 main integrators (Royer, 2011). In

[2] For the USA we can benefit from the extensive data collected by ARMS, in the USDA. ARMS is a complex annual survey applied to a stratified random sample of all U.S. farms gathering information on the financial condition, production practices, resource and contract use, and economic well-being of U.S. farm households.

Table 1.1. Contracting in the agricultural sector in the USA (MacDonald and Korb, 2011).

Item	1991-93	1996-97	2001-02	2005	2008
Share of farms with contracts (%)					
Any contracts	10.1	12.1	11.2	11.1	12.1
Marketing contracts	8.2	10.2	9.0	9.3	10.3
Crop	6.6	8.3	7.4	7.6	9.0
Livestock	1.6	2.0	1.6	1.9	1.4
Production contracts	2.1	2.2	2.6	2.1	2.2
Crop	0.6	0.6	0.5	0.4	0.3
Livestock	1.6	1.6	2.1	1.7	1.8
Share of production under contract (%)					
Any contracts	28.8	32.1	37.7	40.7	38.5
Marketing contracts	17.0	21.5	19.7	22.0	21.7
Crop	11.0	12.2	12.7	13.1	14.9
Livestock	6.0	9.3	7.0	8.9	6.9
Production contracts	11.8	10.6	18.0	18.7	16.8
Crop	0.9	1.0	1.6	0.8	0.5
Livestock	10.9	9.6	16.4	17.9	16.3

France, contracting is mostly developed in specific sub-sectors (e.g. eggs, with 55% of production delivered through contracts; or poultry, with 75% of the total production), and driven mainly by cooperatives. In the Netherlands, 'a large part' of the hog sector operates under contracts, and contracting prevails in the delivery of eggs (over 80%). However, it is very unfortunate that the lack of consolidated data due to the absence of extensive national surveys does not allow having a global picture of the situation.

Notwithstanding this scarcity of aggregated data, numerous studies have substantiated the role of contracts and their characteristics in specific sectors. Pioneering studies are available on broilers and the poultry sector (Goodhue, 2000; Knoeber, 1989; Ménard, 1996; Tsoulouhas and Vukina, 2001), hogs (Key and McBride, 2003; Reimer, 2006; Wang and Jaenicke, 2006), beef packing (Xia and Sexton, 2004), etc. However, we still lack a synthetic view, whether it concerns the data or the analytics of contracts in the agri-food sector (see already Dubois, 2001).

1.2.2 ... with limited impact

This being said, we already know some limitations regarding the usage of contracts in agriculture. First, contracts are far from universal, even in countries where they are most developed. Let me focus again on the example of the USA for which we have data for over two decades. Those data (MacDonald and Korb, 2011; USDA, 2012) show a huge variation across commodities. For example, marketing contracts represent over 65% (in value) in crops such as sugar beets and dairy products. Similarly, production contracts cover up to 90% or more (in value) for livestock such as poultry, egg, hogs. But in field crops such as corn, wheat, soybeans, etc., contracts concern less than 10% in value of the transaction on these commodities.

Second, there are significant variations in contracting practices according to the size of farms, as illustrated by Table 1.2. In some sectors, the number of farms under contract has even declined sharply (by about 6% in fruits and vegetables and even more significantly by about 9% for poultry farms), although the percentage of total production under contracts for these products increased, an indicator of the growing concentration of production (Vavra, 2009: pp. 15-16).

Table 1.2. Contracting practices according to size of farms (USDA, 2012).

Farm sales class (×1000$)	Distribution of farms % of USA farms	Distribution of farm production	Farms with contracts % of sales class	Production under contracts
<25	92.5	29.1	6.2	19.9
250-500	4.1	15.1	43.5	31.3
500-1000	2.1	15.1	59.1	42.6
>1000	1.3	40.7	64.2	53.4

The usage of contracts also varies widely across regions. A striking example is that of rice, with 90% of the production contracted in the 'Fruitful Rim' (California, Arizona), while only 15% is contracted in the 'Mississippi Portal' (Louisiana).

1.2.3 Hence the questions: why to contract and what about this variety?

Notwithstanding their limited application, contracts matter because they provide support to coordination. Indeed, contracts establish common ground among parties, creating regularities that facilitate decision-making and joint-actions. This role of contracts particularly matters for parties to a transaction facing significant asymmetries of information and/or uncertainties affecting the outcome. However, the same factors that contracts intend to overcome are also the very sources of their limits: information asymmetries make most contracts incomplete, and uncertainties challenge the allocation of responsibilities over the resulting outcome as already observed by Stiglitz (1974) in his pioneering paper on sharecropping contracts. This is to say that contracts delineate 'acceptance zones' of action within which coordination develops rather than determining specific actions to be implemented.

There is now quite general acceptance in economic theory of this coordinating function of contracts. However, there are important differences when it comes to the interpretation of their role, their significance, and the potential ambiguities they may generate.

First, contracts can be viewed as tools to manage risks in an uncertain environment (Allen and Lueck, 1999; EC, 2016; Patrick *et al.*, 1998). This is a main point made by agency theory, raising the issue of incentives needed to induce participants to a contract to deliver appropriate actions. The expectation is that in allocating risk among actors, for example in the supply chain system, and in associating adequate incentives to the related transfer of risk, with this adequacy a key to success (hence the search for optimal solutions!), contracts can reduce information asymmetries and the risk that agents derail transactions to their exclusive benefit. In that respect, contracts could be more effective than 'pure' spot markets: for example, contracting farmers would regard cash-forward contracts as effective in reducing risks. However, if it is so, farmers under contract should receive lower average returns since they face comparatively lower risks. But according to some studies, it is not what is observed: farmers under contract tend to benefit from relatively higher prices (MacDonald *et al.*, 2004; Sexton, 2017).

Second, a different perspective building on standard industrial organization suggests that dominant firms (e.g. distributors) may use contracts to extend their control over factors such as quality, thus implementing vertically coordinated chain that consolidates their market power (Hegrenes and Borgen, 2003). Better coordination would then go hand-in-hand with concentration, making independent suppliers captive. Indeed, spot markets becoming very thin, buyers could benefit from their small number to impose lower prices to their (dispersed) suppliers, thus absorbing part of the rent of producers while simultaneously benefiting consumers thanks to the strong competition among these leading buyers. However, there is no available evidence that captive supply procurement would generate low prices for farmers (Schroeter and Azzam, 2004). Sexton (2017) even provocatively argues that the small number of strongly competing firms make them increasingly dependent from farmers to secure provision of required quality, which could explain why market leaders increasingly use contracts that are particularly advantageous to the most efficient farmers.

A third argument is that contracts are transaction enhancing. In easing coordination along the supply chain and securing *ex-post* returns, contracts would provide incentives for *ex-ante* specific investments. In doing so, they would open room for improved product quality and variety, and speed up transfer of technological innovations. Contracts do not need to be extensive to play such role, contrarily to what was suggested by agency theory. All that is needed is that contracts secure investments by delivering a blueprint within which participants can make decision and negotiate adjustments imposed by a changing environment. Note however that this incompleteness may also facilitate opportunistic behaviour, so that there is a delicate equilibrium to find between too much rigidity, embedded into a detailed contract, and too many loose ends, permitted by a relatively vague blueprint. There is another potential drawback coming out of transactional enhancing contracts: the resulting increase in productivity and trade may create environmental damages rarely taken into account and admittedly difficult to assess and to take on board in the contractual process (although this is changing).

Notwithstanding these limitations (and more could be introduced), all arguments above converge in emphasizing the role of contracts as powerful tools to improve coordination and deal with uncertainties, particularly when long term specific investments are at stake. However, there is little in the above arguments to explain why there is such variation in the usage of contracts across sectors, or even within the same sector as illustrated by the rice sector in the USA; and across regions, or even within regions as illustrated by the role of contracts in the dairy industry in the European Community (Vavra, 2009). Two different explanations, that I explore hereafter, can be proposed to this variety of contracts and of the organizational arrangements in which they are embedded. One explanation is rooted in the very characteristics of organizational arrangements and the attributes of transactions they carry; the other, less often explored, is grounded in the properties of the institutional environment within which these organizational arrangements operate.

1.3 The variety of organizational solutions: an extended transaction costs model

Indeed, notwithstanding the path breaking paper from Ronald Coase (1937), the analysis of this variety of solutions and of the forces pushing towards the adoption of a specific arrangement in the set of possibilities as well as the analysis of the exact role of contracts in that context have long remained and still remain underexplored. Agency theory, for example, opened room for the examination of the types of contracts that could incentivize heterogeneous parties to cooperate in an efficient way, that is: at low cost, with high productivity, and with satisfying rewards for members (Bolton and Dewatripont, 2005; Jensen and Meckling, 1976). However, the contributions building on this paradigm almost exclusively focused on the drawing of optimal contracts, that is: contracts that maximize the probability to reach a preassigned goal (Bolton and Dewatripont, 2005; Milgrom and Roberts, 1988). It is striking how little has been done in that framework on issues regarding the conditions of choice among alternative organizational arrangements and about the modalities of governance, which by far exceed the role of contracts. A possible explanation to this bias could be tracked down to the attention paid to the conditions that make a contract as complete as possible (Tirole, 1999).

1.3.1 Contracts as part of a more general framework

Transaction cost economics remains a major source of inspiration when it comes to the exploration of these neglected issues. In its paradigmatic version (Williamson, 1985), the variety and incompleteness of contracts are embedded in the very nature of transaction costs, understood as the costs of organizing the allocation and transfer of rights to use resources, transfer that are made possible because the transactions at stake relate to technologically separable activities. A direct consequence of this conception of transactions is that the contractual dimension must be complemented: contracts must be understood in their interaction with the nature and type of rights to be allocated on the one hand; and the nature and modalities of transactions on the other hand. This configuration defines what has been called the 'golden triangle' of transaction costs economics (Ménard, 2005) within which organizational arrangements are embedded. Figure 1.1 summarizes this conceptual apparatus and suggests how the interactions of these concepts frame the representation of organizational arrangements

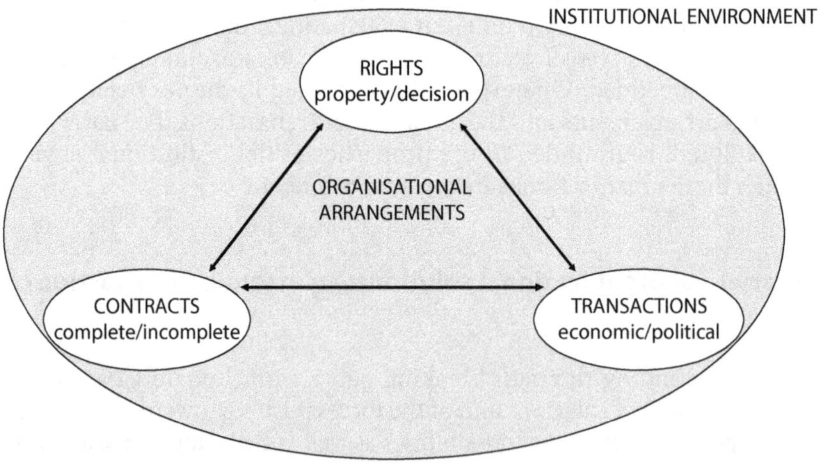

Figure 1.1. The 'Golden Triangle' of transaction cost economics.

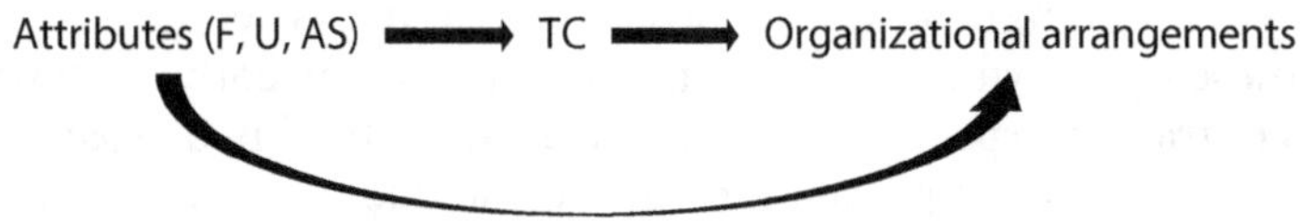

Figure 1.2. From transactional characteristics to organizational arrangements.

The initial version of transaction costs economics viewed contracts as the typical transaction, and read organizational arrangements through their contracting features, characterized by the attributes of the transactions to which contracts and the associated organizational arrangements provide support (Williamson, 1985). These attributes (the frequency of a transaction (F), the uncertainty surrounding a transaction (U), the variable specificity of investments it requires (AS)) determine the cost of that transaction. In a competitive environment decision-makers are pushed towards choosing the organizational arrangement that can meet these attributes at the lowest possible costs. This now well-known approach to organizations is summarized in Figure 1.2, which captures the impact of the attributes on the costs of organizing a specific transaction that in turn frame the conditions under which an organizational arrangement is chosen. The lower arrow, connecting directly the attributes of transactions to the organizational choice, corresponds to the strategy adopted in most empirical studies: facing the difficulty of measuring directly transaction costs, they focus on the direct impact of these attributes on the trade-off between alternative arrangements, implicitly assuming they do so by shaping the associated transaction costs (for a good discussion on this issue, see Masten *et al.*, 1991).

This initial version limited the trade-off to the choice between organizing the targeted transaction through markets versus doing it in-house (the now famous 'make-or-buy' trade-off). It was later extended to include alternative arrangements, identified as 'hybrids' (more on this below), which have characteristics of markets under certain aspects and properties of hierarchical arrangements on other aspects (Makadok and Coff, 2009; Ménard, 2004; Williamson, 1996).

Recent developments have extended and refined the concepts involved. First, following the initial insight from Alchian (1965) and the modern theory of property rights (Arrunada, in press; Baker *et al.*, 2008; Gibbons, 2005; Hart and Moore, 1990; Ménard, in press-b) the nature of rights to be traded has been specified, with a sharper distinction between property rights, which are about who holds (transferable) rights over goods and services to organize transactions and ripe related payoffs; and decision rights, which are about how rights to use can actually be exercised, usually through delegation to 'managers' by holders of property rights. Property rights can be defined De Jure, thus feeding a legalistic approach; or established De Facto, referring to custom and tradition; decision rights depend on the modalities and degree of delegation so that their exercise can be more or less formally determined and more or less centralized. Second, contracts are viewed as defining the relationship of actors to these rights, which operates mainly through incentives. However, in the 'incomplete contract' perspective, incentives are understood in a broader sense than in the 'complete contract' approach in order to include motivations not connected to monetary rewards or penalties, a key issue when it comes to reducing the risk of opportunistic behaviour (Benabou and Tirole, 2011). Third, there have been suggestions, following North (1990b), to extend the concept of transaction cost by differentiating economic transaction costs, coming out of the allocation and transfer of rights as defined above; and political transaction costs, which intend to capture the costs of building the political coalitions that define the rules of the game within which organizations and individuals operate. Political transaction costs thus point out the institutional embeddedness of transactions, an issue discussed in Section 1.4.

1.3.2 The relational dimension of contracts

These developments found a certain achievement in a model developed progressively by Baker *et al.* (2002, 2008) and Gibbons (2005). Building on MacNeil (1974), Goldberg (1980), Williamson (1985) and Hart and Moore (1990), these contributions propose a theory, identified as the 'relational contract' approach, combining transaction cost economics and the 'new' property rights theory.

As suggested by their labelling, these contributions focus on the key role of contracts as a mode of coordination, but in a limited and specific way. In a nutshell, contracts are viewed as most of the time incomplete, due to the presence of 'non-contractibilities' so that contractual clauses leave room for adaptation. The presence of these non-contractibilities also means that relational contracts leave room for opportunistic behaviour: temptation to renege can be highly significant and its prevention requires specific devices to monitor associated risks. Moreover, the need to adapt and to face opportunism imposes quick decisions that most of the time does not allow time for renegotiating the contract (or at costs that are dissuasive). In such an environment, contracts must be complemented by other modalities of coordination. This is typically the role of management (Gibbons and Henderson, 2012), a view that allows reintroducing issues rose by agency theory, but in a very different conceptual context than the one prevailing in contributions inspired by Jensen and Meckling (1976). Moreover, there are organizational arrangements that may require modalities of coordination different from those provided by the management in integrated firms: this is typically the case of hybrid arrangements (see Ménard, 2013a; and the next subsection).

In that perspective, relational contracts can be viewed as playing a substantial role in the allocation and monitoring of the two types of rights defined above, thus being part of the delineation of alternative organizational arrangements. Building on this intuition, Ménard (2013a, in press-c) submitted a synthesis summarized in Figure 1.3.

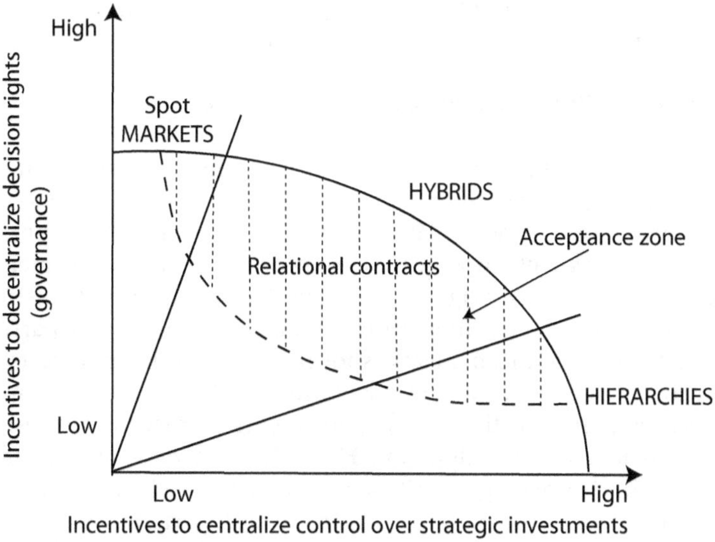

Figure 1.3. A synthetic representation of organizational arrangements Ménard, in press-c).

In this figure, the horizontal axis captures the incentives that parties to a transaction (or contract) have to take control over related assets, depending on how strategic these assets are with respect to the implementation and monitoring of the transaction at stake. In a market economy, control takes the form of holding property rights. The vertical axis relates to the other dimension of rights: it indicates the incentives for holders of property rights to keep more or less tight control over decision rights, thus determining the degree of decentralization that characterizes the modality of governance chosen. The upper bound shows the optimal combination of these two dimensions; the lower bound delineates the area under which the misalignment between decision rights and property rights tends to make transaction costs dissuasive. The unique boundary lines on the extreme left (extreme right, respectively) suggests situations where it is optimal to use 'pure' spot market ('pure' hierarchy, respectively). The lens thus delineated is the domain of relational contracts, which in a sense corresponds to what Simon (1951) identified as the 'acceptance zone' in his pioneering analysis of the employment relationship. The radials are proxies indicating how the combination of the two types of rights and their contractual translation co-determine different organizational arrangements. In that respect this model diverges from the representation of organizations proposed by Williamson (1996), which focused on a single variable (assets specificity, here captured through the more general concept of control of property rights over strategic investments).

Note that there is a missing dimension in this representation, which is about the institutional embeddedness of property as well as decision rights and contracts, an aspect partially captured through the concept of 'political transaction costs' in the golden triangle (more elements on this issue in Section 1.4).

1.3.3 Hybrids: the privileged domain of relational contracts

Figure 1.3 makes explicit the central role of relational contracts in providing modalities of coordination, particularly when it comes to hybrid arrangements. Following Williamson (1996) and Ménard (1996, 2004, 2013a), hybrids are understood as organizational 'arrangements in which two or more partners pool strategic decision rights as well as some property rights, while simultaneously keeping distinct control over key assets'. Hybrids thus differ from arrangements in which parties interact mainly through the price mechanisms (spot markets) so that contracts leave little or no room for mutually negotiated adaptation; and they differ from integrated organizations (hierarchies) within which adjustments are made in last resort through forms of command and subordination. Hybrids are arrangements made among parties that remain distinct legal entities, which matters when it comes to liability issues; and that in last resort maintain control over key rights, which matters in a context in which parties remain competitors. Hence the important role for hybrids of relational contracts through which renegotiations can be conducted and adaptations implemented.

In a sense, this emphasis on the role of hybrids might look paradoxical if one considers some recent evolution in market economies. With respect to production, over the last decades agriculture faced changes similar to that of other sectors, with a strong movement towards consolidation. In the USA, production on large farms increased from 38 to 56% of total production while the total number of small farms went down by 40% over the last 15 years (USDA, 2014). In France, the leading producer in Europe, the number of farms fell by over 50% while the average farm area doubled from 1988 to 2010 (France, 2015). In the distribution of food products the evolution towards concentration is even more spectacular. The top 5 retailers now represent above 90% of total distribution in Canada and Australia and above 50% on average for OECD countries (Vavra, 2009). Last, there have been major changes, still going on, in the relation between production and distribution via the development of centralized procurement through logistic platforms and distribution centres, with a shift towards non-price competition, the implementation of private standards of quality, and the increasing role of third party certification as a central mean for guaranteeing the reputation of the leading firms.

However, these deep changes were also accompanied by the development of hybrid types of relationship, as almost all chapters in this book illustrate. Producers' organizations, cooperatives, franchising, and the now well-known supply chain systems provide important illustrations. Notwithstanding the acknowledgement of their growing importance, supply chain systems are often misunderstood. In a recent document from the European Commission, they are defined as '... a series of consecutive markets involving input providers and producers, then producers and processors, then processors and wholesalers/traders and finally wholesalers/traders and retailers. Each of these markets is shaped by its own specific supply and demand accounting for price formation.' (EC, 2016). This is plainly wrong and misses the main characteristics of supply chain systems that the concept of hybrid helps understanding. Indeed, supply chain systems are network that by far exceed pure market relationship. They are arrangements that operate on a large scale, both vertically from producers to retailers, and horizontally among each layer of the supply chain system. Their development requires tight coordination with respect to quality, quantity, and timing of deliverables. This coordination issue raises specific problems because it must be implemented among legally independent actors that often simultaneously involve cooperation and competition. Besides prices, other mechanisms and devices need being implemented to guarantee this coordination and control the risks of opportunistic behaviour that could destroy the expected value-added of a chain system. The resulting varieties of potential mechanisms and devices that can be chosen to meet these goals translate into differentiated transaction costs that determine the comparative advantages or disadvantages for participants to a specific chain system as well as among different competing systems.

Contracts are definitely part of these mechanisms and associated devices. They are increasingly viewed as a major tool to coordinate and to allow efficient control. Their role is deeply entrenched into the nature of supply chain systems as hybrid arrangements, which are characterized by: (1) pooling resources among otherwise competing entities, which raises problems of allocation of property rights; (2) taking advantage of spillover effects from joint actions, which raises problems of allocation of decision rights; and (3) improving the monitoring of non-contractibilities, which raises the problem of allocation of rents jointly generated. Relational contracts can be viewed as (limited) means to deal with these problems, providing parties a framework within which they can operate safely (that is: limiting the risks of opportunistic behaviour) while keeping the flexibility needed to constantly adjust without having to face costly renegotiations or to rely on prices that would be determined through distinct markets.

However, the other facet of these characteristics of relational contracts is that they cannot do it all. They are only part of complex governance systems required to deliver stability and guarantee resilience (for a detailed empirical example of this imbrication of different mechanisms in hybrids, see Raynaud and Ménard, 2017). Stability is particularly challenging in a context of legally distinct and heterogeneous partners, which for example differ from the heterogeneity of shareholders in a corporation in that hybrids maintain legal as well as economic autonomy of strategic rights among partners who remain competitors, developing a relationship that Brandenburger and Nalebuff (1996) captured as 'co-opetition'. Moreover, resilience is also an issue since there is a variety of hybrid arrangements competing against each other on similar markets, with different transaction costs involved. The Brazilian agri-food sector provides a nice illustration (Ménard, 2013b; Schnaider *et al.*, 2017).

1.3.4 Some fundamental puzzles

The problems raised above translate into several puzzles that organization theory still face in dealing with hybrid arrangements and their contractual practices.

A first puzzle concerns the very existence of these arrangements: why is it that competing economic entities with distinct strategic rights enter into binding relationships, partially formalized in contracts, committing to pool significant parts of their rights and to coordinate their decisions accordingly? The standard answer is that in doing so participating organizations expect the creation of value through interdependence, benefitting from spillover effects (Ghosh and John, 1999), and/or creation of buffer against uncertainties (Carson *et al.*, 2006). However, if markets are efficient (which the above definition of the EC document presumed), why do firms abandon part of their autonomy and do not rely on market mechanisms to deal with uncertainty? And if there are benefits expected from close cooperation, why is it that they do not choose integration, along the line suggested by Grossman and Hart (1986), thus avoiding conflicts over rent sharing?

A second and related puzzle concerns the stability of hybrids. If partners to a binding agreement maintain distinct control over strategic rights and can potentially use them in a way that make them competitors on alternative activities or even on the very activities for which they have an agreement, how come that such arrangement are resilient? What governance structure do they adopt that make the arrangement efficient and resilient in a context of incomplete contracts (Hendrikse and Veerman, 2001)? Numerous empirical studies suggest the complexity and variety of the modalities of governance adopted, as illustrated for different arrangements by Jiang and Hendrikse (2011), Karantininis and Nilsson (2007), Raynaud and Ménard (2017) and Martino *et al.* (2017) among many others. Then, how come that the transaction costs resulting from this complexity are not dissuasive, making hybrid arrangements unstable and transitory (this was the initial view of Williamson, 1975)?

A third puzzle is about what has been identified as 'plural forms' in organizational arrangements (Bradach, 1997; also Lindenberg, 1996; for a discussion, see Ménard, 2013b, Section 1.2). Why is it that some economic entities, usually a firm, decide to not fully integrate nor endorse a well-defined type of agreement with selected partners but rather organize their transactions through different organizational arrangements simultaneously? These combinations have been a major concern for analysts of the so-called 'dual franchising systems', from the pioneering work of Rubin (1978) to more recent developments (Lafontaine and Slade, 2007); and numerous other examples show the generality of this problem, from the automobile industry- (Monteverde and Teece, 1982) to recent studies in the agri-food sector (Lara *et al.*, 2014; Schnaider *et al.*, 2017).

A fourth puzzle concerns organizational innovation. Since transitions most of the times involve high transaction costs, what forces push towards organizational changes, not only within integrated organizations (e.g. a firm) but also along supply chains and other hybrids? The traditional view emphasizes technological pull. Technological changes are of course important factors inducing various organizational arrangements to adapt and/or imposing new modalities of governance, as illustrated by so many examples from the development of biotechnologies to the revolution in the transportation of agricultural products (controlled atmosphere, satellite navigation systems). These changes require major specific investments, which, according to standard transaction cost economics, should induce integration. Several observations noted above (Section 1.3.3) about consolidation and concentration substantiate this analysis. However, supply chain systems, producers' organizations, etc. often adapt well to these technological changes without integration. Moreover, transaction cost economics does not provide a clear explanation when changes go the other way around. Indeed, innovations are often organizational, with technological changes developing *ex-post* to provide adequate support. A formidable example is the development of containers, the box that revolutionized the world (Levinson, 2006). When it comes to the agri-food sector, it is almost impossible to understand the globalization of supply chain systems without the container revolution. So the puzzling question is: considering the transaction costs at stake, what motivates an innovator to engage a complex process of interlocked changes in organization and technology.

1.4 The variety of organizational solutions: institutional embeddedness

All these puzzles and their potential answer raise institutional issues. Developing hybrids requires a redefinition of rights, which by the way challenges the traditional approach that dominates among competition authorities (Ménard, 2005); contracting is embedded in the law; organizational as well as technological innovation impose changes in the rules of the game; and this list could be extended.

1.4.1 Positing the issue at stake

The analysis of institutions went through substantial development over the last decades (e.g. Acemoglu and Robinson, 2012; Aoki, 2001; Greif, 2005; North, 1981, 1990a; among many others). However, most analyses remain at the general level of what can be called the 'macro-institutions'. Macro-institutional rules and norms can be defined at the national level (e.g. Mongolia), the regional level (e.g. the European Union), or even the international level (e.g. the World Trade Organization). However, it is striking how little we know about the transmission mechanisms linking this macro-level and what can be captured as the micro-institutional layer. 'Micro-institutions' correspond to the level at which transactions are actually drafted, negotiated and implemented and within which the different organizational arrangements already mentioned (markets, hybrids, hierarchies) act and interact. These arrangements operate within contract laws, within regulation, and within other rules and norms that establish rights and the modalities of their allocation, thus framing potential usage of these rights. That we know so little about the exact modalities of interaction between these two levels, the macro level and the micro level, can be considered one more puzzle, and a fundamental one.

The challenge it raises should make it high on our research agenda. Recent contributions submit that one way to capture these interactions is by introducing a third, intermediate layer of institutions within which these interactions develop (Kunneke *et al.*, 2010; Ménard, in press-a). This layer can be identified as the domain of 'meso-institutions'. Meso-institutions can be understood as 'the set of mechanisms and devices through which general rules and rights, established at the macro-level, are translated, interpreted, adapted and implemented, thus framing the domain within which alternative organizational arrangements, the micro-institutions, draw and operate transactions and through which they transmit their expectations and requirements to the macro-level'.

As such, meso-institutions provide the actual infrastructure to the actual organization of transactions. For example, macro-institutional rules regarding agriculture are defined at the level of the European Union through the 'Common Agricultural Policy', thereafter embedded into national laws (the 'subsidiarity' principle). However, these rules are then translated, adapted (possibly through negotiations with national or local organizations), and implemented through specific institutional arrangements. Meso-institutions differ from macro-institutions in that they strictly operate within the general rules defined by the latter. Typical examples are regulatory agencies, public bureaus in charge of certifying quality, institutions of arbitrage or specialized courts monitoring conflicts in contractual agreements, etc. Meso-institutions also differ from the micro-layer at which organizational arrangements operate. Indeed, they do not implement actual transactions: they do not produce and deliver actual goods and services that are inputs to other organizations or that are delivered to consumers. What they do is they delineate the specific playing field within which actual transactions are organized. For example, laws (macro-level) may determine a specific fiscal regime for cooperatives (micro-level), but these laws need being translated (for example they may differ according to the type of activity or to the type of cooperatives) and implemented though a public bureau, a professional organization (e.g. accounting organizations that standardize the way cooperatives must respond to the law), etc.

Figure 1.4 visualizes these different institutional layers, also pointing out that all these layers interact with technology and technological changes, an important dimension that is not explored in this chapter (nor in the other chapters of this book in that respect).

1.4.2 Specifying the key role of meso-institutions

Several contributions have already been published on arrangements that correspond to what above proposes to identify as 'meso-institutions'. However, almost all attention regarding these arrangements has focused so far on the characteristics and role of regulations and regulatory devices (agencies, public bureaus) in charge of their

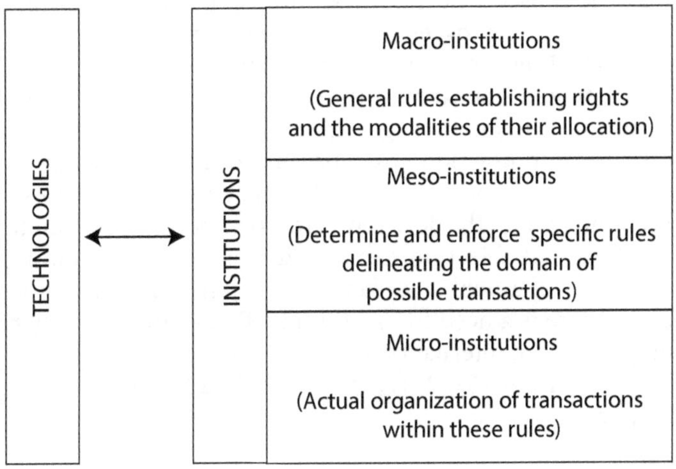

Figure 1.4. Institutional layers (Ménard, in press-a).

implementation (Laffont, 2005, particularly Chapter 7, provides an excellent synthesis). The point I want to make here is that there is more than regulatory agencies to be looked at in order to understand the key role of meso-institutions. Unfortunately, we do not have a general conceptualization of these intermediate mechanisms and devices linking general rules and their actual implementation by actors. For example, what is the theoretical status, beside that of regulatory agencies, of institutional arrangements such as 'interprofessions', which are establishing specific norms and standards; or certifying organizations (whether public, private, or mixed), which are playing an active role in implementing rules of quality established at the macro-level; or public bureaus defining geographical origins within the general framework delineated by the law. With this conceptual issue in mind, let me briefly provide indications on some aspects that I view as deserving special attention in our research agenda.

First, we must find ways to better understand the variety of meso-institutions and the logic behind this diversity. As already mentioned, some of these arrangements are public (e.g. a bureau in the ministry of agriculture; a national or supra-national committee involved in the implementation of a policy); others are private (e.g. an interprofession, a union of producers, a federation of cooperatives); and many are mixed (e.g. certifying organizations, marketing boards). What are the consequences of these different statuses? Can we assess their comparative advantages? For example, when it comes to guaranteeing food safety, what are the advantages and dis7advantages of public certifying organizations operating through command-and-control compared to private ones operating through reputation or to mixed arrangements with representatives of the different stakeholders and operating through consensus building?

Second, what are the consequences of these diverse solutions in terms of transaction costs? Here, the distinction between the two types of transaction costs mentioned in Section 1.3 is relevant. There are economic transaction costs involved in these different institutional arrangements, for example the cost of running a public bureau compared to the costs of running a private arrangement in monitoring quality control. However, there are also political transaction costs, because of the very nature of these meso-institutions, at the intersection between policy-makers and operators. For instance, there are costs coming out of the decision to rely on a committee with representatives of stakeholders (costs of selection process, delays, costs of building a coalition) compared to the costs of a public bureau (costs of accountability, of legitimacy). To assess these costs, there is a need for 'institutional' indicators. A very positive signal is the increasing interest of international organizations in developing such indicators (see for example OECD, 2016a, 2017; World Bank, 2017) but much remains to be done in that respect.

Third, a central dimension of both types of transaction costs relates to the modalities of governance implemented by alternative meso-institutions. In the examples mentioned above, there are issues related to the composition of the board (of an independent agency), of the staff (of a public bureau), or of the representation (of joint committees). The modalities of decision-making, the conditions of implementation of decision made, and their acceptability for operators acting at the micro-level are all important factors and they all involve specific costs. In that respect, it is now quite largely acknowledged among policy-makers that participation of stakeholders is beneficial (OECD, 2016a,b, 2017), but it has also drawbacks, for example what Spiller (2009) and others identified as the risks of third party opportunism. What is the balance between costs and benefits of such arrangements in a democratic regime, e.g. up to what point do they interfere with the responsibilities of elected policy-makers?

Fourth, there is the difficult issue of the overlapping of meso-institutions. Arrangements in charge of interpreting, translating, adapting and implementing general rules into specific ones are unevenly distributed in the vertical hierarchy of institutions. Most of the time, different levels of public decision-making are involved: local authorities, regional or national ones, and increasingly supranational ones. But there are also interferences of different meso-institutions that overlap horizontally because jurisdictions and responsibilities are rarely perfectly defined and assigned. For example, decisions made by a bureau of the Ministry of Agriculture about how to implement the directives of the European Union in the dairy milk industry may well interfere and even be in conflict with the perspective of competition authorities or with the policy of the Ministry of Environment. In sum, the existence of various meso-institutions raises severe problems of coordination, both vertically and horizontally, and the resulting confusion is a source of transaction costs and (often and legitimately) a major source of complaints by operators acting at the micro-institutional level.

Last, but not least, it is necessary to keep in mind that in last resort meso-institutions should ideally be designed so as to establish optimal conditions for an appropriate alignment between organizational choices (e.g. adopting adequate organizational arrangement for guaranteeing food safety at the firm or cooperative level) and rules of the game (e.g. new standard adopted regarding the type and usage of pesticides allowed). Because the possibility and causes of misalignment has been too much neglected so far in the economic literature, adaptations are too often made following misalignments observed *ex-post*, which can significantly increase transaction costs (a good example is provided for fisheries in Haraldsson, 2017).

To sum up, I would argue that we are still in limbo when it comes to understanding the impact of, and the interaction between institutions long assimilated solely to the macro-level at which rules of the game are defined (typically the political and/or legal system) and the organizational level at which transactions are actually made. We need to take on board the more complex setting in which intermediate institutional arrangements play a crucial role in framing, within the context of the general rules and norms, the actual domain and some conditions within which actors and the different organizational arrangements they choose and implement can develop transactions. There is still a long way to go in this direction, which opens exciting perspectives for our research program.

1.5 Conclusions

In this chapter, I have emphasized: (1) the diversity of organizational arrangements in which contracts are embedded, so that we need to dig deeper in the variety of contracts associated to these arrangements; (2) the importance of hybrid arrangements, particularly supply chain systems, their diversity and the key role they play in the context of globalization of economic activities, which is particularly relevant in agriculture; (3) the central status of a specific category of institutions, identified as meso-institutions, that provides the essential intermediation between rules of the game that are too general to be implemented directly, and organizations that are operating within the domain thus delineated; (4) the problems of coordination raised at all these levels, and the role contracts play in that complex environment, an important role for sure, but a limited one, which explains the need to look at their embeddedness into broader governance structures and to look at them in the context of a more general theory.

Although important progress has been made over the recent decades with respect to the nature and role of institutions in economics, a general theory of institutional layers, of which organizational arrangements are an integral part, still remains to be developed. In this chapter, I have emphasized the possible role of a transaction cost approach in that research program. There is of course the need to push the analysis further, which will require the development of new concepts and, ultimately, new theories.

Yes we live in a jungle of organizational arrangements anchored in the broader jungle of institutions. However, a jungle is not as disorganized at it may look and unknown animals are strange only to outsiders' perception. Yes, the knitting of the capitalist fabric is a complex of interwoven institutional layers. However, tools are already available or in development that allow to unfold part of its secrets. What we need above all are young scholars, like those contributing to this book, who are risk-takers and willing to explore new terrains.

References

Acemoglu, D. and Robinson, J., 2012. Why nations fail. The origins of power, prosperity and poverty. Crown Publishers, New York, NY, USA.

Alchian, A., 1965. Some economics of property rights. Il Politico 30(4): 816-819.

Allen, D.W. and Lueck, D., 1999. The role of risk in contract choice. Journal of Law, Economics, and Organization 15: 704-736.

Aoki, M., 2001. Toward a comparative institutional analysis. MIT Press, Cambridge, UK.

Arrunada, B., in press. Property as sequential exchange: the forgotten limits of private contract. Journal of Institutional Economics. DOI: https://doi.org/10.1017/S1744137416000473.

Baker, G., Gibbons, R. and Murphy, J., 2002. Relational contracts and the theory of the firm. Quarterly Journal of Economics 117(1): 39-84.

Baker, G., Gibbons, R. and Murphy, J., 2008. Strategic alliances: bridges between 'islands of conscious power'. Journal of the Japanese and International Economies 22(2): 146-163.

Benabou, R. and Tirole, J., 2011. Identity, morals, and taboos: beliefs as assets. Quarterly Journal of Economics 126: 805-855.

Bolton, P. and Dewatripont, M., 2005. Contract theory. MIT Press, Cambridge, MA, USA.

Bradach, J.L., 1997. Using the plural form in the management of restaurant chains. Administrative Science Quarterly 42: 276-303.

Brandenburger, A.M. and Nalebuff, B.J., 1996. Co-opetititon. Currency/Doubleday, New York, NY, USA.

Carson, S.J., Madhok, A. and Wu, T., 2006. Uncertainty, opportunism, and governance: the effects of volatility and ambiguity on formal and relational contracts. Academy of Management Journal 40(5): 1058-1077.

Coase, R.H., 1937. The nature of the firm. Economica 2(1): 386-405.

Coase, R.H., 1994. Essays on economics and economists. University of Chicago Press, Chicago, IL, USA.

Drescher, K., 2000. Assessing aspects of agricultural contracts: an application to German agriculture. Agribusiness. An International Journal 16(4): 385-398.

Dubois, P., 2001. Contrats agricoles en économie du développement: une revue critique des théories et des tests empiriques, Revue d'Économie du Développement 3: 75-106.

European Commission (EC), 2016. Improving market outcomes: enhancing the position of farmers in the supply chain. Report of the Agricultural Markets Task Force, EC, Brussels, Belgium.

France Ministère de la Transition Écologique et Solidaire (France), 2015. Panorama de l'agriculture. Available at: http://tinyurl.com/y8wl6ezc.

Ghosh, M. and John, G., 1999. Governance value analysis and marketing strategy. Journal of Marketing 63: 131-145.

Gibbons, R. and Henderson, R., 2012. Relational contracts and organizational capabilities. Organization Science 23(5): 1350-1364.

Gibbons, R. and Roberts, J., 2013. The handbook of organizational economics. Princeton University Press, Princeton, NJ, USA.

Gibbons, R., 2005. Four formal(izable) theories of the firm. Journal of Economic Behavior and Organization 58(2): 202-247.

Goldberg, V., 1980. Relational exchange: economics and complex contracts. American Behavioral Scientist 23(3): 337-352.

Goodhue, R., 2000. Broiler production contracts as a multi-agent problem: common risk, incentives and heterogeneity. American Journal of Agricultural Economics 82(3): 606-622.

Greif, A., 2005. Commitment, coercion, and markets: the nature and dynamics of institutions supporting exchange. In: Menard, C. and Shirley, M. (eds.) Handbook of new institutional economics. Springer, New York, NY, USA, pp. 725-786.

Grossmann, S. and Hart, O., 1986. The costs and benefits of ownership: a theory of vertical and lateral integration. Journal of Political Economy 94(4): 691-719.

Haraldsson, G., 2017. The political economy of the ITQ system and resource rent tax in Iceland fisheries. OECD Publishing, Paris, France.

Hart, O. and Moore, J., 1990. Property rights and the nature of the firm. Journal of Political Economy 98(6): 1119-1158.

Hegrenes, A. and Borgen, S.O., 2003. Deregulation and new contractual forms: the case of the Norwegian potato sector. Paper presented at 81st EAAE Seminar. June, 2003. Copenhagen, Denmark.

Hendrikse, G. and Veerman, C.P., 2001. Marketing cooperatives: an incomplete contract perspective. Journal of Agricultural Economics 52(1): 53-64.

Jensen, M. and Meckling, W., 1976. Theory of the firm: managerial behavior, agency costs, and ownership structure. Journal of Financial Economics 3(4): 305-360.

Jiang, T. and Hendrikse, G.W.J., 2011. An incomplete contracting model of dual distribution in franchising. Journal of Retailing 87(3): 332-344.

Karantininis, K. and Nilsson, J., 2007. Vertical markets and cooperative hierarchies. Springer, New York, NY, USA.

Key, N. and McBride, W., 2003. Production contracts and productivity in the U.S. Hog Sector. American Journal of Agricultural Economics 85: 121-133.

Knoeber, C., 1989. A real game of chicken: contracts, tournaments, and the production of broilers. Journal of Law, Economics and Organization 5: 271-292.

Kunneke, R., Groenewegen, J. and Ménard, C., 2010. Aligning modes of organization with technology: critical transactions in the reform of infrastructures. Journal of Economic Behavior and Organization 75(3): 494-505.

Laffont, J.J., 2005. Regulation and Development. Cambridge University Press, Cambridge, UK.

Lafontaine, F. and Slade, M., 2007. Vertical integration and firm boundaries: the evidence. Journal of Economic Literature 45: 629-685.

Lara, V., Ménard, C., Raynaud, E. and Saes, S., 2014. Economia das organizações. Formas plurais e desafios. Atlas, Sao Paulo, Brazil.

Levinson, M., 2006. The box: how the shipping container made the world smaller and the world economy bigger. Princeton University Press, Princeton, NJ, USA.

Lindenberg, S., 1996. Multiple-tie networks, structural dependence and path-dependency: another look at hybrid forms of governance. Journal of Institutional and Theoretical Economics 152(1): 188-196.

MacDonald, J. and Korb, P., 2011. Agricultural contracting update: contracts in 2008. EIB-72. Economic Research Service, USDA, Washington, DC, USA.

MacDonald, J., Perry, J., Ahearn, M., Banker, D., Chambers, W., Dimitri, C., Key, N., Nelson, K. and Southard, L., 2004. Contracts, markets, and prices: organizing the production and use of agricultural commodities. Agricultural Economic Report No. 837. Economic Research Service, USDA, Washington, DC, USA.

MacNeil, I.R., 1974. The many futures of contracts. Southern California Law Review 47: 691-816.

Makadok, R., Coff, R., 2009. Both market and hierarchy: an incentive-systems theory of hybrid governance forms. Academy of Management Review 34(2): 297-319.

Martino, G., Karantininis, K., Pascucci, S., Dries, L. and Codron, J.-M. (eds), 2017. It's a jungle out there – the strange animals of economic organization in agri-food value chains. Wageningen Academic Publishers, Wageningen, the Netherlands, 394pp.

Masten, S.E., Meehan, J.W. and Snyder, E.E., 1991. The costs of organization. Journal of Law, Economics and Organization 7(1): 1-25.

Ménard, C., 1996. On clusters, hybrids and other strange forms: the case of the french poultry industry. Journal of Institutional and Theoretical Economics 152: 154-183.

Ménard, C., 2004. The international library of new institutional economics. Edward Elgar Publisher, Cheltenham, UK.

Ménard, C., 2005. The inadequacy of competition policies: a new institutional approach. In: Oppenheimer, M. and Mercuro, N. (eds.) Law and economics: alternative economic approaches to legal and regulatory issues. Routledge, New York, NY, USA, pp. 27-54.

Ménard, C., 2013a. Hybrid modes of organization. Alliances, joint ventures, networks, and other 'strange' animals. In: Gibbons, R. and Roberts, J. (eds.) The handbook of organizational economics. Princeton University Press, Princeton, NJ, USA, pp. 1066-1108.

Ménard, C., 2013b. Plural forms of organizations: Where do we stand? Managerial and Decision Economics 34(3-5): 124-139.

Ménard, C., in press-a. Meso-institutions: the variety of regulatory arrangements in the water sector. Utilities Policy. DOI: https://doi.org/10.1016/j.jup.2017.05.001.

Ménard, C., in press-b. What approach to property rights? Journal of Institutional Economics DOI: https://doi.org/10.1017/S1744137417000145.

Ménard, C., in press-c. Organization and governance in the agrifood sector: How can we capture their variety? Agribusiness. DOI: https://doi.org/10.1002/agr.21539.

Ménard, C. and Vavra, P., 2008. The role, usage and motivations of contracts in agriculture. OECD Food, Agriculture and Fisheries Working Papers No. 16. OECD Publishing, Paris, France.

Milgrom, P. and Roberts, J., 1988. Economic theories of the firm: past, present, and future. Canadian Journal of Economics 21(3): 444-458.

Monteverde, K.C. and Teece, D.J., 1982. Suppliers switching costs and vertical integration in the automobile industry. Bell Journal of Economics 13(1): 206-213.

North, D.C., 1981. Structure and change in economic history. Norton and Co, New York, NY, USA.

North, D.C., 1990a. Institutions, institutional change and economic performance. Cambridge University Press, Cambridge, UK.

North, D.C., 1990b. A transaction cost theory of politics. Journal of Theoretical Politics 2(4): 355-367.

Organisation for Economic Co-operation and Development (OECD), 2016a. Water governance in cities. OECD, Paris, France. Available at: http://dx.doi.org/10.1787/9789264251090-en.

Organisation for Economic Co-operation and Development (OECD), 2016b. Institutional aspects of governance in fisheries. Five case studies. Report by Claude Ménard of the Trade and Agriculture Directorate. TAD/FISH, Document 2016(5). OECD, Paris, France.

Organisation for Economic Co-operation and Development (OECD), 2017. The political economy of biodiversity policy reform. OECD, Paris, France. Available at: http://tinyurl.com/yc8w3ex2.

Patrick, G.F., Musser, W.N. and Eckman, D.T., 1998. Forward marketing practices and attitudes of large-scale midwestern grain producers. Review of Agricultural Economics 20: 38-53.

Raynaud, E. and Ménard, C., 2017. Ulysses and the sirens: commitment and control in multilateral hybrid organizations. Centre d'Economie de la Sorbonne, Paris, France.

Reimer, J.J., 2006. Vertical integration in the pork industry. American Journal of Agricultural Economics 88(1): 234-248.

Royer, A., 2011. Transaction costs in milk marketing: a comparison between Canada and Great Britain. Agricultural Economics 42(2): 171-182.

Rubin, P., 1978. The theory of the firm and the structure of the franchise contract. Journal of Law and Economics 21(1): 223-233.

Schnaider, P., Ménard, C. and Saes, S., 2017. Heterogeneity of plural forms: a revised transaction cost approach. University of São Paulo, São Paulo, Brazil.

Schroeter, J.R. and Azzam, A., 2004. Captive supplies and cash market prices for fed cattle: the role of delivery timing incentives. Agribusiness 20(4): 347-362.

Sexton, R., 2017. Business-to-business unfair trade practices in the food supply chain: what are they? How common are they? And what should we do about them? Contribution to a symposium on 'Unfair Trading Practices in Supply Chain Systems'. July 17, 2017. European Commission, Brussels, Belgium.

Simon, H., 1951. A formal theory of the employment relationship. Econometrica 19(3): 293-305.

Smith, A., 1776. An inquiry into the nature and causes of the wealth of nations. University of Chicago Press, Chicago, IL, USA (1976).

Spiller, P., 2009. An institutional theory of public contracts: regulatory implications. In: Ménard, C. and Ghertman, M. (eds.) Regulation, deregulation, reregulation: institutional perspectives. Edward Elgar, Cheltenham-Northampton, UK, pp. 45-66.

Stiglitz, J.E., 1974. Incentives and risk sharing in sharecropping. Review of Economic Studies 41: 219-256.

Tirole, J., 1999. Incomplete contracts: where do we stand? Econometrica 67(4): 741-782.

Tsoulouhas, T. and Vukina, T., 2001. Regulating broilers contracts: tournaments versus fixed performance standards. American Journal of Agricultural Economics 83(4): 1062-1073.

United States Department of Agriculture (USDA), 2012. Farms with contracts by farm typology. Economic Research Service, Agricultural Resource Management Survey. USDA, Washington, DC, USA.

United States Department of Agriculture (USDA), 2014. 2012 census. Vol. 1. U.S. national level data. USDA, Washington, DC, USA.

Vavra, P., 2009. Role, usage and motivation for contracting in agriculture. OECD Food, Agriculture and Fisheries Working Papers, No. 16. OECD Publishing, Paris, France.

Wang, Y. and Jaenicke, E.C., 2006. Simulating the impacts of contract supplies in a spot market-contract market equilibrium setting. American Journal of Agricultural Economics 88(4): 1062-1077.

Williamson, O.E., 1975. Markets and hierarchies: analysis and antitrust implications. Free Press, New York, NY, USA.

Williamson, O.E., 1985. The economic institutions of capitalism. Free Press, New York, NY, USA.

Williamson, O.E., 1996. The mechanisms of governance. Oxford University Press, New York, NY, USA.

World Bank, 2017. Doing business: measuring business regulations. Changes to the methodology 2005-2016. World Bank Group, Washington, DC, USA. Available at: http://tinyurl.com/ycwajnd2.

Xia, T. and Sexton, R., 2004. The competitive implications of top of the market pricing and related contract-pricing clauses. American Journal of Agricultural Economics 86: 124-139.

2. Linnaeus in the jungle: configurational lenses for discerning forms of economic organization in agri-business

A. Grandori

Bocconi University, Department of Management and Technology, Via Roberto Sarfatti 25, 20100 Milan, Italy; anna.grandori@unibocconi.it

Abstract

Available frameworks in the theory of economic organization need some extensions for describing, classifying and assessing organization forms in the agri-food sector. Starting from examples of these structural 'anomalies', the paper develops a wider typology of forms, featuring three disentangled dimensions – the allocation of property rights, the organizational coordination mechanisms, and the legal and contractual foundations. Using available studies on 'hybrid' forms, it is shown that the proposed classification systems encompasses a wider variety of forms and permits a more fine-grained comparative assessment among them, which may be useful also beyond the specific sector under analysis.

Keywords: hybrids, organization forms, coordination mechanisms, property rights, contracts, agri-food

2.1 Introduction

The cluster of activities around agriculture is regaining centrality in the economy, and extending its scope in terms of thick relationships with other sectors, not only the traditional up-stream materials and downstream distribution sectors, but also with biotechnologies, food-related services, even entertainment and fashion. Hence the term 'agri-business' has become more appropriate and widely used for indicating this re-emerging set of industries, offering important opportunities for new entrepreneurship and growth.

Together with this regained centrality, new and more complicated problems in the organization of industry surfaced. Quality, traceability and certification along increasingly international chains are difficult and important problems. Traditional organizational arrangements are frequently ineffective and unfair, due to different industry concentration rates in different stages – with the fragmented production-farming stage typically 'squeezed' between the more concentrated stages of chemical/ raw material/ input providers on one side and distribution on the other side.

The design of an effective 'organization of industry' (Richardson, 1972) is therefore of great relevance. At the same time such a task pose challenges to conventional approaches to economic organization design, which have been mostly developed thinking at problems of vertical integration in industrial sectors. The agri-business field is actually astonishingly rich of economic and organizational architectures which are difficult to classify, let alone to assess and improve, using the classic conceptualization schemes as they are.

This paper therefore addresses the task of constructing an extended classification and assessment framework, able to capture the variety of forms observed in agri-business. The framework so produced is an extension and integration of currently available organizational and economic perspectives, such as transaction cost economics, property right theory, and organization design theory. It takes into account transaction cost issues, but also knowledge governance and pooled interdependence issues. Hence, the exercise should be fruitful both in the re-classification and assessment of forms of economic organization in agri-business and in extending theory, with possible applications in other sectors as well.

2.2 'Monsters' in the jungle

'This geometrical figure is a 'monster'! – exclaimed one participant in an imaginary symposium (devised by the great epistemologist Imre Lakatos, 1976), reproducing real debates occurred in the development of thought in mathematics and geometry. What is called a 'monster' is something not 'normal': an 'unclassifiable object', an 'anomaly' with respect to current concepts. But these objects are most helpful in testing and revising the accepted definitions. Hence, let's start with a list of 'monsters' that we are likely to encounter while walking into the agri-business organization jungle.

Some salient anomalies in the agri-business sectors have been already indicated (Grandori, 2015; Ménard, 1996). Let us clarify here why and how they may contribute in developing new dimensions for classifying forms of economic organization.

2.2.1 Non-hierarchical firms

After the rethorically successful opposition coined by Williamson (1975) it has become common to define 'firms' as 'hierarchies', intended as 'classic firms' (Alchian and Demsetz, 1972), characterized by property rights held by only one type of investors (of financial assets) and internal coordination by authority relations. However, the opposition is more rethorically catchy than analytically precise. In fact, in Williamson's papers and other related works, non-hierarchical firms were actually also admitted and analysed. An alternative form of enterprise, is the 'peer group'-like 'collective firm' – exemplified especially by worker cooperatives (Hart and Moore, 1990; Williamson, 1980). Being particularly diffused in the organization of farming activity, cooperative firms are particularly relevant for the present analysis. In addition, as a 'discrete institutional alternative' to market (Simon, 1978), a 'firm' is more proper comparative notion than a 'hierarchy'. Hierarchy is in fact a rather diffused coordination mechanism, that can be and is applied in many systems – from churches (where it comes from: ierai- archein in fact means 'the order of sacred things'), to political parties, to families, to enterprises no matter what their ownership structure is. For example, in cooperative firms, in spite of the sharing of property rights among the workers-owners, hierarchy may be employed as a coordination mechanism especially as firm size grows (Jardim *et al.*, 2013). Conversely, non-hierarchical mechanisms of coordination may be predominant in various types of so called investor-owned firms. Other important alternative coordination mechanisms are the programming and the routinization of activities and knowledge, in stable and known conditions (March and Simon, 1958; Mintzberg, 1979); horizontal coordination via joint team decision making and mutual adjustment in innovative, complex tasks (Burns and Stalker, 1961; Thompson, 1967) and modularized autonomous self-contained units coordinated via the price-like mechanisms of incentives and structured information, in large systems operating in variable environments (Zenger and Hesterley, 1997).

This discussion may already suffice to indicate that the ownership structure of the firm and the type of coordination mechanism used to regulate activities are two different (potentially orthogonal) dimensions. Disentangling them, and reconsidering what the different possible combinations among them are, would be useful for explaining a wider variety of organization forms; especially, albeit not exclusively, in agri-business.

2.2.2 Hierarchical contracts

In addition to be applicable in any type of system, not just in firms, as well in firms with any type of ownership structure, hierarchical coordination can actually be applied among firms and incorporated into inter-firm contracts, as it is actually recognized since a long time (Stinchcombe, 1990). For example, a partnership contract can not only indicate a private arbitrator in case of dispute, but can also stipulate that one firm will have a coordination role, or can establish a central committee vested with authority (i.e. legitimate decision right over the selection of activities to be performed, and/or of control rights over execution). Williamson himself (1991) noticed that hierarchy can be seen as a party-specific, rather than general, court of last resort. That argument can however be applied to inter-organizational contracts as well, that can and do often establish such party-specific courts of arbitration.

This line of argument suggests that another disentanglement of concepts would be useful in further expanding the variety of organization forms that might be interpreted and assessed: separating the notion of contract from that of any particular coordination mechanism. In fact, not only authority relations, but almost any other possible coordination mechanism can be agreed upon and established by contracts among different entities: not only and not even necessarily prices and terms of exchange, but also joint programs, inter-unit committees, liason roles, connected information systems (Grandori and Soda, 1995) and redetermination and renegotiation clauses of all the above (Crocker and Masten, 1991).

2.2.3 Hybrids between markets and democracies

The current characterization of inter-firm networks as 'hybrids' (Ménard, 2004; Williamson, 1991) would also gain in explanatory power, if the continuum of intermediate forms is re-reconceived as originally it was (e.g. Coase, 1937; Richardson, 1972): a polarity between markets and firms, where firms are not necessarily hierarchies. Logically, it would be more coherent, as a market is a complex multi-mechanism institution which might be contrasted with another complex multi-mechanism institution as a firm, not with a single coordination mechanism as hierarchy. But even more importantly for the present argument, extending the pole of firms to a generic integrated entity (that can be coordinated either vertically by hierarchy or horizontally in a peer group mode) is already sufficient to enrich and extend the notion of hybrids significantly: from being limited to a variety of transactional contracts (more or less enriched and relational), to include the full variety of associational, resource pooling contracts among peers (Grandori, 1997b).

Again this extension is particularly important for understanding the organization of the agri-food industry, as these horizontal or associational hybrids are quite common in that field. GIEs ('Groupments d'Interèt Economique'), associations of farmers, and consortia for DOP/DOC products are examples. They most often establish a common central structure, which can be endowed with legal personality, and they may sometimes even own some common assets. The members are legally independent firms, but they may sometimes share gains.

Consortia, in many sectors and types of activity including agri-business, also typically blend a representative democracy decision structure (general assemblies, one-firm-one-vote rights, elected boards) with gain sharing schemes sensitive to the amount of the resources provided: for example, Airlines Consortia (Ménard and Damergy, 2013); R&D consortia (Aldrich and Sasaki, 1995; Ouchi and Bolton, 1988); construction and engineering consortia (Grandori, 1997a); some hotel chains, as Best Western (Grandori and Gaillard, 2011); consortia for joint production, procurement or distribution in agri-business (Berardino, 2013; Martino, 2016).

The puzzling feature of those arrangements with respect to the market and hierarchy continuum is their team-like, 'inter-organizational democracy' (Lammers, 1993), governance structure. Therefore, if hybrids, they may be more properly conceived as hybrids between markets and democracies (Grandori, 1997a) rather than between markets and hierarchies.

2.2.4 'Networks' that are actually firms

Agriculture is particularly rich in cooperative firms, not only at the level of basic production units, but also at the higher level of groups of firms. In particular farmers, individually structured in cooperatives or entrepreneurial firms, often associate into a 'second-tier' cooperative firm, who is able to integrate activities along the value chain – i.e. with downstream food processing and distribution firms (Hendrikse, 2013) and/or with up-stream input (e.g. seeds, chemicals) providers, or with other related activities (as R&D) (Grandori, 2015).

Even if it seems obvious, it is worthwhile pointing that that when an agreement among a set of firms implies the establishment of a new firm…well, the resulting structure is a firm, not a hybrid or a network. The observation is worthwhile because, surprisingly, firms constituted by multiple firms as partners are often misclassified as networks, generating confusion about their governance properties. This confusion occurred about joint venture firms, as the leading transaction cost economist of joint ventures have repeatedly pointed out (Hennart, 2013). And it occurred also for cooperative firms whose members are firms rather than individuals (Grandori, 2015).[1]

To the purpose of classifying a wider variety of governance forms, though, it is also worth noticing that the sharing of property rights is a matter of degree. Before culminating in the establishment of that legal entity endowed with full property rights called a firm, there are proprietary agreements sharing some but not all property

[1] To say that entities that legally are firms are actually inter-firm networks or hybrids is so strange that one may perhaps provocatively speculate that a reason why these multi-tiered firms are misclassified as networks lies somewhere in the identification between firms and hierarchies, so that when firm governance is multi-actor and peer based it is concluded that the entity is not a firm…(?!).

rights and, to the extent this sharing does not imply the constitution of a new firm, they can still be classified as inter-firm networks (Daems, 1983). For example, there are forms of association (as the 'association in participation' under Italian law, born and used precisely in the agri-food sector, such as the production of milk) under which one party provides a technical asset or tool to another firm in exchange of becoming residual claimant to a share of profits. Or there are federations of entrepreneurs sharing the ownership and right of use of an asset (such a brand of cheese, or a mill or a piece of land) but maintaining all gains separate.

In conclusion, the general lesson that can be learned by encountering and studying all these 'monsters' is that there is a need for disentangling the dimensions we use for classifying them. Multiple mechanisms can co-occur to define a much wider variety of internal and external governance and organizational configurations than commonly admitted (Grandori, 1997a). Here, we attempt to work out a way of 'dimensionalizing governance' useful for classifying and assessing the wide variety of governance forms present in agri-business. As in Williamson (1991), we use the term governance as a broad concept encompassing a variety of mechanisms, including the allocation of all classes of property rights, the type of contract used, and the key coordination mechanisms employed. Precisely for this breadth, we propose that these different components of governance should be analysed as conceptually distinct dimensions, and their possible combinations should not be taken for granted, but theoretically and empirically explored. Once governance forms have been reconstructed in this way, an exercise in comparatively assessing their performance properties can be conducted over a wider set of more fine-grained 'structural alternatives'.

2.3 From one to many *continua*: an extended classification system of governance forms

Research on hybrid forms has taken some steps in identifying sub-types or forms within that vast category. The usual definitions refer to forms as they became institutionalized in practice and law: the 'objects' are things like licensing, sub-contracting, franchising, consortia, joint ventures – roughly ordered from more market-like to more firm-like form of economic organization. This ordering, however, at a closer scrutiny, is quite imprecise. Consider the following empirical results on the content and shape of a variety of those hybrid forms in a variety of sectors.

- Licencing is usually considered a form of hybrid relatively close to the market pole of the continuum. As a counter-example, consider licencing contract in high tech and R&D: they often include property rights over the licensee firm hold by the licensor institution (Jensen and Thursby, 2001). Such licencing contracts, then, are closer to an equity alliance, that is something closer to the integrated firm pole than to the market pole of the traditional continuum. In outsourced R&D, 'property right sharing' – typically the main contractor holding stock in research firms – is more the rule than the exception (Carson and John, 2013).

- Sub-contracting has been initially conceived as a combination between market contracting and hierarchical coordination that have been diffused in textiles in early times, as 'putting out' systems (Williamson, 1980), and reappeared in the modern organization of fashion, construction, automotive and electronics industries. In those settings, in addition to traditional vertical forms, horizontally organized, co-makership relations were found out (Dyer and Chu, 2003; Eccles, 1981; Suarez-Villa and Rama, 1996). In agri-business, a diffused form, arguably similar to sub-contracting, is 'contract farming'. However, also in this case, quite different real arrangements may be at work behind the legal form. The contract farming legal notion prospects a collaborative, decision right sharing notion, and some realities may well conform to it. However, in practice, contract farming often reduces to very asymmetrical arrangements, similar to the 'squeezing sub-contracting' found also in other industries (as automotive). For example, a case described in Grandori (2015) reports that the growers of asparagus in Australia lament that buyers, to whom they are linked by contract farming agreements, have too many control rights on quality, and can easily manipulate quality assessment, adversely affecting their income.; that the buyers do not provide the technical assistance and know-how transfers that may sustain quality; and that prices are determined too unilaterally and communicated too late. Farmers also typically lament the absence of renegotiaton clauses, with the consequence that they bear excessive risk for natural exogenous events and variations (Cembalo *et al.*, 2014).
- Franchising contracts can also range from quasi-market licenses of using a brand against a fee, to quasi-firm arrangements including property right sharing clauses, defining how to share residual reward and decision rights, asset ownership and other ownership 'surrogates' (e.g. exclusivity clauses) (Windsperger, 2013).
- Joint Venture and Consortia are respectively considered a type of hybrid. However, there are variants that cannot be claimed to have the same governance properties at all, as they can range from thin joint action contracts (e.g. Joint venture contracts used by drilling companies for joint bidding, or consortia used only for common branding and DOP/DOC functions) to the constitution of a common company as a joint venture firm or a consortiated company (Grandori, 1997b).

These results indicate that each and every type of inter-firm agreement usually considered as 'a form' of network (e.g. licencing, sub-contracting, franchising, consortia, joint venture contracts, etc.), actually encompasses many forms that are likely to have different governance properties, as they range from market-like to firm-like arrangements. In other terms, the legal/institutionalized form adopted in a hybrid does not qualify its organization form, as much as the legal/institutionalized form adopted by a firm does not qualify its internal organization (e.g. a corporation, or a LLC, can design its internal organization in many ways). Legal forms set some constraint but a lot can still vary, including property right allocations and coordination mechanisms. Therefore, one single continuum for ordering those forms is too narrow a cage, subject to many counterexamples (unclassifiable objects).

A possible response is to envisage many *continua*. For example, for each juridical or otherwise institutionalized form of hybrid (say franchising or consortia) we can specify a spectrum of forms positioned on a continuum ranging from an integrated entity to a set of disconnected entities. However, this linear framework would be unable to capture all the qualitatively different combinations among the modes of connecting the entities that we have shown to exist.

Hence, we opt for another possible response: to leave aside how an arrangement is juridically defined (something that by the way varies across countries) and to construct a typology of forms based only on the dimensions of governance structures. According to the argument conducted so far, at least three dimensions of governance deserve to be distinguished: the type of contract, the allocation of property rights and the organizational coordination mechanisms employed. This approach will also help, as it is going to be shown, to evaluate when which combination of these elements is going be effective, as the efficiency and effectiveness of different types of contract, organization mechanism, and ownership structure under different circumstances have been widely studied.

As a first step, the profile of any real arrangement can be empirically studied and positioned on the three dimensions through the following types of operational question.

- Property rights
  - Which property rights (on inputs and/or on outputs; residual rewards; residual decision rights) are shared or separated? To what extent?
  - If shared, are they distributed among different actors (physical or juridical persons) or concentrated ? To what extent?
  - Are those partners all of the same kind (e.g. all farmers, all financial investors, all professionals) or of different kind?
  - Do different types of partners hold different types of property rights, i.e. are property rights 'unbundled'?
- Coordination mechanisms
  - Are prices and fees present? Procedures for (re-)determining action and terms of exchange or association?
  - Is a common administrative apparatus/structure for ad hoc decision, control and coordination instituted?
  - Is coordination entrusted to a central actor or performed multilaterally by actors through mutual adjustment, eventually through representatives?
- Types of contract
  - Is the arrangement based on (a set of) bilateral contracts of exchange? Or on a multilateral contract of association?
  - Does the arrangement establish an independent legal entity, like an association or a company?
  - Are the parties of the contract physical persons, or firms (first-tier, second-tier, n-tiered contract)?
  - Is the contract informal and self-enforcing ('relational') or formal and rich of enforceable clauses ('obligational')?

The result would be a three-dimensional typology of the type represented in Figure 2.1.

With such lenses it is possible to classify many familiar types of agreements as particular cases, but also to conceive possibilities and forms that are usually not seen for lack of appropriate glasses. Let's illustrate both familiar and unfamiliar but possibly interesting examples.

The south-west corner is where classic market governance would fall. That's the familiar form. However, a market is not the only type of governance systems based on autonomous decentralized decisions coordinated by codified information. If not only transactional but also pooled interdependence situations are considered, we can also see crowds (in the sense of sets of independent actors pooled together by some purpose or constraint, as in crowd-sourcing) and 'commons' (in the sense of sets of actors in a commensality relation, using the same resources) as alternatives here. From a governance point of view the important features are that each actor is a separate entity and decides autonomously, using just a common language and information system (like the price system, or a website for casting votes or micro-investments as in crowd-funding). Those features also help in clarifying the possible independent variables or requirements for the viability of those arrangements: units should be separable (not linked by complex technical and task interdependence), they should hold clear knowledge locally of what the alternatives are, and information for aggregating choices should be codifiable.

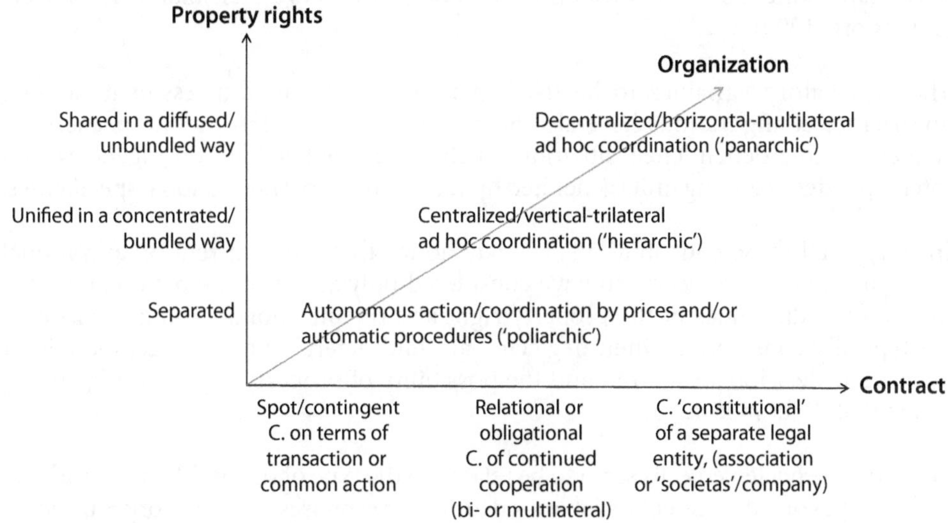

Figure 2.1. From one to three continua: a configurational view of governance forms.

In a general abstract way, Sah and Stiglitz (1988) define this class of decision architectures 'poliarchies', and notice that markets 'roughly correspond' to this archetype, but other real institutions may too. Using the combinative framework for example, we may ask: can a poliarchic mode of coordination be applied also within a single entity, like a company? Or in combination with relational or obligational contracting? If we conceive such unfamiliar alternatives, and look for them, we can find examples. An instance of 'poliarchic companies' are large 'disaggregated' firms, composed by 'molecular units' deciding autonomously and coordinated by 'highly powered incentives' (Zenger and Hesterley, 1997) or just by overarching rules (Hedlund, 1986) (that in fact have sometimes been called 'heterarchies'). Or a licencing contract including only fees for a right of using a brand name (like producing accessories, or now wines, marked with the name of a fashion house) but nothing else may be an example of a 'poliarchic contract' (rather than of a 'hierarchical contract' in which the production is also coordinated by one party – in the example, the fashion house).

Back to the familiar forms, let's observe that the focus of current organizational economics has typically been on comparisons between the base value of each axis to its intermediate value: e.g. between bilateral spot contracting and more continuous relational contracts; between autonomous 'poliarchic' coordination and 'trilateral governance' / 'hierarchical' coordination among and within firms; and between separate and unified property rights (where 'unified' ownership is intended as meaning that 'one party owns the means of production' (Williamson, 1980); and 'one party', if not intended strictly as 'one person', is at least conceived as a set of persons of the same kind, i.e. all financial investors, or all workers) (Hansmann, 1996; Hart and Moore, 1990).

The explanatory variables to be used in those comparative assessments among governance configurations have also been clarified in that tradition, especially for sequential interdependence situation, such as unpredictable contingencies and interdependence among units generated by various types of transactional specificities.

In spite of all those contributions, we may notice that most current organizational economic approaches seems to have considered only a part of the picture: only the low and middle values of the property right and organizational arrangements axes are typically considered. Often neglected are multilateral contracts, democratic or 'panarchic' decision structures, and the possibility of property right sharing among actors that differ in kind.

Why so? The neglected cells seem to be related to the type of the problems considered and the types of independent variable used to devise responses. The main organizational economic models have been developed as a response to problems of asset specificity and costs of writing and enforcing contracts across stages of production. Hence, transactions, i.e. exchanges, i.e. sequential interdependence is the focus, and how much it is complicated by uncertain contingencies and specific investments the core independent variables. What remains out of sight in such a frame are conditions of uncertainty calling for knowledge generation and validation rather than just for adaptation (Grandori, 2013); and conditions of pooled rather than sequential interdependence (Ostrom, 1990; Vanberg, 1994).

Organization theory, on this ground, is perfectly complementary, since it has been focusing a lot on what types of coordination mechanisms may improve the quality and innovativeness of solutions, rather than just how costly those coordination mechanisms are (e.g. Van der Ven *et al.*, 1976). Of course a system in which each actor decides for himself, or one in which one actor decides for everybody, are always less costly than any form of joint decision making, but no one would conclude, to put it provocatively, that therefore anarchy or dictatorship are always better than democracy.... It is also a truism that if everybody entitled to make decision have similar interests and knowledge the cost of decision is lower (another often mentioned reason given in organizational economics for not 'unbundling' decision rights and assigning them to different types of actors) (Hansmann, 1996). The problem is that single actors and highly homogenous teams are much more likely to be in error and to produce low quality decision in any problem of significant complexity. In fact, both in theory and practice, decision bodies composed by actors bringing differentiated knowledge and interests at one table are supposed to be necessary when problems are complex and the risk of making mistakes is high.

Are there examples of arrangements featuring property rights shared and diffused among different types of actors, and by a team-like, horizontal, democratic coordination mechanisms? For the sake of terminological consistency, next to 'poliarchy' and 'hierarchy', this third abstract coordination mode could be and has been called a 'panarchy', i.e. the governance of the many together (De Bellis, 2009; Grandori, 2013).

Let's give some descriptive evidence in the agri-food sector of governance structures that may be located in the property right sharing, panarchic coordination cells of Figure 2.1. Consider the following cases.

Ménard and Reynaud (2009) described the interesting hybrid governance structure of a mill chain, born as a GIE among flour producers (mills) in France, and later become a corporation. The governance structure of that corporation though, combined residual reward rights distributed according to shares, with a one-mill-one vote association-like (and cooperative-like) representative democracy decision structure. In addition, the associated farmers expanded in the distribution stage through another 'strange' hybrid: the associated mills own the central firm, the central firm owns the brand, the central firm acts as a franchisor of the brand, and reaches hundreds of bakeries (franchisees) via Tier-1 master franchises who are the owners-mills themselves.

Another, even 'stranger' entity is the Mercato Metropolitano (MM) of Milan. Established in a dismissed urban area, obtained from the Municipality thank to the

EXPO opening for food-related initiatives, by an entrepreneur with former activities in the agri-food sector and high environmental sensibility. The 'Market' is a life and entertainment centre, with reminiscences of ancient markets, where food is entirely coming from small farmers and producers of Italian excellence. They are pooled together in an Association in which they are the 'socii' and principals, providing agricultural items. The MM Brand is instead owned by a firm registered in London, with subsidiary in Milan, founded by an entrepreneur, which gradually expanded through associating a team of partners (about 20), who are shareholders and at the same time the main workers, providing know how, assistance, and their skills and capabilities in initiatives such as the Milano market, with prospects of starting other projects in other towns around the world based on mobilizing local craftsmanship or other small high quality businesses in need of association and outlet. The link between MM and the Associated producers is not a simple sale contract for the certified, high quality goods sent to the market; but is going to be a profit sharing one, whereas, in addition to the price received *ex-ante* for the material, they are entitled to a share of revenues according to how much is sold to the public and used for restaurant activities in the market.

The governance of those initiatives is interestingly 'hybrid within hybrids' (to use an interesting expression coined by Ménard and Damergy, 2013). In fact, more than one governance configuration (more than one Cell in our Figure) should be used for characterizing the overall governance solution. In the MM case, first of all two types of 'societas' are established: an association of land owners and agri-food good providers, and a company of human capital providers. Both are peer-group like and democratic in their internal decision and ownership structure. The two 'societas' are further linked together by contractual agreements involving some further property right sharing, but not any further entity. The actors figuring as 'socii' are differentiated in kind: financial capital has been raised at the outset in the typical entrepreneurial way (the '3 F': family, friends and fools) and a few other sources; financial assets in any case played a minor role with respect to the technical resources and materials from agriculture, the human resources from the entrepreneurial team, and the urban area provided by the Municipality of Milan. All these resource providers are represented in the ruling boards (as it also typically occurs in innovative entrepreneurial firms).

In the French mill chain, under the early GIE form, the ownership of assets was separated, but that type of association contract did institute a common legal entity (the 'Groupment') and the decision structure was democratic. When the ownership of assets became shared under a corporate form, residual reward rights became proportional to the value of conferred assets, but decision rights remained allocated as in the association (one firm one vote) and coordination remained based on joint decision making. At the same time, an obligational contracting form of franchising was employed to govern the relation with distributors, and an associational contracting form of franchising constituted a first-tier belt of 'agents' who are in fact 'the principals' themselves (the mills).

Those nexuses of contracts are by no means all 'market-like' (as in the famous Jensen and Meckling's notion): some are contracts of exchange, some are contracts of association and some are company establishing contracts. They can be argued to follow effectively the different features of the interdependence relations, the criticality of the various assets, and the complexity of the various tasks to be performed, thereby contributing to explain the impressive performances observed in those experiences.

2.4 Summary and conclusions

This paper started from the observation that the agri-business field is rich in economic and organizational architectures which are difficult to classify, let alone to comparatively assess and improve, using the conceptualization schemes of institutional and organizational economics as they currently are. Analysing a series of counter-examples and researches on those 'strange forms' of organizing in agri-business, the paper then proposes to extend those frameworks. In particular, the usually conceived continuum between 'markets' and 'hierarchies' is expanded and disentangled into a three-dimensional classification framework. Those three dimensions of governance are: the type of contract, the allocation of property rights and the organizational coordination mechanisms employed. The paper then shows how this extended framework helps in characterizing and evaluating the properties of many strange forms otherwise difficult to interpret, such as organizational democracies, hierarchical and poliarchical contracts, property rights sharing licencing and multi-tiered governance structures that make use of even more than one combination of the three dimensions for governing different relations.

In perspective, therefore, if a number of cases can be made available including descriptions of the main three attributes illustrated above, configurational analysis could be performed and yield interesting results on the possibility (or impossibility) of combinations among dimensions, the relative diffusion of those combinations, and their effectiveness in relation to different types of conditions and results. At the very least, a repertory of forms and a reasoned map for figuring out a path in the jungle of forms in the agri-food sector can be constructed.

References

Alchian, A.A. and Demsetz, H., 1972. Production, information costs and economic organization. American Economic Review 62: 777-795.

Aldrich, H. and Sasaki, T., 1995. R&D consortia in the United States and Japan. Research Policy 24: 301-316.

Berardino, C., 2013. Le reti di impresa nell'agribusiness: caratteristiche, finalità e limiti delle reti nel settore oleario. MSc-thesis, Bocconi University, Milan, Italy.

Burns, T. and Stalker, G.M., 1961. The management of innovation. Tavistock Publications, London, UK, 269 pp.

Carson, S.J. and John, G., 2013. A theoretical and empirical investigation of property rights sharing in outsourced research, development and engineering relationships. Strategic Management Journal 34: 1065-1085.

Cembalo, L., Pascucci, S., Tagliafierro, C. and Caracciolo, F., 2014. Development and management of a bio-energy supply chain through contract farming. International Food and Agribusiness Management Review 17(3): 33-51.

Coase, R., 1937. The nature of the firm. Economica 4: 386-405.

Crocker, K.J. and Masten, S.E., 1991. Pretia ex machina? Prices and process in long-term contracts. Journal of Law and Economics 34: 69-99.

Daems, H., 1983. The determinant of hierarchical organization of industry. In: Francis, A., Turk, J. and Willman, P. (eds.) Power efficiency and institutions. Heinemann, London, UK, pp. 35-53.

De Bellis, G.P., 2009. On panarchy: a brief review and a personal view. Available at: http://tinyurl.com/3ey3ym7.

Dyer, J.H. and Chu, W., 2003. The role of trustworthiness in reducing transaction costs and improving performance: empirical evidence from the United States, Japan, and Korea. Organization Science 14: 57-68.

Eccles, R.J., 1981. The quasi firm in the construction industry. Journal of Economic Behaviour and Organizations 2: 335-357.

Grandori, A., 1997a. Governance structures, coordination mechanisms and cognitive models. Journal of Management and Governance 1: 29-47.

Grandori, A., 1997b. An organizational assessment of interfirm coordination modes. Organization Studies 18: 897-925.

Grandori, A., 2013. Epistemic economics and organization. Forms of rationality and governance for a wiser economy. Routledge, London, UK, 156 pp.

Grandori, A., 2015. Improving organization forms in the agri-food industry. British Food Journal 117: 1-18.

Grandori, A. and Gaillard Giordani, L., 2011. Organizing entrepreneurship. Routledge, London, UK, 320 pp.

Grandori, A. and Soda, G., 1995. Inter-firm networks: antecedents, mechanisms and forms. Organization Studies 16: 183-214.

Hansmann, H., 1996. The ownership of enterprise. Harvard University Press, Cambridge, MA, USA, 384 pp.

Hart, O. and Moore, J., 1990. Property rights and the nature of the firm. Journal of Political Economy 98: 1119-1158.

Hedlund, G., 1986. The hypermodern corporation: a heterarchy? Human Resource Management 25: 9-35.

Hendrikse, G. and Feng, L., 2013. Inter-firm cooperatives. In: Grandori, A. (ed.) Handbook of economic organization. Integrating economic and organization theory. Edward Elgar, London, UK, pp. 501-520.

Hennart, J.F., 2013. Internal and external hybrids and the nature of joint ventures. In: Grandori, A. (ed.) Handbook of economic organization. Integrating economic and organization theory. Edward Elgar, London, UK, pp. 481-499.

Jardim, G.F., Saes, M.S. and Ferraz de Mesquita, L., 2013. Governance structures and innovation: a case of the Brazilian coffee roasting and grinding industry. International Conference on Economics and Management EMNet Conference. November 21-23, 2013. Agadir, Morocco.

Jensen, R. and Thursby, M., 2001. Proofs and prototypes for sale: the licensing of university inventions. American Economic Review 91: 240-259.

Lakatos, I., 1976. Proofs and refutations. In: Worral, J. and Currie, G. (eds.) The logic of mathematical discovery. Cambridge University Press, Cambridge, UK, 183 pp.

Lammers, C., 1993. Interorganizational Ddmocracy. In Lindenberg, S. and Schreuder, H. (eds.) Interdisciplinary perspectives on organization studies. Pergamon Press, Oxford, UK, pp. 323-337.

Lerner, J. and Merges, R.P., 1998. The Control of Technology Alliances: an Empirical Analysis of the Biotechnology Industry. Journal of Industrial Economics 46: 125-156.

March, J.G. and Simon, H.A., 1958. Organizations. Wiley, New York, NY, USA, 262 pp.

Martino, G., Rossetti, E. and Giacchè, G., 2016. Organizing the co-production of health and environmental values in food production: the constitutional processes in the relationships between Italian solidarity purchasing groups and farmers. Sustainability 8: 316.

Ménard, C., 1996. On clusters, hybrids, and other strange forms. The case of the French poultry industry. Journal of Institutional and Theoretical Economics 152: 154-183.

Ménard, C., 2004. The economics of hybrid organizations. Journal of Institutional and Theoretical Economics 160: 345-376.

Ménard, C. and Damergy, G., 2013. Hybrids within hybrids. International Conference on Economics and Management EMNet Conference. November 21-23, 2013. Agadir, Morocco.

Ménard, C. and Raynaud, E., 2009 Ulysses and the sirens: relational commitment and control in hybrid organizations. 13th Annual Conference of the International Society for New Institutional Economics. June 18-20, 2009. Berkeley, CA, USA.

Mintzberg, H., 1979. The structuring of organizations. Prentice-Hall, Englewood Cliffs, NJ, USA, 509 pp.

Ostrom, E., 1990. Governing the commons: the evolution of institutions for collective action. Cambridge University Press, Cambridge, UK, 294 pp.

Ouchi, W.G. and Bolton, M.K., 1988. The logic of joint research and development. California Management Review 30: 9-33.

Richardson, G.B., 1972, The organization of industry. Economic Journal 82: 883-896.

Sah, R.K. and Stiglitz, J.E., 1988. Committees, hierarchies and polyarchies. Economic Journal 98: 451-470.

Simon, H.A., 1978. Rationality as process and as product of thought. American Economic Review 68: 1-16.

Stinchcombe, A.L., 1990. Organizing information outside the firm: contracts as hierarchical documents. In: Stinchcombe, A. (ed.) Information and organizations. University of California Press, Berkeley, CA, USA, pp. 194-235.

Suarez-Villa, L. and Rama, R., 1996. Outsourcing, R&D and the pattern of intra-metropolitan location: the electronics industries of Madrid. Urban Studies 33: 1155-1197.

Thompson, J.D., 1967. Organization in action. McGraw-Hill, New York, NY, USA, 192 pp.

Van der Ven, A.H., Delbecq, A.L. and Koenig, R., 1976. Determinants of coordination modes within organization, American Sociology Review 41: 322-338.

Vanberg, V.J., 1994. Rules and choice in economics. Routledge, London, UK, 482 pp.

Williamson, O.E., 1975. Markets and hierarchies: analysis and antitrust implications. Free Press, New York, NY, USA, 286 pp.

Williamson, O.E., 1980. The organization of work. A comparative institutional assessment. Journal of Economics Behavior and Organization 1: 5-38.
Williamson, O.E., 1991. Comparative economic organization: the analysis of discrete structural alternatives. Administrative Science Quarterly 36: 269-296.
Windsperger, J., 2013. The governance of franchising networks. In: Grandori, A. (ed.) Handbook of economic organization. Integrating economic and organization theory. Edward Elgar, London, UK, 522-538.
Zenger, T.R. and Hesterley, W.S., 1997. The disaggregation of corporations: selective intervention, high powered incentives and molecular units. Organization Science 8: 209-222.

3. A three-tiered approach to the economics of hybrids: an application to the agricultural innovation sector

J.E. Hobbs

Department of Agricultural and Resource Economics, University of Saskatchewan, 51 Campus Drive, Saskatoon, SK, S7N 5A8, Canada; jill.hobbs@usask.ca

Abstract

Occupying the middle ground between spot markets and full vertical integration, lies a diverse array of inter-firm relationships and organisational forms in the agri-food sector, so-called 'hybrids', from informal trust-based relational contracts, to explicit contracts, cross-licencing agreements, strategic alliances, networks, clusters, franchising agreements and joint ventures. The chapter offers reflections on the economics of hybrids and the challenges inherent in categorising these organisational forms into a workable typology as the basis for explaining their adoption and evolution. A three-tiered approach to the analysis of hybrid modes of organisation is discussed: analysis which examines the influence of the external (institutional) environment; analysis which examines the inter-firm environment, drawing upon insights from transaction cost economics, agency theory, and the property rights approach; and analysis which considers the intra-firm (managerial) environment, including insights from the resource-based view of the firm and elements of strategic management theory. The use of hybrid modes of organisation to co-ordinate inter-firm relationships in the agricultural biotechnology innovation sector is discussed as an illustrative example of the three-tiered analytical approach. Three challenges are identified for future research: the sheer diversity of hybrid organisational forms; the development of theoretical models of organisational forms amenable to empirical validation; and moving beyond static models to a dynamic perspective on the adoption, development and diffusion of hybrid organisational forms.

Keywords: institutions, transaction costs, agency theory, property rights, biotechnology

3.1 Introduction

Supply chains in the agri-food sector are characterized by a complex array of organisational forms and inter-firm linkages. Spot market transactions lie at one bookend of the vertical co-ordination spectrum, while vertically integrated firms (within-firm transactions) occupy the other end. Much of the initial literature in this field focused on these two alternatives, beginning with the rationale for the existence of a firm first espoused by Coase (1937) in his seminal article, and continuing with extensive literature in economics, management and organisational sciences examining a firm's 'make-or-buy' decision. Since that time there has been widespread recognition of the diversity of organisational forms that lie between the bookends of spot markets and within-firm transactions, and extensive analysis has delved into the determinants, organisation and outcomes of so-called 'hybrid forms', from relational contracts with an emphasis on informal trust-based relationships, to formal written contracts of various types, strategic alliances, networks, clusters, franchising agreements, licensing arrangements, joint ventures, and many more.

This chapter offers reflections on the diversity of organisational forms in the agri-food sector and discusses insights from various strands of literature that enhance our understanding of these supply chain relationships. The chapter also offers thoughts on the challenges inherent in developing a conceptual model to categorise hybrids into a workable typology as the basis for explaining the adoption and evolution of these organisational forms. The chapter is organised as follows: the next section reviews types and definitions of organisational forms within an agri-food context. The core of the chapter discusses insights from several strands of literature grouped into three tiers: the over-arching (external) institutional environment, the inter-firm environment, and the intra-firm/managerial environment. As an illustrative example, these insights are applied to inter-firm relationships in the biotechnology innovation sector. The chapter closes with a discussion of three challenges for researchers in developing a typology of organisational forms: the sheer diversity of organisational forms; developing 'predictive assertions' amenable to empirical validation; and incorporating a dynamic vs static perspective.

3.2 Types and definitions of organisational forms

In his seminal contribution 'The Nature of the Firm' (1937) Ronald Coase asked a deceptively simple question that was to become the basis of a rich literature in institutional and organisational economics: why do firms exist – why do some transactions occur within firms while others occur across a market interface? The answer he gave, as is well known, is that there are costs to using the market mechanism: costs of discovering prices, of negotiating and specifying contracts, of monitoring and enforcing contractual agreements; these costs can be avoided (reduced), or so the story goes, if firms vertically integrate, directing economic activity through within-firm managerial orders. Yet there are also internal costs of co-ordination within a firm. Markets and firms are positioned as alternative modes of organising economic transactions. Fast-forward eighty years and this relatively simple question, with the juxtaposition of markets and firms as alternatives, has multiple dimensions and nuanced layers of complexity. Yes, (spot) markets and firms (hierarchies) are alternative modes of organisation but between these two polar opposites there exists a wide array of other organisational forms, exhibiting varying degrees of co-ordination, control and shared rights.

Figure 3.1 offers a simple stylised depiction of organisational forms across the vertical co-ordination spectrum, with more formalised methods of co-ordination located to the right of the spectrum. As Mighell and Jones (1963) first observed, there is always some form of 'co-ordination' within an agri-food supply chain. In the case of spot markets, such as auctions, characterised by one-off transactions between multiple buyers and sellers in the current time period, there is very little room for negotiation of the terms of the sale, products move been buyer and seller in response to price signals, hence, price is the co-ordinating mechanism. In contrast, in a vertical integrated firm, products move through successive stages of production, processing or distribution in response to within-firm managerial decisions.

The middle ground of hybrids is diverse, and one could debate the positioning of these organisational forms across the co-ordination spectrum depicted in Figure 3.1. Implicit or relational contracts can be considered relatively loose forms of co-ordination, where trust and reputation play central roles. In a pan-European study Fischer *et al.* (2009) examine factors affecting the choice of implicit Vs explicit contracts and relationship sustainability across producer-processor and processor-retailer relationships in the pork, beef and cereals sectors of selected European countries. Of note are marked differences in the prevalence of implicit (relational) vs explicit written contracts across sectors, between countries, and at different stages of the supply chain. Market, industry and firm-specific characteristics are identified as potential explanatory factors.

Strategic alliances between firms, or between networks of firms, can take a wide variety of forms and may have strong elements of relational contracting but typically are centred around a clearly defined common strategic goal with an emphasis on information sharing among partners. In an empirical treatment, Oxley (1997) examines the use of strategic alliances in technology transfer relationships where firms face significant appropriability hazards. Differentiating types of alliance by the degree of hierarchical control, she finds that more hierarchical governance structures are chosen when appropriability hazards are more severe.

Explicit contracts represent a broad umbrella, and in reality many types of contractual relationships between firms exist, exhibiting differing degrees of formality, co-ordination and control (indeed, licensing and franchising arrangements are types of contract). In an agricultural commodity marketing context, scholars typically differentiate between marketing contracts and production contacts. Commonly

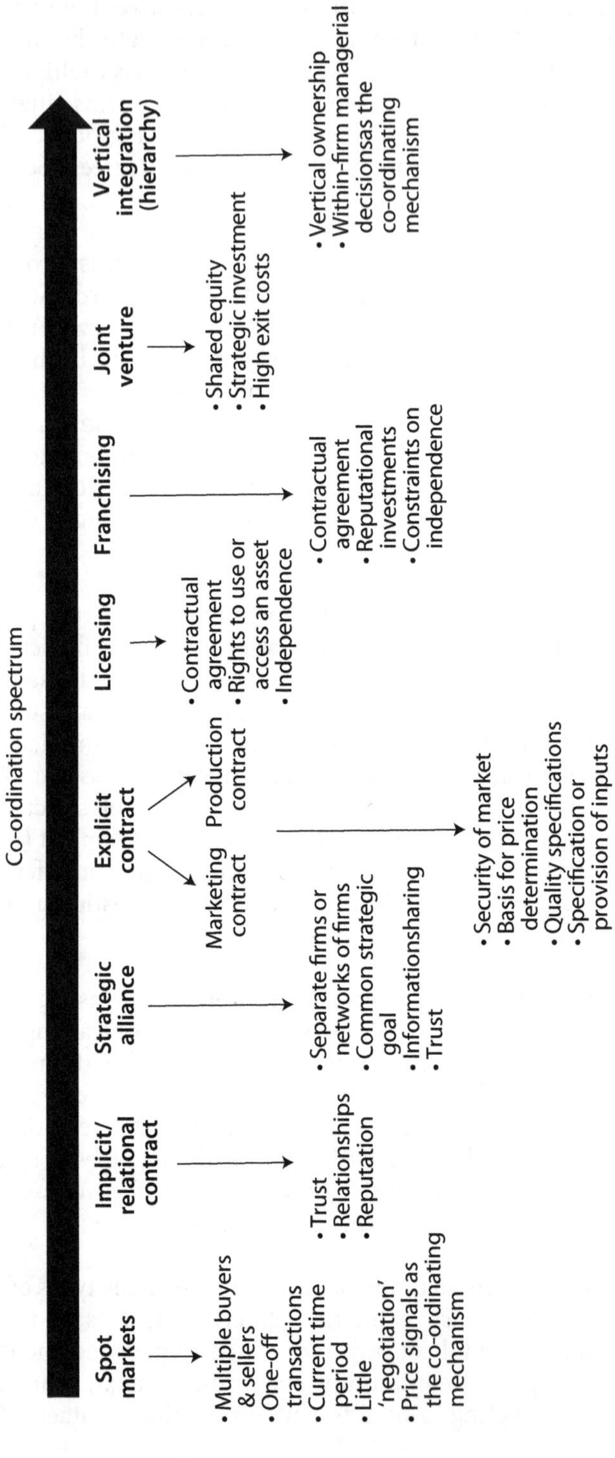

Figure 3.1. Types and key features of organisational forms.

used in the marketing of grain, marketing contracts are wide-ranging in nature (for example, forward contracts, basis contracts, deferred delivery contracts) and typically cover one crop season wherein the buyer agrees to provide a market for the seller's output, and price or the basis of price determination (often tied to quality) is specified in the contract. Production contracts invest more control with the buyer; in addition to providing a market for the seller's output, buyers specify (or in some cases, provide) key inputs, for example, contracts to source identity preserved grains with specific output traits or contracts between producers and processors in the US hog sector (Hobbs and Young, 2000).

Licensing and franchising arrangements specify the rights to use or access an asset. Significant reputational investments (e.g. in a brand name) are an important component of franchising arrangements which often feature contractual limits on independence to protect those reputational investments. Joint ventures usually imply a strategic investment by two firms featuring shared equity ownership. Collaboration and cooperation are central components of a joint venture investment, with high exit costs creating an incentive for partners to take a long-term view.

Figure 3.1 is intended to be illustrative rather than exhaustive, the key point being to highlight the diverse array of organisational relationships and some of the ways in which hybrid forms differ. In a detailed treatment of the topic, Menard (2012) defines hybrids as arrangements where two or more partners have pooled their strategic decision rights, in addition to some property rights, establishing joint mechanisms of governance to co-ordinate activities and determine payoffs. His categorization of hybrids delineates these organisational forms on two dimensions: the degree of decentralization of co-ordination or control, and the extent to which strategic resources or rights are pooled among partners. The resulting spectrum of hybrid governance arrangements stretches from information-based networks with distinctly separate ownership of assets and decision rights and with co-ordination managed relatively informally through information platforms, to a central ground of collaborative arrangements with relationship-specific investments protected by third party verification or arbitration, to a reliance on strategic centres with formal authority where partners pool rights and forego some independence through contractual limits.

In whatever form we choose to characterize these governance arrangements, the result will be broad umbrella groupings of organisational types which often overlap but which differ in recognisable dimensions with respect to the degree of co-ordination and control, the allocation of property rights, and the formality of the arrangement. Beyond recognising that different organisational forms exist, a rich literature has emerged that attempts to tease apart the determinants of different inter-firm relationships, their evolution and their outcomes. The next section offers insights from selected strands of this literature at three levels of analysis.

3.3 A three-tiered analytical approach

The ruling economic paradigm for the analysis of markets, industries and firms at the time of Coase`s original insight into the nature of the firm was the traditional neoclassical economic model enshrined in the theoretically neat, mathematically tractable world of perfect competition. It is worth pausing for a moment to consider the central tenets of the neoclassical economic transaction: a perfectly competitive industry with a large number of competing firms producing the same (a single) product under the same cost conditions and facing the same market demand curve. The exchange of a homogenous product removes the possibility of quality variations, assuming away measurement costs and negating the need for quality measurement institutions. Economic agents are assumed to have perfect, costless information such that there is no uncertainty over prices, product characteristics or the behaviour of competitors or exchange partners. Monitoring and enforcement of transactions are not needed. Transactions occur in a single time period between multiple buyers and sellers; there is no market power because alternative buyers and sellers are always available.

The neoclassical model concentrates on equilibrium outcomes, with no real explanation of how or why transactions occur and how equilibriums are reached. In essence, it ignores, or takes as given, the institutional environment within which transactions occur. Transactions are implicitly treated as though they arise in a frictionless economic vacuum. Somewhat ironically, the 'Theory of the Firm' has very little to say about firms, why they exist, or the conditions under which they are an efficient mode of governance. It also has very little to say about the institutions necessary for markets to function effectively. In fact, many of the game-changing issues in the agri-food sector in recent years have involved quality variability and uncertainty (genetically modified organisms, food safety, food quality), asymmetric information (food labelling, certification), market power, and the emergence of transformative new institutions: all issues conveniently assumed away within a strictly neoclassical framing. The perceived shortcomings of the traditional neoclassical economic model led economists to search for more realistic explanations of firm and market behaviour. Numerous contributions followed including, among others, transaction cost economics, property rights theory, agency theory, strategic management theory and the resource-based theory of the firm. These approaches are grounded in more realistic behavioural assumptions, such as the recognition that agents are boundedly rational and have the propensity to act opportunistically, and explicitly recognise that markets (and, by extension, inter-firm relationships) suffer from imperfect information.

This chapter proposes that the analysis of hybrid modes of organisation – the 'strange animals' to which Menard (2012) refers – is informed by a multi-tiered approach that considers: (1) the external (institutional) environment; (2) the inter-firm environment; and (3) the intra-firm/managerial environment. Figure 3.2 depicts three tiers of analysis which offer a suite of theoretical lenses through which to consider hybrids.

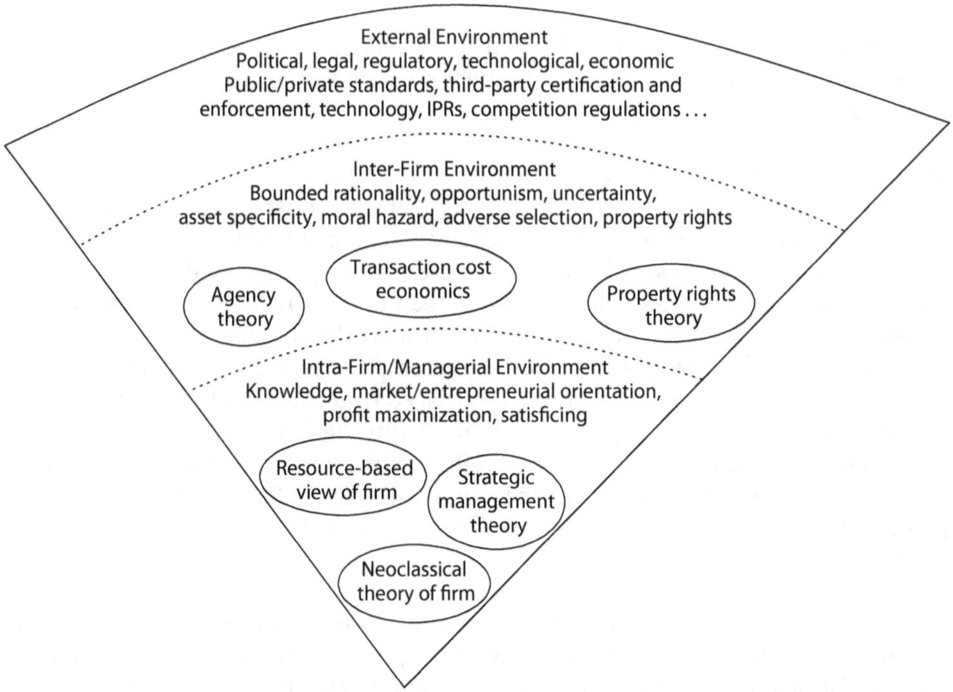

Figure 3.2. Three-tiered analytical approach.

The external environment reflects the set of institutions, norms or regulations within which firms operate and which influence the relationships between firms, including aspects of the political, legal, regulatory, technological and economic environment. Examples include the existence of public (regulatory) standards governing food safety, animal welfare, environmental protection or mandatory traceability requirements; competition regulations; the existence of private standards for food safety and quality; third party certification, monitoring and enforcement; technological change; intellectual property rights (IPR) protection; security of property rights in general, and so on. In unpacking motivations for the adoption of hybrid modes of governance and their evolution over time, consideration of the over-arching institutional environment within which firms operate deserves close attention.

The second layer of analysis, and the focus of much of the attention in the analysis of hybrid modes of governance, is the inter-firm environment, with notable contributions from transaction cost economics, agency theory and property rights theory. The recognition that bounded rationality characterises economic agents is an important departure from the neoclassical theory of the firm, allowing for explicit consideration of the incomplete contracting problem, of hold-up and of transaction costs. Opportunism, or 'self-interest seeking with guile' (Williamson, 1979) is given particular emphasis in the transaction cost approach and, when combined with bounded rationality (we can't know which agents will act opportunistically and when), is a source of uncertainty in inter-firm relationships and, by extension, a source of transaction costs.

The extent of interdependence between economic agents is a central feature of the inter-firm environment, with the presence of asset (relationship)-specific assets given particular emphasis in empirical analyses of the choice of contract type or alliance relationship in agri-food supply chains (see for example, Boger, 2001; Franken *et al.*, 2009; Vukina and Leegomonchai, 2006; Weseen *et al.*, 2014). There is also explicit recognition that economic agents operate in a world characterised by imperfect information such that asymmetric information complicates the inter-firm environment, whether it be the adverse selection problem faced by a principal in selecting reliable agents, or the moral hazard problem arising from the inability to monitor agents' behaviours. Previous research has shown that hybrid modes of governance emerge to reduce the monitoring costs arising from asymmetric information, while creating pressure for institutional adaptation to redress information impactedness (Hobbs, 2003, 2004).

Various strands of literature embedded in the 'inter-firm environment' offer insights into how and why hybrid modes of governance emerge (or wane). Building upon the foundational work of Williamson (1979, 1986, 1991), transaction cost economics examines how the dimensions or characteristics of transactions (uncertainty, asset specificity, frequency, complexity) affect governance outcomes. Transaction costs (search, negotiation, monitoring and enforcement costs) are a consequence of the characteristics of transactions within a given institutional environment.

While the transaction cost approach yields relatively clear predictions for the use of spot markets versus hierarchy (vertical integration) and explains the use of contractual safeguards, distinguishing between organisational forms within the 'fuzzy' middle ground of hybrids in discrete structural ways has proven more challenging (Williamson, 1991). Nevertheless, the transaction cost approach remains a core element of the theoretical toolkit available to organisational researchers. That toolkit also includes agency theory, with its focus on moral hazard, risk aversion and risk sharing, which has implications for the design of efficient, incentive compatible contracts in agri-food markets (Sheldon, 1996), as well as the property rights approach (Grossman and Hart, 1986), with its emphasis on incomplete contracts, investment distortions and hold-up in the presence of poorly defined property rights.

The final suite of theoretical lenses offering insights into hybrid organisational forms focuses on the internal (intra-) firm or managerial environment. Considerations of the managerial (firm) objective function, including (profit) maximising versus satisficing behaviour, are pertinent. The neoclassical theory of the firm, while in fact having little to say about the intra-firm environment per se, offers insights pertaining to profit maximising objectives and exchange at the margin that remain relevant.

Other managerial theories of the firm add considerable richness and depth to an analysis of hybrid forms. Strategic management theory takes a pragmatic, functionalist view of the firm as a collection of functions, with an emphasis on internal firm strategies to gain competitive advantage. Porter (1980) suggested that the source of competitive advantage for industries and the development of competitive strategy within firms lay in understanding rivalry among existing firms, the threat of new entrants, the threat of substitutes, the bargaining power of suppliers and the bargaining power of buyers. Viewed through this lens, the use of hybrid forms of governance to manage inter-firm relationships may be one element of a broader set of internal firm strategies to attain (or retain) a competitive advantage; for example, the development of joint venture or strategic alliance partnerships to create barriers to entry or to counteract the bargaining power of suppliers (buyers).

Strategic management theory also recognises the importance of understanding the cultural and organisational norms within a firm. A 'market orientation' relates to the values, beliefs or actions that a firm exhibits in a demand-pull focus, while an 'entrepreneurial orientation' emphasises innovation, risk taking and a proactive outlook. Viewed through these perspectives, the tendency to seek out collaborative business relationships that enable adaptation to market demands, or that facilitate new innovations, may be an outcome of a market or an entrepreneurial orientation of the firm.

The resource-based view of the firm offers an additional theoretical lens through which to view the use of hybrid governance within a supply chain. Firms are conceptualised as a set of resource endowments which form the basis of their competitive advantage and inform strategic decisions (Wernerfelt, 1984). The core competencies or capabilities of a firm represent a particularly salient resource in explaining the boundaries of a firm and the incentives for firms to adopt hybrid modes of governance as a means of managing knowledge assets and creating access to human capital. Teece *et al.* (1994) describe a firm's competence as '...a set of differentiated technological skills, complementary assets, and organisational routines and capacities that provide the basis for a firm's competitive capacities in one or more businesses' (p.18). Just as with individuals, firms have different skills sets and aptitudes and are necessarily limited in what they know how to do well, providing motivations for supply chain relationships that maximise the outputs of a firm's own competencies or that expand upon its competencies through co-operation or collaboration with other agents.

Knowledge and the nature of knowledge assets are central to this view of the firm. While codified knowledge can be specified in formulae and designs and may be patentable, tacit knowledge is acquired through learning-by-doing, while distributed knowledge is only valuable if used in conjunction with other economic agents, requiring co-operation across firms. Transferring tacit knowledge between firms is complex, subject to uncertainty and the source of transaction costs, providing a motivation for closer co-ordination through strategic alliances (Oxley, 1997). The core competencies or capabilities of a firm therefore set limits on its boundaries or create incentives for firms to engage in hybrid modes of governance to access (or protect) technological know-how. This is especially pertinent in the innovation and commercialisation processes within high-tech sectors, as indicated by the complex and evolving array of inter-firm relationships in the agricultural biotechnology sector. An application of the three-tiered analytical approach to understanding the diversity of organisational forms in the agricultural biotechnology sector is presented in the next section.

3.4 An application: alliances and networks in the biotechnology innovation system

Agricultural biotechnology has been at the forefront of technological change and innovation in the agricultural sector through the development and adoption of transformative genomic technologies. Large multinational enterprises dominate the sector and garner most of the attention, however, small and medium-sized firms have played pivotal roles with respect to the generation of ideas and the development of explorative technologies. Access to these technologies is managed through a complex array of inter-firm relationships – both vertically along agri-food supply chains and horizontally across innovating firms – from strategic alliances to joint ventures, licensing agreements, networks, clusters, and contracts (Hobbs, 2014). The three-tiered analytical approach depicted in Figure 3.2 provides a basis from which to analyse the use of hybrid organisational forms in the biotechnology innovation system.

The external environment for innovating firms in the agricultural biotechnology sector is characterised by technological and institutional complexity. Platform technologies, including the sequencing of genomes, have formed the basis of countless subsequent innovations (Pray and Naseem, 2005). Gene stacking (where plants contain several patentable traits) often means that one patentable trait is developed on the basis of other patentable traits or processes, with the need for complex inter-firm agreements to allocate the rents that arise from this technological complexity (Hobbs, 2014). The spectre of overlapping patents and multiple pieces of intellectual property incorporated within a single crop variety complicates the commercialisation of new innovations; for example, Smyth and Gray (2011) report that new crop varieties developed through the application of biotechnology may sometimes incorporate up to 30 separate pieces of intellectual property. They note that inter-firm agreements on over 70 pieces of intellectual property were needed prior to the initiation of research that led to the development of Golden Rice. The regulatory environment governing intellectual property rights, the security of those property rights, and the speed and efficacy with which patent disputes are resolved, are important influences on the nature of contractual agreements and technology-based alliances between firms.

The agricultural biotechnology sector has been the scene of extensive merger and acquisition activity in the past, and large mergers continue to be contemplated.

In March 2016, for example, the Chinese firm ChemChina announced a takeover bid for Syngenta, prompting calls for the US Department of Treasury to review the proposed acquisition (Hopkins, 2016), while in September 2016 Monsanto and Bayer announced a merger agreement under which Bayer will acquire Monsanto (Bayer, 2016). In cases where the regulatory environment, as enshrined in competition and anti-trust rules, limits consolidation, firms may have stronger incentives to pursue cross-licensing agreements and alliances as a more flexible alternative to vertical or horizontal integration.

Technological complexity within agricultural biotechnology supply chains means that the inter-firm environment is characterised by high levels of uncertainty, complexity and asset specificity (Gaisford *et al.*, 2001). Uncertainty takes several forms: uncertainty over the ownership of intellectual property, as well as uncertainty arising from the process of scientific discovery given often lengthy time lags between the genesis of a research idea and eventual commercialisation of a technology. Regulatory delays and uncertainty, particularly in the face of vocal public opposition to agricultural biotechnology in some jurisdictions, compounds the inherent technological uncertainty. An incomplete contracting problem arises due to challenges in writing a fully contingent contract to co-ordinate the development of an agricultural biotechnology innovation years before a product might reach commercialisation in an unpredictable market environment. Stronger inter-firm linkages or vertical integration offer possible solutions to the incomplete contracting problem (Gaisford *et al.*, 2001; Hobbs, 2014).

Asset specific investments arise if an innovation is dependent upon a platform technology, a patentable trait, or is specific to the germplasm through which the innovation will be delivered. While vertical integration offers one solution (the spate of mergers and acquisitions in the agricultural biotechnology sector in the 1990s are a case in point), hybrid governance modes present another route. Cross-licensing agreements between agricultural biotechnology firms have become commonplace and take the form of *ex-post* agreements allowing for shared access to existing technology platforms, as well as *ex-ante* agreements to foster the development of new crop traits (Hobbs, 2014; Smyth and Gray, 2011).

The third tier of analysis in Figure 3.2 points to the need to consider intra-firm factors and managerial decision-making processes in analysing the use of hybrid modes of governance. In this vein, the relationship between firms in the biotechnology sector can also be examined from the perspective of the management of knowledge assets and the existence of knowledge spillovers between firms. The resource-based view of the firm characterises the firm as a collection of resources, including competencies. In the biotechnology sector knowledge resources, or those related to technological know-how, are particularly pertinent, including human capital, knowledge management and intellectual property rights. As noted in Hobbs (2014), viewed through this lens, firms engage in strategic alliances, joint ventures, licensing agreements or vertical integration to expand access to core competencies including knowledge resources, since transferring tacit knowledge between independent firms across a market interface may be ineffective. While a firm's particular set of competencies allow it to create new products and technologies in response to market signals, these competencies also set limits on the boundaries of the firm and provide an incentive for mergers, acquisitions, strategic alliances and other forms of inter-firm linkages to expand that knowledge resource base.

As Hobbs (2014) notes 'the sources of innovation rarely lie exclusively within company boundaries, nor are new ideas always traded in markets' (p. 65). Instead, innovations occur within networks often comprising a combination of high-tech firms, downstream users and scientists in public sector research institutions. This phenomenon is not new. Business networks among knowledge-intensive companies became commonplace across a number of high-tech sectors in the 1990s, including semiconductors, nanotechnology and medical biotechnology (Hobbs, 2014; Niosi and Queenton, 2010; Peña, 2009). Research suggests that proximity matters within biotechnology networks, whether it be proximity to 'star scientists' (Niosi and Queenton, 2010), or geographical clusters of firms, institutions and individuals (Phillips and Ryan, 2007). The potential for positive knowledge spillovers along supply chains, within networks, or across clusters provides a compelling motivation for collaborative hybrid modes of governance as a component of successful innovation strategies in an increasingly complex regulatory and technological environment.

The concept of entrepreneurship, and the entrepreneurial orientation of a firm, have been examined extensively in the strategic management literature. In this vein, the concept of 'intrapreneurship' refers to entrepreneurial and innovation processes within a firm and the extent to which internal organisational norms and orientation encourage innovation (Hobbs, 2014). In an extension to that approach, recent research has recognised that innovation processes are often an outcome of *inter-firm* collaboration through so-called 'interpreneurial networks' (Bonnafous-Boucher and Lamy, 2010; Peña, 2009). In the agricultural biotechnology sector, firms may lack all of the necessary knowledge resources to develop their technologies fully in-house. Interpreneurial networks or strategic alliances can be an effective means of pooling complementary knowledge assets to enhance innovative capacity along the supply chain (Hobbs, 2014; Peña, 2009). In the presence of significant technological or institutional uncertainty (for example, with respect to IPRs or novel food approval regulations), gradual investment through interpreneurial R&D alliances may offer a more cautious market entry strategy than outright acquisition of technologies and firms during the technology development phase (Hobbs, 2014). Indeed, Peña (2009) find that major US and European firms in the life science and medical biotechnology sectors who opted for interpreneurial R&D alliances rather than making large in-house R&D investments tended to exhibit stronger financial performance.

While vertical and horizontal integration through mergers and acquisitions remain an ongoing feature of the agricultural biotechnology sector, hybrid modes of governance nevertheless abound, from alliances, to simple licensing agreements, cross-licensing agreements, R&D contracts, clusters and networks. Taken together, the three tiers of analysis depicted in Figure 3.2 allow us to view the use of hybrid modes of governance from various perspectives: the influence of the external environment (institutional, regulatory, technological), inter-firm factors (transaction costs, property rights, agency relationships), and the intra-firm/managerial environment (knowledge resources, spillovers, and managerial orientation). Much like the parable of the blind men and the elephant, at present, these remain separate lenses through which to view individual pieces of the elephant, and efforts to integrate these perspectives represent a fruitful avenue for further research.

3.5 Looking ahead: challenges and opportunities

In deepening our analysis of hybrid modes of governance in the agri-food sector, there is merit in continuing to develop typologies of hybrid forms to allow these organisational modes to be categorised in discrete structural ways. This will have stronger traction as a component of a broader research strategy to develop a more comprehensive theoretical basis for explaining when, why and how hybrid forms emerge and evolve. In doing so, three challenges are evident.

First, the sheer diversity of organisational forms and the dimensions by which these could be described makes this a complicated endeavour. At a fairly aggregate level, progress has been made in this regard, for example, grouping hybrid organisational forms on the basis of the degree of co-ordination/control and the extent to which strategic resources or property rights are pooled versus remain independent, yielding a trade-off between the benefits of closer co-ordination/control and the costs of governance (Menard, 2010, 2012). Further refinement of existing typologies is warranted but primarily as an input to a broader understanding of the adoption and evolution of hybrid organisational forms, rather than for its own sake.

A second challenge lies in developing typologies and theoretical frameworks to explain hybrids that are amenable to empirical validation and application. The original (Old) Institutional Economics approach of the 1920s and 1930s floundered because it was largely descriptive, categorising and describing institutions but lacking a firm theoretical basis from which to explain the use of specific governance structures. In developing the transaction cost approach, Williamson (1986) stressed the importance that the theory generate what he termed 'refutable implications' or 'predictive assertions'. Indeed, testable hypotheses are the bread and butter of applied economics. In developing conceptual frameworks to advance our understanding of hybrid organisational forms, therefore, the capacity for empirical validation is an ongoing consideration. Are the dimensions by which we describe (categorise) hybrids identifiable and measurable in a discrete fashion? Are the deterministic factors hypothesised to be driving the establishment of hybrid organisational forms observable and measurable? If not, do reliable proxy measures exist? Can insights from richly detailed case study analyses be extrapolated beyond the context of the case study, or do these form the basis for the development of comparative empirical models? The agri-food sector presents a rich ground upon which to develop empirical applications.

The third challenge lies in incorporating a dynamic element to an analysis of hybrid organisational forms. Typically, organisational theories have tended to be static in nature, indeed this has been a criticism of the transaction cost approach. In an application to contract farming and vertical integration in U.S. agriculture, for example, James *et al.* (2011) argue that in order to explain the adoption, diffusion and evolution of organisational forms, a more dynamic theoretical framework is required that draws upon theories of organisational complementarities and technology diffusion. While much valuable work has been done to observe and document the existence of hybrids, by nature this is usually a snapshot view of hybrids at a given point in time. How do we incorporate dynamics into a conceptual framework of hybrids? How (and why) do hybrids evolve over time? Data challenges notwithstanding, considerable scope exists for empirical applications that examine these questions across sets of firms over time (a panel data-type approach), as well as comparative analyses of the adoption of hybrid organisational forms across agri-food sectors and across geographical regions.

Hybrid modes of organisation within the agri-food sector continue to offer a rich ground for applied research. From analyses of the adoption and diffusion of marketing versus production contracts between farmers and processors, to the development of strategic alliances, networks, clusters and cross-licencing agreements within the innovation sector, to the use of co-branding, third party certification and collaborative collective reputation signals for food quality, the agri-food sector presents significant opportunities for new research. In exploring these applications, approaches which consider the combined ramifications of the external (institutional) environment, the inter-firm environment and the intra-firm (managerial) environment enrich our understanding of the diversity of economic organisation in agri-food value chains.

References

Bayer, 2016. Bayer and Monsanto to create a global leader in agriculture. Available at: http://tinyurl.com/ybkla296.

Boger, S., 2001. Quality and contractual choice: a transaction cost approach to the polish hog market. European Review of Agricultural Economics 28(3): 241-261.

Bonnafous-Boucher, M. and Lamy, E., 2010. Hybridisation of the profiles of entrepreneurs and intrapreneurs: the case of French biotechnology SMEs. International Journal of Entrepreneurship and Small Business 10(3): 410-427.

Coase, R.H., 1937. The nature of the firm. Economica 4: 386-405.

Fisher, C., Hartmann, M., Reynolds, N., Leat, P., Revoredo-Giba, C., Henchion, M., Albisu, L.M. and Gracia, A., 2009. Factors influencing contractual choice and sustainable relationships in European agri-food supply chains. European Review of Agricultural Economics 36(4): 541-569.

Franken, J.R.V., Pennings, J.M.E. and Garcia, P., 2009. Do transaction costs and risk preferences influence marketing arrangements in the Illinois hog industry? Journal of Agricultural and Resource Economics 34(2): 297-315.

Gaisford, J.D., Hobbs, J.E., Kerr, W.A., Perdikis, N. and Plunkett, M., 2001. The economics of biotechnology. Edward Elgar Press, Cheltenham, UK, 246 pp.

Grossman, S. and Hart, O., 1986. The costs and benefits of ownership: a theory of vertical and lateral integration. Journal of Political Economy 94: 691-719.

Hobbs, J.E., 2003. Information, incentives and institutions in the agri-food sector. Canadian Journal of Agricultural Economics 51(3): 413-429.

Hobbs, J.E., 2004. Information asymmetry and the role of traceability systems. Agribusiness 20(4): 397-415.

Hobbs, J.E., 2014. The private sector: MNEs and SMEs. In: Smyth, S.J., Phillips, P.W.B. and Castle, D. (eds.) Handbook on agriculture, biotechnology and development. Edward Elgar Publishing Ltd., Cheltenham, UK, pp. 56-70.

Hobbs, J.E. and Young, L., 2000. Closer vertical co-ordination in agrifood supply chains: a conceptual framework and some preliminary evidence. Supply Chain Management 5(3): 131-142.

Hopkins, M., 2016. Senators call on treasury department to review ChemChina's acquisition of Syngenta. CropLife. Available at: http://tinyurl.com/y7bjnspu.

James, H.J., Klein, P.G. and Sykuta, M.E., 2011. The adoption, diffusion, and evolution of organizational form: insights from the agrifood sector. Managerial Decision Economics 32(1): 243-259.

Menard, C., 2010. Hybrid organizations. In: Klein, P.G. and Sykuta, M.E. (eds.) The Elgar companion to transaction cost economics. Edward Elgar Publishing Ltd., Cheltenham, UK, pp. 176-184.

Menard, C., 2012. Hybrid modes of organization: alliances, joint ventures, networks, and other 'strange' animals. In: Gibbons, R. and Roberts, J. (eds.) The handbook of organizational economics. Princeton University Press, Princeton, NJ, USA, pp. 1066-1108.

Mighell, R.L. and Jones, L.A., 1963. Vertical coordination in agriculture. USDA-ERS, Agricultural Economics Report No. 19. USDA, Washington, USA.

Niosi, J. and Queenton, J., 2010. Knowledge capital in biotechnology industry: impacts on Canadian firm performance. International Journal of Knowledge-Based Development 1: 136-151.

Oxley, J.E., 1997. Appropriability hazards and governance in strategic alliances: a transaction cost approach. Journal of Law, Economics and Organization 13(2): 387-409.

Peña, I., 2009. Interpreneurial strategic behaviour and business performance. International Journal of Entrepreneurship and Innovation Management 10(1): 3-19.

Phillips, P.W.B. and Ryan, C., 2007. The role of clusters in driving innovation. In: Krattiger, A., Mahoney, R.T. and Nelson, L. (eds.) Intellectual property management in health and agricultural innovation: a handbook of best practices. MIHR, Oxford, UK and PIPRA, Davis, CA, USA, pp. 281-294. Available at: www.ipHandbook.org.

Porter, E.M., 1980. Competitive strategy. Free Press/Macmillan, New York, NY, USA.

Pray. C. and Naseem, A., 2005. Intellectual property rights on research tools: incentives or barriers to innovation? Case studies of rice genomics and plant transformation technologies. Agbioforum 8(2-3): 108-117.

Sheldon, I., 1996. Contracting, imperfect information and the food system. Review of Agricultural Economics 18(1): 7-19.

Smyth, S.J. and Gray, R., 2011. Intellectual property sharing agreements in gene technology: implications for research and commercialisation. International Journal of Intellectual Property Management 4(3): 179-190.
Teece, D., Rumelt, R., Dosi, G. and Winter, S., 1994. Understanding corporate coherence: theory and evidence. Journal of Economic Behavior and Organization 23(1): 1-30.
Vukina, T. and Leegomonchai, P., 2006. Oligopsony power, asset specificity, and hold-up: evidence from the broiler industry. American Journal of Agricultural Economics 88(3): 589-605.
Wernerfelt, B., 1984. A resource-based view of the firm. Strategic Management Journal 5(2): 171-180.
Weseen, S., Hobbs, J.E. and Kerr, W.A., 2014. Reducing hold-up risks in ethanol supply chains: a transaction cost perspective. International Journal of Food and Agribusiness Management 17(2): 83-106.
Williamson, O.E., 1979. Transaction cost economics: the governance of contractual relations. Journal of Law and Economics 22: 233-262.
Williamson, O.E., 1986. Economic organization: firms, markets and policy control. Harvester Wheatsheaf, Hemel Hempstead, UK, 310 pp.
Williamson, O.E., 1991. Comparative economic organisation: the analysis of discrete structural alternatives. Administrative Science Quarterly 36(2): 269-296.

Part II – **Case studies on organizational diversity**

4. Horizontal integration between farmers – governing cooperation through different enforcement mechanisms

A. Chlebicka[1], J. Fałkowski[2] and B. Łopaciuk-Gonczaryk[2]*

[1]Warsaw University of Life Sciences (SGGW), Faculty of Economic Sciences, Chair of Economics and Organisation of Enterprises, Nowoursynowska St. 166, 02-787 Warsaw, Poland; [2]University of Warsaw & CEAPS, Faculty of Economic Sciences, Chair of Political Economy, 44/50 Długa St. 44/50, 00-241 Warsaw, Poland; aleksandra_chlebicka@sggw.pl

Abstract

Cooperation between farmers has often been advocated as a potential stimulus to increase efficiency of agricultural production and strengthen farmer's bargaining power vis-a-vis representatives of other segments in the agro-food supply chain. We propose an explorative research in an attempt to provide some evidence linking the process of establishing agricultural producer groups with the issue of trust. Using micro level data from 260 interviews with Polish farmers, we investigate to what extent farmers' willingness to involve in formal cooperation depends on the level of general trust and interpersonal relationships. The former is associated with more formal governing mechanisms based on a third party enforcer. Instead, interpersonal relationships are related to more informal enforcement mechanism, based on reputation and mutual trust. First, we look at the extent to which participation in producer groups is driven by kinship or friendship to other group members. Second, we explore to what extent farmers being involved in producer groups differ from farmers acting on their own in their opinions on the general trust and willingness to cooperate in their neighbourhood. Finally, we also show some tentative evidence to link different types of trust to different cooperative equilibria emerging at the regional level.

Keywords: producer groups, trust, interpersonal relationships

DOI 10.3920/978-90-8686-844-5_4, © Wageningen Academic Publishers 2017

4.1 Introduction

Cooperation between farmers has often been advocated as a potential stimulus to increase efficiency of agricultural production and a tool to increase farmer's bargaining power vis-a-vis representatives of other segments in the agro-food supply chain. The relevant literature provides at least several arguments to support such claims (Bijman *et al.*, 2012; Hendrikse and Veerman, 2001; Menard, 2007). Most often however it is assumed that acting together may allow farmers to take advantage of the economies of scale and to minimise transaction costs which are related to production processes on the one hand, and to organising sales of agricultural products on the other. In line with this view, starting a collective action in the form of producer groups should enable farmers to concentrate the supply of agricultural products. This in turn should make them more essential in production of processing industry and thus facilitate negotiations of the conditions of their sales. Similarly, by opening the possibility for planning joint production and concentrating demand for agricultural inputs, it should strengthen farmers' bargaining power vis-a-vis the suppliers of agricultural inputs.

High hopes of the positive effects coming from farmers' collaborative actions have been frequently expressed not only in academic but also in political circles. This is clearly reflected in several political actions initiated at the national and international level. This could be seen, for example, in the European Union where several member states adopted explicit measures aimed at supporting horizontal integration at the farm level. The recommendations of the High Level Group on the Competitiveness of the Agro-Food Industry which acknowledge the role of producer groups and agro-food cooperatives in influencing the competitiveness of the European food industry can serve as another illustration.[1]

Interestingly though, despite the fact that the benefits which are likely to arise from farmers' collective action are widely recognised, the scale of farmers' horizontal integration in Europe varies to a large extent by countries and sectors. In fact, as reported by Bijman *et al.* (2012), in France, Germany, United Kingdom, Spain, Denmark, depending on a sector, the share of cooperatives/producer groups in the market ranges from 25 to 95%. In yet another countries, for example Poland, which is the focus of this study, this share is often much lower and does not exceed 10% (Chlebicka *et al.*, 2014).

This clearly calls for better understanding of why cooperative behaviour between farmers may (not) take place and what potential stimulus one can use to foster collective actions. Improving our knowledge on mechanisms that can sustain cooperative equilibria seem to be equally important. This is because coordination problems very often are characterised by different equilibrium configurations, some

[1] The Group, working under the chairmanship of the European Commission, was launched in 2008. In 2010 it was dissolved and the High Level Forum for a Better Functioning Food Supply Chain was established to follow the implementation of recommendations formulated by the Group. These final recommendations are available at: http://tinyurl.com/y9t6tjfs.

of them involving close cooperation and some of them involving no cooperation at all (e.g. Acemoglu and Jackson, 2015). This is important as we may find different patterns of cooperative behaviour in similar socio-economic environments.

There is an extensive literature studying various determinants of cooperative behaviour among farmers (e.g. Hendrikse and Bijman, 2002; Lopez and Spreen, 2008; Menard and Valceschini, 2005; or Pascucci *et al.*, 2012). A factor which has received a special attention in this context is trust. As commonly argued, trust reduces the level of agency costs in collective action and thus makes the emergence of cooperative equilibria more likely (for the recent contributions see, e.g. Martino, 2010; Nilsson *et al.*, 2009; or Szabo, 2010). Our research adopts a similar point of view and tries to further explore to what extent the lack of trust can be perceived as an important obstacle to setting up of producer groups. In contrast to the existing literature however, we focus our discussion on distinguishing between two different enforcement mechanisms which can be used to start and govern cooperation and which are likely to depend on different types of trust. On the one hand, we look at mechanisms based on informal institutions and mutual trust between individuals that interact with each other. On the other hand, we consider enforcement mechanisms based on formal institutions, by which a more general notion of trust (to people in general, to the state, etc.) would be of special importance. In this context, we attempt to see how does the process of establishing agricultural producer groups relate to resources stemming from farmers' interpersonal relationships on the one hand, and to general trust to people on the other.

To do so, we use a unique micro data from Poland, which combines information on farmers being involved in a collective action of participating in producer groups with the information on farmers who are not affiliated to those groups. More specifically, we check whether interpersonal relationships were important for the first type of farmers when they were making their decision to join a producer group and whether lack of such relationships could explain why other farmers choose not to be involved in a formal cooperation. We also explore whether this difference in behaviour could be attributed to the fact that both types of farmers live in neighbourhoods characterised by a different level of generalised trust towards other people. We refer this discussion to different types of enforcement mechanisms which can be used to govern collaborative actions.

Clearly, the analysis which we present is only a first step on the way to improve our understanding of the issues in question. In that sense, the insights which we present can be at best indicative of a potential impact that general trust and interpersonal relationships, via their linkages to different enforcement mechanisms governing collective action, may have for the process of establishing producer group. Thus the results we present should be seen only in a correlational sense and we do not claim to capture any causal relationships. That said, we nonetheless believe that this kind of evidence could be a good starting point for future research.

There is also another point which should be mentioned here. The focus of this study is on Poland. What follows, a justified question arises to what extent the discussion we present here can be generalised on a broader scale. That said, there are some arguments pointing to the fact that the insights gained from our investigation might be of broader interest and provide useful information also for other contexts. This is because, as argued by several authors, the situation in Poland is similar in some respects to that observed in other Central and Eastern European countries which have undergone the transition from a centrally planned to a market economy. In particular, it has been argued that, due to negative experiences with collective action under the communist dictatorship, farmers in Central and Eastern Europe have a relatively low level of social capital and are generally unwilling to cooperate (e.g. Chloupkova and Bjornskov, 2002; Csaki and Forgacs, 2008; Paldam and Svendsen, 2001; Tisenkopfs *et al.*, 2011). This reasoning is quite intuitive. In addition, it could be supported on theoretical grounds by arguments illustrating a negative relationship between totalitarian rule and the level of social capital within the society (Putnam, 1993; Wintrobe, 1998). Interestingly though, there is still no systematic evidence which would prove that this intuitive explanation for the slow development of horizontal integration between farmers in former communist countries finds support in the data (Wolz *et al.*, 2004). In this context therefore, we also hope to contribute to this part of the debate and provide some insights on the role that resources embedded in farmers' various relationships might have played in forming their attitude to cooperate.

The reminder of the paper is organised as follows. In the next section we briefly sketch the theoretical framework which we use to motivate our analysis. Further, we provide some background information about the development of producer groups in Poland. The next section presents our data and main insights from our empirical investigation. Finally, the last section concludes.

4.2 Theoretical considerations

In order to highlight some key aspects determining the process of establishing formal cooperation between farmers, it is useful to note that every cooperation must be supported by an effective enforcement mechanism (Greif, 2006). Otherwise, there will be no credible way to punish opportunistic behaviour from individuals participating in a collective action. In such case it will be very difficult to discourage such behaviour. Consequently, individuals would withdraw from cooperation at all.

In principle we can think of two kinds of enforcement mechanisms which could be put in place here (e.g. Greif, 2006). The former one rests on community based mechanisms (e.g. Kandori, 1992). In this case, cooperative behaviour is assured using informal institutions (such as reputation) among those involved in a collective action. The second enforcement mechanism on the other hand rests on formal institutions. In this case a third party is responsible for monitoring individuals' actions and for rewarding or punishing their behaviour.

That said, regardless of the enforcement mechanism which can be used, cooperation can be established only if individuals trust that their collective action can be efficiently sustained and that they can benefit from it. This means that appropriate incentives for group members (if it is the mechanism based on informal institutions) or for a third party (if it is the mechanism based on formal institutions) have to be put in place so that individuals who deviate from a cooperative behaviour are punished. The very different nature of these two enforcement mechanisms suggests however that for each of them a different type of trust may be of special importance. Mutual trust and interpersonal relationships within the group of individuals involved in cooperation seem to be of key importance for establishing collaborative actions based on informal institutions. Setting up collective action based on formal institutions in turn seems to crucially depend on trust in the third party which is supposed to monitor and punish potential deviations from the expected course of behaviour. While in our empirical analysis we do not directly measure the latter type of trust, in what follows, we proxy for it with trust in people in general and their general willingness to cooperate. This approach can come for criticism and thus should be taken into account when interpreting our findings. Yet, as shown by several studies, trust in people in general and their general willingness to cooperate are highly correlated with trust in formal institutions and government which gives credence to our strategy (e.g. Brehm and Rahn, 1997; Keele, 2007).

Recognising the differences between these two types of enforcement mechanisms and, consequently, acknowledging that depending on circumstances different types of trust might be needed for setting up farmers' collaborative actions seem to have important implications for the policy making. This is because it clearly indicates that various measures encouraging formal cooperation between agricultural producers may bring different effects in different contexts. For instance, it is plausible to assume that groups which are governed by informal institutions will be of relatively small size. The reason for this is the fact that with this type of enforcement mechanism monitoring and punishing is done by the group members themselves. This in turn is possible only in groups of limited size as in larger groups information about individuals' behaviour is not so easily available. As a result, our efforts to increase the scale of cooperation when collective actions are mainly based on informal institutions would be successful only if we can generate individuals' trust towards a third party enforcer. In effect, in regions/ sectors where there is no such trust, projects promoting small-scale cooperation based on interpersonal relationships could be more efficacious than that aimed at stimulating cooperation on a large scale, especially in the short-run.

Second hypothesis which we may pose in relation to these considerations allows us to assume that the choice of different types of enforcement mechanism will be linked to groups' specialisation patterns (compare with Acemoglu and Wolitzky, 2015). This argument is based on the observation that governance mechanisms using third party enforcers should be called for more often, the more technologically advanced is the production process. The reason for this is that with more advanced technologies it is more difficult to accurately observe individual's behaviour. In addition, processes using more advanced technologies are usually associated with a production of more complex goods. This is often linked to producing goods of higher quality. The latter in turn needs to be accurately assessed which is normally done by a third party. Taken together these observations suggest that formal enforcement (rather than informal enforcement) should form the basis of cooperation particularly among farmers specialising in animal production. This is because animal production is relatively more constrained by various production standards than crop production. Further, this is predominantly the former type of production which is dependent on third parties as regards the monitoring of quality issues. What follows, one may assume that establishing a producer group specialised in animal production requires a given level of general trust towards this third party, which is not necessarily the case with producer groups in crop sector.

This line of reasoning gets further support by argumentation based on popular distinction between bonding social capital (involving strong ties and mutual trust) and bridging social capital (often indicated by generalised trust) (Putnam, 2000). As it is commonly argued, in case of complex goods, which are associated with the need of learning new practises and dealing with external parties, it is bridging rather than bonding social capital which is required. The latter type of social capital can be even counterproductive for starting and expanding collaborative actions among producers of such goods (compare it with the discussion in e.g. Burt, 2005).

Below we try to explore these issues using micro data from Poland. Before moving to empirical analysis however, we first present some information on the current state of development of horizontal integration of Polish farmers. This is to motivate the choice of our case study and to build the background for further analysis.

4.3 Producer groups in Poland – some background information

Given the socio-economic changes that Poland experienced with the overthrowing of the communist dictatorship in 1989, the process of creating formal structures of cooperation among farmers in Poland had, in fact, to begin anew. What we describe here therefore is a relatively recent phenomenon, although long traditions of cooperative movements dating back to the 19th century have to be acknowledged (e.g. Chloupkova *et al.*, 2003).[2]

It should be recognised however that potential advantages of agricultural producer groups were discussed in political and academic circles already in the 1990-ties (see, among others, Majewski and Dalton, 2000). Furthermore, and importantly from our perspective, the issues related to horizontal integration at the farm level constitute a

[2] While we do not provide a general literature review of the restructuring of Polish agriculture in the 1990s, we refer the reader to existing reviews. Studies analysing the collapse of state-owned and collective farms in Poland include, among others, those of Hunek (1994), Maliszewski (1995) or Milczarek (2002). An excellent survey of the literature on agricultural transition in general is that of Rozelle and Swinnen (2004).

frequent reference point also in current debates (e.g. RDP, 2010). Producer groups are often presented as a potential tool to achieve, for example, greater stability on agricultural markets or to increase the transparency within agro-food supply chains (e.g. Bondyra, 2003; or Chlebicka, 2011; Krzyżanowska, 2003). Moreover, given a slow pace of land concentration processes, farmers' horizontal integration is presented as the only real possibility of reinforcing their market position (Karasiewicz, 2001). As such the issue of farmers' cooperation is frequently mentioned also in a broader context of rural development and the future of the Polish farming sector. Below we briefly report the main statistics which allow to illustrate the current state of farmers' horizontal integration in Poland.

Our focus is on agricultural producer groups. As a consequence, we do not cover here those cooperatives which are not recognized as producer groups, that is: about 700 agricultural producer co-operatives (for more information see Dzun, 2009), 120 apiculture co-ops and 90 dairy co-operatives (data for 2014). A comprehensive description of the current level of development of farmer cooperatives in Poland can be found, for example, in Matczak (2012). A historical view on rural cooperatives in Poland is presented, among others, by Brodziński (2014).

Under the definition laid down in Polish law established in 2000,[3] a producer group is identified based on the specification of its functions and goals. According to the Act on agricultural producer group and associations, each legal person is regarded as a producer group provided that it has been established on agricultural producers' own initiative and its aim is to primarily improve economic effectiveness of its member farms, mainly by adapting production and sales to the market requirements. What should be noted, the term agricultural producer group does not imply any specific legal form.[4] In particular, the intention of a legislator was not to impose a co-operative as an obligatory legal form for a producer group, because of its bad reputation caused by the communist legacy. This approach was later indicated as a recommendation for the policy-makers in other transitional and developing countries where cooperatives are negatively perceived (Beckmann *et al.*, 2015). In effect, agricultural producer group refers simply to the organization whose major objective is to commercialise commodities produced by group members, thereby ensuring them maximum benefits, proportionately to the quantity of products sold by the group.

Starting from the passing of the relevant legislation and membership to the European Union, we have been observing a continuous development of formal cooperation between farmers in Poland. In fact, as mentioned by several authors, legislation on the functioning of producer groups established in 2000 and financial support offered under Common Agricultural Policy were an essential stimulus for setting up new

[3] Act on agricultural producer group and associations of 15 September 2000, Dz.U. (Official Journal) No 88 from 2000.

[4] Factors affecting the choice of various legal forms of agricultural producer groups are discussed e.g. by Beckmann *et al.* (2015) or Lemanowicz (2005).

producer groups (e.g. Banaszak and Beckmann, 2008). Below we report some basic statistics to illustrate both the current development of horizontal integration of farmers in Poland and its dynamism in recent years.

At the end of 2015, 1,338 agricultural producer groups were registered in Marshal's (region governors) registers, of which 824 were formed in 2007-2012 and around 330 between 2012 and 2015. In addition to agricultural producer groups, at the end of 2015, there were 310 fruit and vegetable producer groups (236 in 2012). The latter were subject to separate regulations before 2014 and were set up and provided with support under the common organization of the market in fruit and vegetables (the so-called Pillar I of the Common Agriculture Policy). Thus, overall, around 1,650 entities may be regarded as producer groups (there were around 1,300 of them in 2012).

The very number of producer groups, however, does not prove that farmers are well organized. If we compare the number of group members – approximately 35,000 (according to unpublished data of the Ministry of Agriculture and Rural Development) and the number of farmers who are CAP beneficiaries, it transpires that in 2015 as little as 2.5% of agricultural producers were associated in producer groups. Therefore, even though the number of group members has importantly increased between 2012 and 2015 (by roughly 9,000 members), the share of farmers formally cooperating with others is still very low. Furthermore, unpublished data of the Agency for Restructuring and Modernization of Agriculture on trade of agricultural producer groups indicate that the value of products sold by such groups (excluding fruit and vegetable groups) was in 2012 about one billion euro, which made up as little as approximately 6% of agricultural commodity production in that year, in current prices. A relatively modest share of producer groups in supplies of agricultural products on the market in Poland is clearly seen also in international comparisons (recall the statistics reported in the introductory part of the text). That said, it is worthwhile to notice here that the degree of farmers' horizontal integration varies according to the sector and region (Matczak, 2012). In terms of value, the highest sales of products through producer groups were observed in the poultry, oilseeds, pig production as well as fruit and vegetable sectors.

Taken together these statistics tend to suggest that regardless of the noticeable acceleration of the process of setting up agricultural producer groups in recent years, the level of formal cooperation among farmers in Poland is relatively low. In line with the theoretical considerations presented earlier, we would like to look for the causes of such situation by focusing on the issue of trust and, closely related to it, social capital. As argued above, these two factors are likely to affect, to a considerable extent, both the speed of the development, and the stability of, cooperation between agricultural producers (Kenis and Oerlemans, 2008).

Adopting this perspective in our context seems to be appropriate as many concerns have been expressed that lack of trust/social capital is among the main reasons for relatively low level of farmers' cooperation in Poland. Indeed, as argued in numerous studies, the level of social capital in Poland is low and this concerns not only urban areas, but also rural areas (e.g. Czapiński, 2008). In their studies concerning the latter, various scholars emphasize in unison that rural inhabitants are characterized by a relatively low level of trust in people, including local residents (Fedyszak-Radziejowska, 2006; Wieruszewska, 2002). They also point out to the problem of relativism of values and selective conformity to social norms (Halamska, 2011) which in turn leads to an over-politicisation of local authorities and corruption. Further, academics point out to the weak status of civil society, which, if exists at all, has a familial-egoistic nature (Halamska, 2011) and functions despite the lack of, not based on, mutual trust (Fedyszak-Radziejowska, 2006). Given these accounts, the lack of trust and low level of social capital naturally come to mind as plausible explanations for a relatively slow pace of the development of formal cooperation between Polish farmers, mentioned above. Below we try to see to what extent this is really the case.

4.4 Empirical analysis

The data that we use in our analysis come from two surveys. The first survey was done with 200 farmers who already have been members of producer groups. These farmers come from 10 regions (out of 16 in the whole country), which yields 20 observations per region. They are members of 50 producer groups specialised in selling various products from poultry, pigs, cereals, and cattle, to fruits and vegetables. The number of respondents per group is four, except for two groups for which it equals to three and five.

What should be stressed here is that all farmers from this survey were participants in one of the trainings organised by the National Cooperative Council. What follows, the sample analysed cannot be treated as representative for the country. Further, our respondents constitute only a fraction (sometimes larger, sometimes smaller) of all members of agricultural producer groups included in our sample. Therefore, we cannot exclude that information from members missing from our sample would change the picture we present. Finally, due to the nature of our data just described, what we show here are correlations and we are not pretending to estimate any sort of causal relationships. These caveats should be kept in mind while interpreting our results.

To have a group of controls, we also run a second survey, this time with 60 farmers who have not been participating in any producer group. These farmers come from three regions (again 20 per region) which were also included in the survey with group members. They were selected to resemble as much as it was possible our respondents who were members of producer groups (in terms of scale of production and technology used). In contrast to respondents questioned in the first survey, these farmers were not taking part in the training organised by the National Cooperative Council. What should be stressed though is that they were also approached by this organisation. Again, this should be taken into account when reading our results.

The focus of both surveys was on eliciting information which could be helpful in establishing the factors that are crucial for setting up producer groups in Poland. Most importantly from our perspective, both surveys included questions capturing farmers' attitude towards cooperation, their interpersonal relationships to people belonging to agricultural producer groups, or their opinion on the general level of trust of their neighbours. Finally, both questionnaires were designed in a way that would make comparing the two groups of respondents possible. All the data were collected between January and June 2014.

With that in hand we now move to the analysis of the data. It seems useful to start with investigating to what extent the common opinion arguing that a lack of trust hinders setting up of producer groups is confirmed by our respondents. To answer that, we analyse two questions from our survey. First, we look at the closed question: what barriers in your opinion hinder setting up producer groups? When answering this question farmers could have chosen more than one answer and one of them was 'lack of trust'. Among the respondents who were members of producer groups this answer was chosen by 70% of the sample (four missing observations not included) and this was far the most often chosen alternative. Among non-members respondents this reason was mentioned by 50% of those who responded. In this sample however, except for this reason, farmers equally often mentioned also lack of leaders (who could initiate the process) and lack of other strong producer groups that could serve as an example. The second question which we use to analyse the role of trust in establishing producer groups is an open question in which farmers were asked to mention the most important barriers which they think hinder setting up producer groups. Among group members lack of trust appears in 44% of answers (19 missing obs. not included), whereas among non-members this reason is mentioned by 31% of our respondents (9 missing obs. not included). By this question, this answer was the most often chosen in both samples.

Taken together, these statistics seem to confirm that lack of trust is indeed an important obstacle to establish producer groups in Poland. Our respondents who are involved in collective action even perceive it as the most important barrier. These findings seem to be consistent with those coming from other studies related to agricultural producer groups in Poland. For example, Banaszak (2008), based on a survey with leaders of producer groups which were closed, reports that problems with commitment, loyalty and trust in the leader and other members were among the most important factors which resulted in groups finally splitting up. Similarly, lack of trust in potential partners is also argued to importantly impede the strengthening of cooperatives/ producer groups' position by Matczak (2012).

With that in hand, it is interesting to see what type of trust (the one present in interpersonal relationships or a more general one) is mainly missing. To do so we analyse whether having a relative or a friend in a producer group was an important determinant for our member-respondents to join the group, whether our non-member respondents can take advantage of such social relationships, and to what extent our two types of respondents differ in terms of their opinion on the general level of trust of their neighbours and their willingness to cooperate.

We start with the analysis of the importance of close relationships of those who have already been involved in a producer group. More specifically, we investigate our respondents' answers to the question 'what was decisive for your participation in a producer group?'. As above, it was a closed question and the surveyed farmers could have selected more than one answer. Interestingly, 50% of them mentioned that an important reason was having a friend among other members. Another 23% mentioned that they joined the producer group because of their relatives being members as well. The remaining (roughly) 27% of our sample did not mention having friends or relatives to be an important reason for their participation in a producer group.[5] This in turn suggests that for a majority of our respondents who are involved in a producer group having close interpersonal relationships with other group members was important. These findings correspond well with that reported by Banaszak and Beckmann (2006) who investigated the role of leadership in the process of establishing and the functioning of producer groups in one of the Polish provinces. As the authors show, interpersonal relationships characterised most of the groups which were included in the study. More specifically, interviewed leaders agreed that they knew most of the members of their group (almost 80% respondents) and were involved in some business with most of their groups' members (about 40%) before the group was established.

These results may suggest that for farmers, for whom interpersonal relationships with other group members are important, an opportunity to rely on informal enforcement mechanisms to establish and govern cooperation could be a key issue. To some extent, this view can be supported by statistics which arise from collating the reasons for joining a group with the opinion on the role of trust in encouraging development of producer groups. Interestingly, for those mentioning relatives or friends as an important reason to join a producer group the share of responses indicating lack of trust as an important barrier to setting up a producer group was 77 and 74% respectively. Instead, for those who did not refer to interpersonal relationships this share was significantly lower and accounted for 57% of responses. What follows, these results may suggest that those who perceive trust to be the most important barrier for successful cooperation join producer groups only if they have friends or relatives being a member as well. In that sense, friendship or kinship relations may serve for them as an insurance against distrust to cooperate from others they do not know personally.

To have further insights on that, we look at the size of producer groups to which our respondents belong. We take advantage of the fact that respondents mentioning having relatives as an important reason to join a producer group are not uniformly

[5] Four farmers did not answer this question and thus we have four missing observations. We do not include them when calculating these statistics.

distributed across 50 groups included in our sample but are present in only 20 of them. Thanks to this we can compare the size of these 20 groups with the size of the remaining 30 groups, for which we have no evidence that kinship relationships were important. Interestingly, in line with our theoretical predictions, the former groups are on average smaller than the latter (32 vs 45 members respectively). This difference seems to be robust to excluding groups with more than 100 members, or analysing only groups with less than 25 members. While this cannot be treated as a definite evidence, these observations again suggest that groups where family relationships were important, could use more informal enforcement mechanisms to govern cooperation. This is because the latter, as argued above, can be efficiently used only in groups of a relatively small size.

This last conclusion is further supported by the fact that farmers belonging to larger groups seem to have a better opinion about the generalised trust of people in their neighbourhood or their general willingness to cooperate. Interestingly, for those agreeing or rather agreeing with the statement that 'in general people in the neighbourhood trust each other' the median size of a producer group is 9 members and the mean is 41.5 members.[6] For those who disagree or rather disagree with this statement the respective statistics were 6 and 26.1 members. The same pattern could be observed when we look at the statement: 'in general people in my neighbourhood are willing to cooperate'. Among those (rather) agreeing with it the median and mean size of the group was 12 and 43.8 members respectively, whereas for those who (rather) disagree with it the respective numbers were 6 and 15.5 members. If we assume that a general trust could be used as a proxy for farmers' willingness to cooperate under more formal enforcement mechanism, these statistics show that indeed this seems more likely to happen in larger groups.

The question obviously arises whether a more general trust and mutual trust which we capture by the importance of close interpersonal relationships, can be indeed used as proxies for individual's perceived possibility of relying on different enforcement mechanisms. The situation in which these two types of trust are not positively correlated may warrant this interpretation. While this issue certainly needs further investigation, our data provide some evidence which justifies taking such assumption. More specifically, when we look at those who mentioned having a family member in a producer group as an important reason for getting involved in a collective action, the share of those positively assessing neighbours' willingness to cooperate (general trust) is 52% (60%). For those referring to having a friend or to other reason not related to interpersonal relationships these numbers are 73% (77%) and 71% (68%) respectively. Taken together, these statistics show that, especially for those for whom kinship relationship are important, mutual trust and generalised trust can indeed

[6] When analysing farmers' answers to this question, we look specifically at the share of those that either agree or rather agree with this statement in contrast to the share of those that either disagree or rather disagree with it. Answers 'hard to say' were treated as missing observations. The same rules apply to the analysis of farmer's opinion about the statement: 'in general, people in my neighbourhood are willing to cooperate'.

capture different things. For those mentioning friendship relationships as important the picture is not that clear though. We return to this issue below.

Let us now have a look at respondents who act on their own and do not belong to any producer group. If the assumption that farmers for some reasons would like to use informal enforcement mechanisms as a basis for cooperation is correct, then we should observe among this type of respondents hardly any who have family members being involved in producer groups. Interestingly, this is precisely what we observe in the data. Among 58 farmers who answered the relevant question only seven said that they have relatives being involved in a producer group (six mentioned to have only a few of them, whereas one mentioned having many of them). Similarly, only seven farmers (out of 57 who answered the relevant question) mention having relatives who have similar farms to theirs (in terms of both size and production type). 19 respondents say that they have relatives running a farm of the same production type but of different size, whereas 31 report not having relatives being involved in agricultural activities similar to them. In consequence, the potential circle of relatives with whom they could set up a producer group is relatively limited.

As far as friendship relationships are concerned, as above, the picture is less clear. This is because 32 farmers who are not involved in formal cooperation mention having a friend who is a member of a producer group (10 of them claim to have many such friends). Further, 23 farmers (out of 57 who answered the relevant question) mention having friends who have a farm similar to theirs (in terms of size and production type). In addition 26 respondents claim to have friends running a farm of the same production type but of different size and only seven report not having friends being involved in agricultural activities similar to them. Therefore, the group of friends with whom they could potentially involve in collective action is relatively large. Overall, with statistics referred to earlier, these observations may suggest that friendship relationships could be insufficient to create efficient enforcement mechanisms to start cooperation using informal institutions. This conclusion though needs further research.

Next, let us have a look whether the two types of respondents (members vs non-members) differ in terms of their opinion on the level of general trust and willingness to cooperate in their neighbourhood. In principle, if producer groups' development is hindered mainly by farmers' attachment to informal enforcement mechanisms then the two types of respondents may not differ in the way they answer to these questions. Instead if farmers believe that formal institutions can effectively sustain cooperation, we should expect that the assessment of general trust/willingness to cooperate in their neighbourhood among member-respondents should be higher than that observed among non-members.

If we look at simple averages the share of those agreeing or rather agreeing that in general people in the neighbourhood are willing to cooperate is slightly higher among group members than among non-members (70 vs 63% respectively). The same result we get when we look at the share of those agreeing or rather agreeing with the statement that, in general, in the neighbourhood people trust each other. In that case the respective shares are 67 and 65, respectively. Based on that, we do not observe a radical difference between members and non-members respondents. An interesting insights emerge however, when we analyse these data at regional level. As mentioned before, non-members were surveyed in three regions, for which we also have data for members respondents (in each case 20 observations per region). When we collate the abovementioned data only for these three regions it appears that in one of them (*Mazowieckie* region), the share of those (rather) agreeing with the two statements mentioned above is much higher for farmers being involved in producer groups than for those acting on their own. Instead, for two other regions (*Podkarpackie* region and *Kujawsko-Pomorskie* region) the situation is reversed and we observe a much higher share of those (rather) agreeing with the two statements among non-members respondents.[7] The relevant statistics are presented in Tables 4.1. and 4.2.

This clearly shows that there might be considerable differences across regions in terms of how people perceive different enforcement mechanisms governing collective action. Thus, it may provide some explanation for the emergence of different regional cooperative equilibria suggested by other studies focusing on farmers' cooperation in Poland (e.g. Matczak, 2012). What might account for these differences? As shown above, (rather) agreeing with the abovementioned statements is positively related to the size of a producer group. This in turn is consistent with the observation presented earlier that smaller groups tend to be more dependent on informal enforcement mechanisms based on interpersonal relationships. Put it in a different way, problems with general trust may explain why these farmers look for opportunities to form a producer group based on mutual trust. This interpretation is consistent with the fact that in *Podkarpackie* region 90% of the surveyed farmers report that either relatives' or friends' membership in a producer group was important for their involvement. As regards *Kujawsko-Pomorskie* region however, this share is much lower and accounts for 30%. At the same time though, 50% of our respondents from this region report that there was a family member in the group they joined (although did not mention it to be an important reason for starting formal cooperation). It is also interesting to note that in *Podkarpackie* region, in comparison to *Kujawsko-Pomorskie* region, average size of a producer group is smaller. This in turn may suggest that in case of producer groups in the latter region there is a problem with generalised trust but at the same time relying with a collective action on mutual trust and informal enforcement is not that often chosen.

A supplementary analysis shows also that relatively larger problems with general willingness to cooperate and general trust are reported by farmers in groups specialised in crops rather than in animal production. This again is in line with the

[7] For all other regions for which we have data only for members respondents the share of those (rather) agreeing with the two statements is always higher than the share of those (rather) disagreeing and it ranges from 63 to 100%.

Table 4.1. In my neighbourhood people are generally willing to cooperate.

	Farmers belonging to producer groups ('agree' and 'rather agree')	Farmers not belonging to producer groups ('agree' and 'rather agree')
Kujawsko-pomorskie region	38%	53%
Mazowieckie region	81%	59%
Podkarpackie region	32%	76%

Table 4.2. In my neighbourhood people generally trust each other.

	Farmers belonging to producer groups ('agree' and 'rather agree')	Farmers not belonging to producer groups ('agree' and 'rather agree')
Kujawsko-Pomorskie region	40%	55%
Mazowieckie region	75%	54%
Podkarpackie region	41%	83%

hypothesis that third party enforcers should be more often used when more advanced technologies are applied. In that sense, our results showing that farmers involved in animal production report less problems with general willingness to cooperate and general trust is not that surprising. Coming back to the analyses exploring regional differences, our respondents from *Mazowieckie* region specialise in animal production, while in *Podkarpackie* and *Kujawsko-Pomorskie* regions farmers specialised in crop production outnumber those specialised in animal production. This is fully consistent with the statistics reported in Tables 4.1. and 4.2. indicating that in *Mazowieckie* region the perceived level of general trust/general willingness to cooperate is much higher than that observed in *Podkarpackie* and *Kujawsko-Pomorskie* regions. Concerning the two latter regions, this specialisation in crops is more evident in Podkarpackie than *Kujawsko-Pomorskie* region. Given the potential linkage between production type and mechanisms governing cooperation, this may provide additional explanation for why in the latter region relatively less farmers consider to rely with a collective action on informal coordination mechanisms and, consequently, why regional differences in collaboration equilibria appear. While we are not able to draw here any definite conclusions, these observations clearly show that this issue needs further in-depth research.

4.5 Conclusions

In this study we tried to provide some evidence linking the process of establishing agricultural producer groups with the issue of trust. Thus we aimed at contributing to an extensive literature which suggests that trust is necessary for efficient horizontal integration among farmers. The channel through which we attempted to link trust and setting up farmers' collective actions relates to different enforcement mechanisms that are used to start and govern cooperation. More specifically, we investigated to what extent farmers' willingness to involve in formal cooperation with each other depends on the level of general trust in the neighbourhood and interpersonal relationships. The former was associated with more formal governing mechanisms, based on a third party enforcer. Instead, interpersonal relationships and mutual trust related to them were associated with more informal enforcement mechanism, based on individuals' interactions with each other.

Using micro level data from 260 interviews with Polish farmers, we tried to explore the issues in question in two ways. First we looked at the extent to which farmers' participation in producer groups is driven by interpersonal relationships (kinship or friendship) to other group members. Second, we explored to what extent farmers being involved in collective action in the form of a producer group differ from farmers acting on their own in their opinions on the general trust and willingness to cooperate in their neighbourhood.

The explorative nature of our research determines that our results can be treated as only tentative. Thus they should be interpreted with caution. Nevertheless they do suggest interesting points for future research. Our findings tend to show that lack of trust is indeed a barrier to the producer groups' development in Poland. That said, this general remark while important is not too informative. This is because, our analysis suggests that for farmers in smaller groups, in particular those specialised in crop production, the problem might be predominantly with mutual trust between group members. As regards larger groups, in particular those specialised in animal production, this can mainly concern trust to a third party enforcer, external to the group. These results seem to be in line with two testable hypotheses which we derive from theoretical considerations. Further, the data we have analysed suggest that the patterns of cooperative behaviour may significantly differ across regions. Future work, which would try to verify these findings with more robust methods and/or data for other countries, could be a very promising line of research. We hope therefore that the presented analysis will help, at least to some extent, move this debate further.

Acknowledgements

The authors would like to thank Marcin Martynowski for his help in collecting the data. Financial support from the National Science Centre, Poland (grant no. DEC-2011/01/B/HS4/04733) is gratefully acknowledged. The authors would like to thank an anonymous referee for providing helpful comments on earlier drafts of the manuscript.

References

Acemoglu, D. and Jackson, M.O., 2015. History, expectations, and leadership in the evolution of social norms. Review of Economic Studies 82(1): 1-34.

Acemoglu, D. and Wolitzky, A., 2015. Sustaining cooperation: community enforcement vs. specialised enforcement. Available at: http://tinyurl.com/y83tp86w.

Banaszak, I., 2008. Determinants of successful cooperation in agricultural markets: evidence from producer groups in Poland. In: Hendrikse, G., Cliquet, G., Tuunanen, M. and Windsperger, J. (eds.) Strategy and governance of networks. Physica-Verlag, Heidelberg, Germany, pp. 27-46.

Banaszak, I. and Beckmann, V., 2006. The role of leadership in the process of establishing and sustaining cooperation. Available at: http://tinyurl.com/y7qagsox.

Banaszak, I. and Beckmann, V., 2008. The incentive structure in public policies - the example of the law on producer groups in Poland. Society and Economy 30(1): 79-92.

Beckmann, V., Otto, I.M. and Tan, R., 2015. Overcoming the legacy of the past? Analyzing the modes of governance used by the Polish agricultural producer groups. Agricultural Economics - Czech 61(5): 222-233.

Bijman, J., Iliopoulos, C., Poppe, K.J., Gijselinckx, C., Hagedorn, K., Hanisch, M., Hendrikse, G.W.J., Kühl, R., Ollila, P., Pyykkönen, P. and Van der Sangen, G., 2012. Support for farmers' cooperatives. Final report. Wageningen University, Wageningen, the Netherlands. Available at: http://tinyurl.com/y8zrqox8.

Bondyra, K., 2003. Kooperacja a przemiany na wsi (Co-operation and changes in rural communities). Wieś i Rolnictwo 3(120): 350-356.

Brehm, J. and Rahn, W., 1997. Individual-level evidence for the causes and consequences of social capital. American Journal of Political Science 41(3): 999-1023.

Brodziński, M.G., 2014. Oblicza polskiej spółdzielczości wiejskiej: geneza, rozwój, przyszłość. (Faces of Polish rural cooperatives: genesis, development, the future). Warszawa: Wydawnictwo Frel, Warsaw, Poland, 514 pp.

Burt, R.S., 2005. Brokerage and closure. An introduction to social capital. Oxford University Press, Oxford, UK, 279 pp.

Chlebicka, A., 2011. Czynniki wpływające na sukces grup producentów rolnych (Factors affecting success of agricultural producers groups). Journal of Agribusiness and Rural Development 4(22): 31-39.

Chlebicka, A., Fałkowski, J. and Łopaciuk-Gonczaryk, B., 2014. Grupy producentów rolnych a kapitał społeczny - potencjalne zależności w krótkim zarysie (Agricultural producer groups and social capital - the potential dependence). Wieś i Rolnictwo 3(164): 137-156.

Chloupkova, J. and Bjornskov, C., 2002. Could social capital help Czech agriculture? Agricultural Economics 48(6): 245-248.

Chloupkova, J., Svendsen, G.L.H. and Svendsen, G.T., 2003. Building and destroying social capital: the case of cooperative movements in Denmark and Poland. Agriculture and Human Values 20: 241-252.

Csaki, C. and Forgacs, C., 2008. Observations on regional level. In: Csaki, C., Forgacs, Milczarek-Andrzejewska, D. and Wilkin, J. (eds.) Restructuring market relations in food and agriculture of Central and Eastern Europe - impacts upon small farmers. Agroinform, Budapest, Hungary, pp. 29-51.

Czapiński, J., 2008. Kapitał ludzki i kapitał społeczny a dobrobyt materialny: polski paradoks (Human and social capitals as sources of economic well-being. Polish paradox). Zarządzanie publiczne 2: 5-28.

Dzun, W., 2009. Agricultural production cooperatives in the process of system transformations. Zagadnienia Ekonomiki Rolnej 4(321): 67-89.

Fedyszak-Radziejowska, B., 2006. Kapitał społeczny wsi – w poszukiwaniu utraconego zaufania (Rural social capital – in search of trust). In: Szafraniec, K. (eds.) Human capital and rural social resources. People – the local community – education. IRWiR PAN, Warsaw, Poland, pp. 71-120.

Greif, A., 2006. Institutions and the path to the modern economy: lessons from medieval trade. Cambridge University Press, New York, NY, USA, 526 pp.

Halamska, M., 2011. Społeczna kondycja polskiej wsi. Struktura, kapitały i społeczeństwo obywatelskie w I dekadzie XXI wieku (The social condition of the Polish countryside. Structure, capital and civil society in the first decade of the twenty-first century). In: Nurzyńska, I. and Drygas, M. (eds.) Rozwój obszarów wiejskich w Polsce. Diagnozy, strategie, koncepcje polityki (The development of rural areas in Poland. Diagnosis, strategies, policy concepts). IRWiR PAN, Warsaw, Poland, pp. 39-55.

Hendrikse, G.W.J. and Bijman, J., 2002. On the emergence of new growers' associations: self-selection versus market power. European Review of Agricultural Economics 29(2): 255-269.

Hendrikse, G.W.J. and Veerman, C.P., 2001. Marketing cooperatives and financial structure: a transaction costs economics analysis. Agricultural Economics 26: 205-216.

Hunek, T., 1994. Reorienting the cooperative structure in selected Eastern European countries: case-study on Poland. Central and Eastern Europe – Agriculture in Transition 5. Food and Agriculture Organization, Rome, Italy.

Kandori, M., 1992. Social norms and community enforcement. Review of Economic Studies 59(1): 63-80.

Karasiewicz, G., 2001. Systemy dystrybucji artykułów rolno-spożywczych na rynku polskim. Diagnoza i koncepcja zmian (Distribution systems of agro-food products on the Polish market. Diagnosis and the concept of changes). Wydawnictwa Naukowe Wydziału Zarządzania Uniwersytetu Warszawskiego, Warsaw, Poland.

Keele, L., 2007. Social capital and the dynamics of trust in government. American Journal of Political Science 51(2): 241-254.

Kenis, P. and Oerlemans, L., 2008. The social network perspective: understanding the structure of cooperation in: inter-organizational relations. In: Cropper, S., Ebers, M., Huxham, Ch. and Ring, P. (eds.) The Oxford handbook of inter-organizational relations. Oxford University Press, Oxford, UK, pp. 289-313.

Krzyżanowska, K., 2003. Przedsiębiorczość zespołowa rolników i jej uwarunkowania (Team entrepreneurship of farmers and its determinants). Wydawnictwo SGGW, Wydawnictwo SGGW, Warsaw, Poland.

Lemanowicz, M., 2005. Organizacje i Grupy Producentów Szansą na Poprawę Konkurencyjności Polskiego Rolnictwa (Producer organizations and groups as a chance for improving the competitiveness in Polish agriculture). Wydawnictwo SGGW, Warsaw, Poland.

Lopez, R.A. and Spreen, T.H., 2008. Co-ordination strategies and non-members' trade in processing co-operatives. Journal of Agricultural Economics 36(3): 385-396.

Majewski, E. and Dalton, G., 2000. Strategiczne opcje dla polskiego sektora agrobiznesu w świetle analiz ekonomicznych (Strategic options for the Polish agro-business sector in the light of economic analyses). Wieś Jutra, Warsaw, Poland, 588 pp.

Maliszewski, A., 1995. The fall of the co-operative movement in Poland: causes and consequences. ICA Review 88(1): 88-95.

Martino, G., 2010. Trust, contracting, and adaptation in agri-food hybrid structures. International Journal of Food System Dynamics 4: 305-317.

Matczak, P., 2012. Support for farmers' cooperatives: country report Poland. Wageningen University, Wageningen, the Netherlands. Available at: http://tinyurl.com/y8zrqox8.

Menard, C. and Valceschini, E., 2005. New institutions for governing the agri-food industry. European Review of Agricultural Economics 32(3): 421-440.

Menard, C., 2007. Cooperatives: hierarchies or hybrids? In: Karantininis, K. and Nilsson, J. (eds.) Vertical markets and cooperative hierarchies. Springer, New York, NY, USA, pp. 1-17.

Milczarek, D., 2002. Privatization as a process of institutional change. The Case of State Farms in Poland. Institutional Change in Agriculture and Natural Resources No 11. Shaker Verlag, Aachen, Germany.

Nilsson, J., Kihlen, A. and Norell, L., 2009. Are traditional cooperatives an endangered species? About shrinking satisfaction, involvement and trust. International Food and Agribusiness Management Review 12(4): 101-121.

Paldam, M. and Svendsen, G.T., 2001. Missing social capital and the transition in Eastern Europe. Journal for Institutional Innovation, Development and Transition 5: 21-34.

Pascucci, S., Gardebroek, C. and Dries, L., 2012. Some like to join, others to deliver: an econometric analysis of farmers' relationships with agricultural co-operatives. European Review of Agricultural Economics 39(1): 51-74.

Putnam, R.D., 1993. Making democracy work: civic traditions in modern Italy. Princeton University Press, Princeton, NJ, USA, 280 pp.

Putnam, R.D., 2000. Bowling alone: the collapse and the revival of American community. Simon and Schuster, New York, NY, USA, 544 pp.

Rural Development Program (RDP), 2010. Polish rural development plan 2007-2013. March 2010 version. Available at: http://tinyurl.com/ydclllda.

Rozelle, S. and Swinnen, J.F.M., 2004. Success and failure of reform: insights from the transition of agriculture. Journal of Economic Literature 42(2): 404-456.

Szabo, G.G., 2010. The importance and role of trust in agricultural marketing co-operatives. Studies in Agricultural Economics 112: 5-22.

Tisenkopfs, T., Kovach, I., Lostak, M. and Sumane, S., 2011. Rebuilding and failing collectivity: specific challenges for collective farmers marketing initiatives in post-socialist countries. International Journal of Sociology of Agriculture and Food 18(1): 70-88.

Wieruszewska, M., 2002. Społeczność wiejska – podstawy samoorganizacji (Rural community – the basics of self-organization). In: Wieruszewska, M. (eds.) Samoorganizacja w społecznościach wiejskich: przejawy, struktury, zróżnicowania (Self-organization in rural communities – manifestations, structure, diversity). IRWiR PAN, Warsaw, Poland, pp. 12-90.

Wintrobe, R., 1998. The political economy of dictatorship. Cambridge University Press, Cambridge, UK, 404 pp.

Wolz, A., Fiege, U. and Reisenberg, K., 2004. The role of social capital in promoting institutional changes in transitional agriculture. In: Van Huylenbroeck, G., Verbeke, W. and Lauwers, L. (eds.) Role of institutions in rural policies and agricultural markets. Elsevier, Amsterdam, the Netherlands, pp. 413-114.

5. Italian network contracts viewed from the perspective of transaction cost economics

F. Ventura[1], M.C. D'Arienzo[2] and G. Martino[3]*

[1]Department of Civil and Environmental Engineering, University of Perugia, Piazza Università 1, 06123 Perugia, Italy; [2]Confagricoltura, C.so Vittorio Emanuele II 101, 00186 Roma, Italy; [3]Department of Agricultural, Food and Environmental Sciences, University of Perugia, Borgo XX Giugno 74, 06100 Perugia, Italy; flaminia.ventura@unipg.it

Abstract

This chapter explores the network contract, a type of contract recently codified within Italian law. It creates a formal structure within which farmers can pool resources and information and undertake joint activities in order to increase their own innovative and competitive capabilities. Greater coordination among farmers also gives them leverage in integrated food chains. This increases the potential of primary producers to retain more value-added, although such horizontal networking also increases the complexity of food chains, since farmers need to coordinate themselves in order to reap these benefits. This paper elaborates a conceptual interpretation of the network contract from the perspective of transaction cost economics (TCE). We interpret network contracts in terms of the theoretical categories proposed by TCE (particularly focusing on the role of uncertainty and the characteristics of the governance mechanisms). On the basis of this conceptual framework we then examine network contracts and discuss their potential role in helping farmers to cope with market and technological uncertainties.

Keywords: network contract, transaction costs, coordination

5.1 Introduction

This study considers inter-firm networking in agribusiness value chains (AVC) focusing on the network contract, a type of contract recently codified in Italian law. The network contract allows farmers and other actors within the AVC to pursue common goals, mainly in relation innovation and competitiveness. This paper delineates the main characteristics of this type of contract and provides a tentative interpretation of it in the light of transaction cost economics (TCE).

The network contract is but one many of the organizational agreements that exist within the AVC. Scholars who have investigated vertical linkages within AVC recognize the increasing organizational complexity of such chains. (AVCs take a wide variety of organizational forms. Agricultural economists conceptualize this evidence as a continuum of forms and as the result of the adaptation and innovation process of the farm fostering new organizational arrangements and exchanges governance structures. (Peterson *et al.*, 2001; Saccomandi, 1998). According to Williamson (1985), exchange on the spot market is just one of many possible governance structures. The influence of property rights allocation models (Grossman and Hart, 1986) on the modelling of the organizational adaptation is also recognized (Saccomandi, 1998). Transaction costs are a significant influence on the organizational forms adopted within agribusiness supply chains (Ménard, 2005; Peterson *et al.*, 2001; Sauveé, 1998). Much research has been done, and much more is needed, to empirically validate the theoretical insights provided by TCE and the Theory of Organization in relation to AIVCs.. These disciplines provide useful theoretical insights to explain why AVCs contains a wide variety of organization forms (Grandori, 2017; Mènard, 2015, 2017). Networks are becoming increasingly widespread as an organizational form within AVCs and TCE offers a useful conceptual framework to investigate this phenomenon. Omta *et al.* (2001) highlighted the relevance of the concept of networks in analysing AVCs. Lazzarini *et al.* (2001) introduced the concept of the 'netchain', explaining how agribusiness firms address the problem of vertical coordination by cooperating with each other through establishing effective horizontal connections This insight pushed networks to the fore of conceptualizing the economics and governance of AIVC (Fritz and Schiefer, 2008). In this paper we address the legal and governance aspects of netchains, how their legal characteristics can be interpreted in terms of governance and the management possibilities (and constraints) that these give rise to.

The focus of this study is on the contract network as an instrument of the organization setting up in AVC, where both horizontal and vertical ties among the firms can be effective. The paper is structured as follows. Section 5.2 outlines the main characteristics of the network contract, as recently specified in Italian legislation. Section 5.3 introduces some key elements of transaction cost economics, which provide an analytical basis for our interpretation of the new legislation, developed in Section5.4. In Section 5.5 we devise a functional typology of contract networks and make some concluding remarks.

5.2 The general characteristics of the network contract and the opportunities it offers for farm networks

5.2.1 Network contracts and inter-firm collaboration

A *network contract* is an agreement through which two or more entrepreneurs, who may be spread across more than one sector, commit themselves, within a pre-established legal framework, to work together in specified areas in order to improve their innovative capacity and/or their competitiveness. This type of contract was initially defined in Italy by law n. 33/2009 which provides a basic regulatory framework. The law sets no dimensional or territorial boundaries, so micro, small, medium or large enterprises, and even foreign companies operating in Italy, can participate in such a network, even if they are located in different parts of the country. Moreover, companies operating in different sectors (production processing, logistic, etc.) can also participate in the same network.. To be eligible to participate actors must be registered in the Italian Register of Enterprises and be considered to be entrepreneurs, according to the Italian Civil Code. The basic requirements for the latter are: (1) to be engaged in an economic activity; (2) to produce and/or exchange goods and/or services; and (3) be appropriately organized and pursue their economic activity in a professional manner.

This new aggregation tool is designed to promote collaboration between enterprises, encourage them to innovate and enhance their competitiveness. It is specifically aimed at enterprises that are too small to compete meaningfully on markets that are increasingly globalized. The regulation establishes the network contract as a new governance model for inter-firm collaboration. The network contract allows firms to realize shared projects and objectives with a view to increasing their innovative capacity and competitiveness, while maintaining their independence, autonomy and specialization. Thus, entrepreneurs can enter into networks without giving up their legal and managerial autonomy. This can be done in order to improve their research and marketing activities, to expand their market scope, or to develop specific product or process-related know-how, in order to achieve business growth goals, which would be difficult to attain by operating individually.

5.2.2 The content and form of the network contract

The network contract aims to enhance the innovative and competitive capacity of firms who wish to collaborate through a network. This collaboration is usually based on a collectively designed strategy which sets out common objectives and seeks to trigger different partners' business The basic terms of the contract normally cover six areas:

- The definition of a common network program, including the rights and duties of each participating entrepreneur, the network's common objectives and the strategies to be pursued.

- The expected results, in terms of enhancing the participants' competitive capacity and the indicators for measuring progress towards these goals.
- The duration of the contract, the procedures for recruiting other entrepreneurs and, conditions regarding early withdrawal from the network.
- Where necessary, setting up a common fund, with clearly defined rules setting out the contributions to be made by participating companies, the activities within the common network program to be financed, the assets(soft or hard) to be acquired, and the contributions to be made by any new companies joining the network.
- Establishing a joint board to manage and, unless otherwise specified in the contract, to represent the network in negotiating with public authorities, accessing credit, developing common strategies in the internationalization and innovation processes and, developing and using promotional and protective instruments for products (such as quality marks and designations of origin that guarantee the genuineness and/or authenticity of the product(s).
- The rules concerning decision-making on any matter of common interest that lies beyond the powers conferred on the joint board, and the conditions under which the network's commonly-agreed program can be modified.
- Member farms are obliged to participate in common activities, and the benefits of so doing need to be sufficiently attractive to motivate farmers to join the network.

5.2.3 Farms networks and joint assumptions

The network contract is constantly evolving. The latest regulatory innovations, introduced by the Decrees Law 76/2013 and 91/2014, focus on network contracts in the agricultural sector. These provisions, create the possibility to develop a new organization formula, which allows new aggregated types of transaction for farms that can now formally participate in networks aimed at enhancing their innovativeness and competitiveness, while still maintaining their legal and fiscal autonomy. In the agricultural sector, a network contract may be signed by the parties with the assistance of one or more national agricultural organizations. This procedural simplification aims to reduce the economic burden and administrative complexity of making such agreements.

This legislation, whose introduction was strongly advocated by one of the main farmers organizations (Confagricoltura), allows individual participating farms to reorganize their production activities,. Through allowing them to use production factors. A distinctive feature of the farms network contract is the possibility for the farmers to pool resources and manage them collectively, including using the labour resources of a given farm within the network. The possibility to engage in joint recruitment, established by the Employment Package (D.L. 76/2013) has been developed to meet the needs of farms looking for new development opportunities. The new network contract allows a network of farms to employ one or more employees who can be assigned to carry out activities in the participating farms, with the wages and social security costs being (proportionately) shared among the participants of the network. This opportunity can be used by farms linked to one another by an Agriculture network contract or when at least 40% of the firms participating in the contract are farms. This new job sharing tool is particularly useful for networks going through a phase of strong innovation, giving the enterprises more flexibility (and less risk) to hire labour and engage the skills it needs.

A network contract establishes a stable and organized collaborative structure that allows farms to cooperate with each other through pooling their own production factors (funds, tools, equipment, know how, etc.) to produce a 'common' agricultural production, which is attributed to each participating farm. This common production can be distributed among the participants according to quotas determined by the network contract. In this way, the 'farm network' allows farms to pool their assets and resources in order to improve their businesses and strengthen their competitiveness. The medium to long term effects should be a reduction in costs and/or an increase in revenues.

These important regulatory innovations allow us to consider farm networks as a new organizational form for the agricultural sector with considerable potential to help micro-enterprises and SMEs to increase their productivity, reduce their costs, introduce innovations and access new markets without merging or losing their autonomy.

5.2.4 The advantages of farm networks

The benefits of a network contract can stem from different sources, some may be contractual, while others may derive from incentives provided by public authorities and/or financial institutions. Together, they improve the performance of affiliated firms through a number of means.

- *Increased market access*: a network allows members to develop a more complex and/or larger scale package of offers, and the potential to generate new sales opportunities on existing and/or new markets.
- *Reduction in production costs*: a network can help members reduce both variable and fixed production costs. Variable costs can be reduced by enhancing the network's negotiating stance with suppliers and fixed ones and by enabling firms to have a greater saturation of their production capacity, relying on the possible production collaboration of other network companies.
- *Reduced supply times*: by establishing stable relationships with other firms in the network, participating enterprises can create, develop and consolidate sub-supply networks among themselves, reducing supply times and optimizing inventories, and as a consequence, production times and the costs of maintaining excessive stock.
- *Fiscal advantages*: a single tax credit of 40% on the enterprise income per year is available to individual farms and cooperatives, grouped together in these consortia, for the development of new products, practices, processes and technologies (up to four hundred thousand euros of detax).

- *Easier public funding access*: contract networks get priority access to European Rural Development Funds.
- *Possibility to participate in tenders*: contract networks are eligible to participate in tenders.
- *Easier access to private credit*: contract networks receive advantageous conditions from banks, as a growing number of lenders have a positive attitude towards aggregated business structures and the guarantees of solidarity that they offer. In addition, the European Union and national and local public administrations are paying more attention to networks, which makes it likely that more incentives will be available to establish and develop networks over the coming years.
- *Increased innovative capacities*: affiliated companies can develop their innovative capacity through exchanging know-how and skills with other network companies.
- *Human resource development*: owners of, and workers within, networked businesses benefit from a more stimulating professional environment, multiplying their contacts with other businesses and becoming familiar with their realities, thereby enriching their work experience.
- *Easier access to international markets and business ties*: very small firms or those lacking resources struggle to enter, or remain successful in, international markets. Collaborating with other businesses gives such companies more access to financial resources (credit and public contributions), human resources and knowledge and allows them to share costs and organizational burdens. All these factors can help overcome this problem and enable smaller businesses to take advantage of the benefits of internationalization.

5.2.5 Defining the problem

The network contract is a tool that allows firms to participate in a structure that enables them to achieve common economic benefits while maintaining their legal identity and autonomy. The network is an organizational form that can successfully support participating firms (and farms) which transfer (limited) decision making powers, over some aspects of their business, to a joint board. In this section we explore the characteristics of this model, explain its economic nature and its potential for developing more effective management strategies.

The network form is already well-researched within economic and organizational studies. Grandori and Soda (1995, p. 1885) state that:

> An inter-firm network is a mode of regulating the interdependence between firms, which is different from the aggregation of these units within a single firm and from coordination through market signals (prices, strategic moves, tacit collusion, etc.) and which is based on a cooperative games with partner-specific communication.

Based on this definition they go on to provide a classification of networks on the basis of coordination modes which emphasize the specificity of the network. Powell (2003) previously argued that a network is based on organizational principles that differ from both the market and the firm. He recognized the specificity of networks as a tool for allocating resources in order to organize transactions in a reciprocal, preferential, and mutually supportive, way (cf. p. 303). Both these approaches emphasize the cooperative aspects of networks and consider conflicts and competition among participants to be less relevant in analytical terms. Transaction cost economics (TCE) classifies network as a hybrid mode: which has features of both cooperation and competition (Mènard, 2004). We elaborate on this idea in this paper, focusing on the problems of governance that network contracts raise. In the following section we summarize some basic elements of the TCE approach to governance, and then apply this set of concepts to network contracts.

5.3 Theoretical framework: drawing on transaction cost economics

TCE provides a useful theoretical structure to investigate the economic and organizational nature of network contracts. Mènard (2013) provides a comprehensive analytical representation of governance structures in which it is possible to position network contracts. He also identifies three basic problems associated with the management of hybrid structures. We summarize his work and then move, in Section 5.4, to discuss the characteristics of network contracts.

TCE states that governance is based upon devices that instil order in joint activities by allocating assets and rights in order redistributes benefits from mutual gains while mitigating any conflicts that may arise (Ménard, 2013; Williamson, 1996). Governance can be achieved through different alternative structures (Williamson, 1985, 1991). Ménard (2013, p. 1076-1704) invites us to consider two firms, 1 and 2, and four assets (A, a; B, b): a and b can only be jointly used, while A and B are used within each firm's

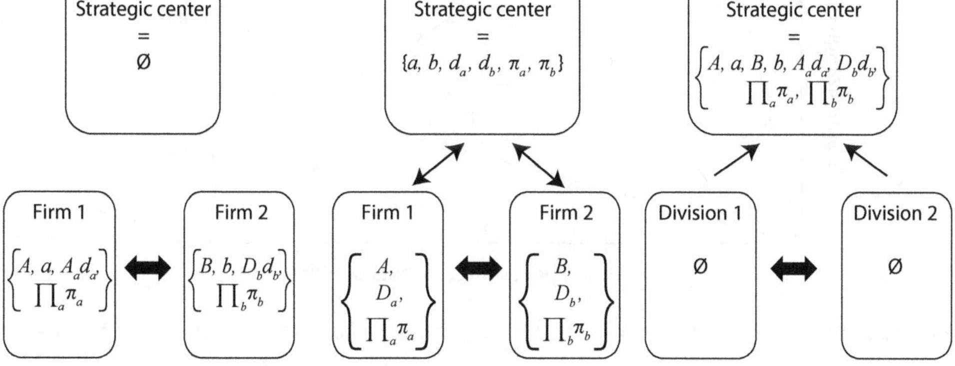

Figure 5.1. Modes of organization (Ménard, 2013, p. 1074).

boundaries. As a consequence, the two firms have full decision rights over 0 and B (D_A and D_B), but decision rights over a and b (d_a and d_b) require the firms to cooperate. The resulting payoffs are π_A, π_B, π_a π_b (the π_a last two are only generated if the relevant assets are jointly used). The strategic centre is the institutional entity under which transactions are initiated, negotiated, monitored, adopted, enforced and terminated. The decision-making rights allocated to the strategic centre will differ within each governance structure. Figure 5.1 compares the alternative modes of governance: market, hybrid and firm, in terms of the allocation of decision-making rights.

Mènard (2013) specifically states that the forces that induce the choice of a hybrid governance structure and the strategic choices that define the degree of centralization give rise to various types of governance structures. This results in a variety of governance forms (Figure 5.2). One of the key motivations for adopting a hybrid structure is the need to cope with uncertainty, both in terms of markets and technological developments (Robertson and Gatignon, 1998; Walker and Weber, 1984) which drives individual actors to create a cohesive governance structure based on shared rules and to pool resources and rights (Ménard, 2013, p. 1094). In so doing they need to strike a balance between their desire to preserve their autonomy and co-operating and sharing control over strategic issues to a limited and predefined extent (Ménard, 2013, p. 1094).

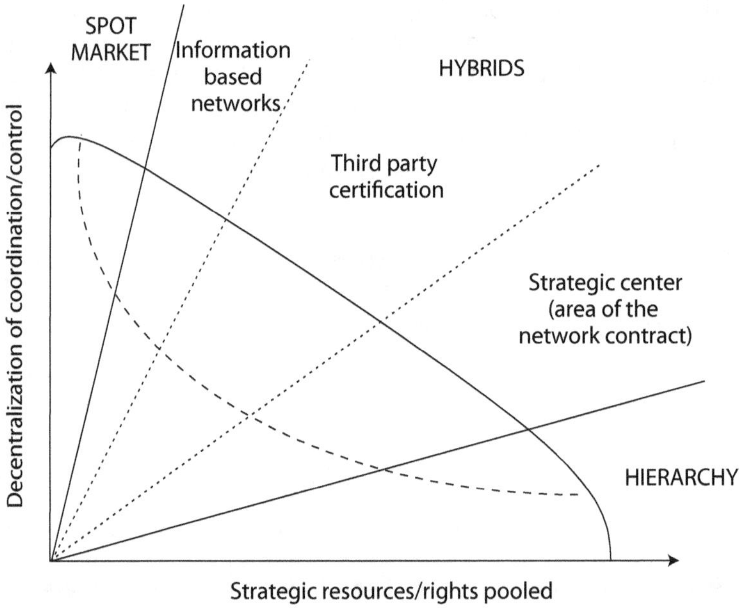

Figure 5.2. Network contract in the Ménard's hybrid typology (elaborated from Mènard, 2013, p. 1096).

Ménard (2004, p. 351) identified three main common features of hybrid governance structures:

- a tendency for independent firms to share (some of) their resources (pooling);
- the regulation of relationships between actors (contracting);
- the importance of competitive pressures in shaping the specific governance form (competing).

When pooling resources one first needs to identify the resources that need to be pooled (Ménard, 2013, p. 1087). Pooling can be achieved in a number of ways, such as developing a subset of shared resources from which each actor can draw, by sharing resources sequentially, or by building and maintaining joint assets. Pooling has three main practical implications: (1) it highlights the centrality of selecting the right potential partners, since common investments can give rise to opportunistic behaviour; (2) common planning systems must be put in place, and; (3) it requires the establishment and ongoing management of the implementation of a proper information system among participants. In essence, pooling requires an ongoing relationship which, in turn, implies the need to cooperate and to coordinate. This is the core of the first problem characterising hybrid forms: 'how to carry out low-cost coordination among agents without losing the advantages of decentralised decision-making'.

Contracting provides, a more or less formal means to regulate relationships and transactions among the independent partners within the hybrid structure. Contracts are often intrinsically incomplete. This is partly an expression of uncertainty, but can also leave room for opportunistic behaviour as asset specificity grows. On the other hand, it implies that the contract is often seen as a simple framework intended to leave more or less room for manoeuvre. A second typical problem that emerges here is 'the need to identify efficient mechanisms which are consistent (aligned) with the nature of the transactions' (e.g. in terms of the parties' ability to monitor the organisation and to search for solutions to emerging questions).

Competition influences hybrid forms because partners may also be competing against each other (e.g. while they may cooperate in some fields – typically R&D – they might also be competing for the same target market), and also because these organisations may be competing against other arrangements, including other hybrids. Moreover, if the asset specificity is low or moderate, participants may be encouraged to move into alternative and competing organisations, rendering the organisation unstable. So the third typical problem concerns 'the design of a mechanism that allows the organisation to make joint decisions in order to discipline partners, resolve conflicts, and prevent free-riding'.

These three features influence the extent to which the coexistence of cooperation and competition patterns the hybrid arrangements, and the extent to which their efficiency and workability depend upon mechanisms designed to reconcile legal autonomy and interdependence.

5.4 The network contract in the light of a transaction cost perspective

The network contract should be considered as tool to establish a hybrid governance mode in which participants maintain their autonomy, while defining an area of joint activity, based on allocating a subset of decision-making rights to a joint board, which manages the network. The subset of assets, rights and payoffs are chosen by the participant firms on the basis of their expectations in relation to their objectives in cooperating.

Identifying the partners to be invited to be involved in a network contract is the first step in building up the network. This step imposes 'search costs' on each partner (and particularly the initiator(s)). Potential partners have to meet basic criteria. They have to have resources (skills, knowledge, contacts, finance, land, etc.) that will aid the network to achieve its goals and need to be willing to cooperate with the network and to be monitored for opportunistic behaviour.

Secondly, the firms/farms setting up the network contract have to take the strategic decision over which resources and rights to pool. This requires the partners to identify the type and the amount of resources they are willing to pool and is related to the types of decision-making rights they want to share. This decision allows partners to coordinate their activities at low-cost, on a decentralized basis, without losing their autonomy. Pooling decision rights involves establishing a joint board, although the law gives networks the freedom to establish type of the board it deems to be suitable. This will depend greatly on market or technological uncertainties. As the network is likely to be involved in one or both of these areas, the effects of a critical level of uncertainty may reduce the capacity of the network in achieving the expected payoffs of the joint activities. The level of decision-making powers allocated to the joint board plays a key role in defining the scope of the network's activities, and its future performance.

The level of decision-making rights allocated to the joint board will, of course, vary according to the type(s) of activities the network plans to carry out. If they are seeking to innovate the partners will have to organize the technological creation and the design and implementation of the new process or product. While the parties will need to manage specific appropriability hazards (Oxley, 1997; Teece, 1996), the allocation of decision-making rights to the joint board will also need to bear in mind the need to minimize transaction costs related (Martino *et al.*, 2017) and the complementarity among alternative patterns.

In the case of marketing oriented networks the partners need to identify the effective equilibrium between the benefits of cooperating and the reality of competing for the same, or similar, market segment(s) (Ménard, 2004). Quality issues become prevalent, especially in the AIVC (Raynaud *et al.*, 2005) while the network's position within the value chain may influence the allocation of decision-making rights with respect to transactions in the chain (Raynaud *et al.*, 2009). A third issue is the necessity of the network managing specific market functions on behalf of the participant partners (Royer *et al.*, 2016).

Thirdly, the partners have to set up efficient mechanisms to monitor their behaviour. The complexity and cost of monitoring will vary according to the number of partners and the number of activities jointly carried out. This may influence the size of the network and, also will add to the negotiation costs when setting-up the network. Moreover the network needs to set up well-defined procedures to resolve potential conflicts among partners. These type of procedures are normally defined when establishing the network and often involve a form of arbitrage to resolve future (and possibly unforeseen) conflicts. While it is in the interest of the network reduce the costs of the conflict resolution, the law may limit the possibility of arbitrage application and a further set of procedures may be required to prevent free-riding behaviour. This kind of behaviour may increase the appropriability hazards as well as possibly exacerbating competition among the partners in the same market segment. Figure 5.3 summarizes the analysis illustrated above.

5.5 Conclusions

The network contract is an innovative model, intended to enhance cooperation between companies. To understand how it might work requires a precise understanding of the instrument and its likely effects. One also needs to bear in mind the attitudes of those likely to be drawn to this type of arrangement. Business relationships within a network contract need to be based on loyalty and trust: whereas, not being governed by the Civil Code, the contractual provisions must be clear and specific.

Farms and firms participating in network contracts may benefit unequally. For example, some companies in a network that helps its members participate at trade or food fairs are likely to sell more produce than others. And, in networks that pool production factors, some firms may benefit more than others.

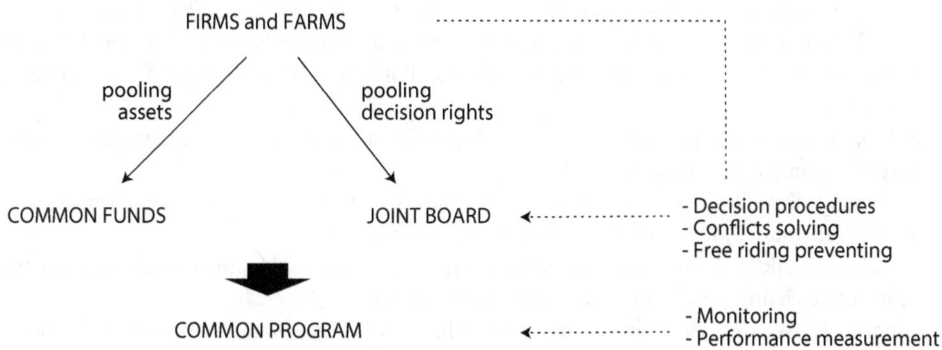

Figure 5.3. Network contract in a transaction cost economics perspective.

If we try to classify networks on the basis of their inter-company organizational structure, it is possible to recognize two main categories: (1) vertical or branch networks, in which network members contribute to the formation of the whole chain (purchase of raw materials, processing, marketing); and (2) horizontal networks: in which affiliated companies may carry out the same or different activities but collaborate on the joint realization of one or more business functions that are part of their business.

In terms of functions networks can be divided into several categories (which may sometimes overlap):
- supplying networks;
- innovation networks;
- production networks;
- networks for the acquisition and/or delivery of common goods and services; and
- marketing networks.

It is also important to point out that, as far as the merely contractual aspect is concerned, it is not possible to construct or use a business network using a uniform, template, network contract, since each network will have its own specific set of objectives, activities, and organization. As such each network contract needs to be tailor-made to ensure maximum compliance with the needs of participating firms and the network. However the regulatory guidelines do allow for the use of 'blueprint' network contracts, which can be adapted to specific circumstances and requirements.

References

Fritz, M. and Schiefer, G., 2008. Food chain management for sustainable food system development: a European research agenda. Agribusiness 24(4): 440-452.

Grandori, A., 2017. Linnaeus in the jungle: configurational lenses for discerning forms of economic organization in agri-business. Chapter 2. In: Martino, G., Karantininis, K., Pascucci, S., Dries, L. and Codron, J.-M. (eds.) It's a jungle out there – the strange animals of economic organization in agri-food value chains. Wageningen Academic Publishers, Wageningen, the Netherlands, pp 51-66.

Grandori, A. and Soda, G., 1995. Inter-firm networks: antecedents, mechanisms and forms. Organization Studies 16(2): 183-214.

Grossman, S.J. and Hart, O.D., 1986. The costs and benefits of ownership: a theory of vertical and lateral integration. Journal of Political Economy 94: 691-719.

Lazzarini, S., Chaddad, F. and Cook, M., 2001. Integrating supply chain and network analyses: the study of netchains. Journal on Chain and Network Science 1(1): 7-22.

Martino, G., Rossetti, E., Marchini, A. and Frascarelli, A., 2017. Process innovation in milling stage in olive oil sector: evidence from an empirical analysis in Umbria (Italy). British Food Journal 119: 1748-1765.

Ménard, C., 2004. The economics of hybrid organizations. Journal of Institutional and Theoretical Economics 160: 345-376.

Ménard, C., 2005. A new institutional approach to organization. In: Mènard, C. and Shirley, M.M. (eds.) Handbook of new institutional economics. Springer, New York, NY, USA, 281-318.

Mènard, C., 2013. Hybrid modes of organization. In: Gibbons, R. and Roberts, J. (eds.) Handbook of organizational economics. Princeton University Press, Princeton, USA, pp. 1066-1106.

Mènard, C., 2015. Do contracts solve it all? Keynote Lecture at the International Seminar 'Trough the Lens of the Contract: Regulating AgriFood System in the Future of Europe'. March 5, 2015. Department of Agricultural, Food and Environmental Sciences, University of Perugia, Perugia, Italy.

Ménard, C., 2017. Finding our way in the jungle: insights from organization theory. Chapter 1. In: Martino, G., Karantininis, K., Pascucci, S., Dries, L. and Codron, J.-M. (eds.) It's a jungle out there – the strange animals of economic organization in agri-food value chains. Wageningen Academic Publishers, Wageningen, the Netherlands, pp 27-50.

Omta, S.W.F., Trienekens, J.H. and Beers, G., 2001. Chain and network science: a research framework. Journal on Chain and Network Science 1: 1-6.

Oxley, E., 1997. Appropriability hazards and governance in strategic alliances: a transaction cost approach. Journal of Law, Economics and Organization 13(2): 387-409.

Peterson, H.C., Wysocki, A. and Harsh, S.B., 2001. Strategic choice along the vertical coordination continuum. International Food and Agribusiness Management Review 4(2): 149-166.

Powell, W.W., 2003. Neither market nor hierarchy: network forms of organization. In: Handel, M.J. (ed.) The sociology of organizations: classic, contemporary, and critical readings. SAGE, Thousand Oaks, CA, USA, pp. 315-330.

Raynaud, E., Lioc, S. and Valceschini, E., 2005. Alignment between quality enforcement devices and governance structures in the agro-food vertical chains. Journal of Management and Governance 9: 47-77.

Raynaud, E., Sauvée, L. and Valceschini, E., 2009. Aligning branding strategies and governance of vertical transactions in agri-food chains. Industrial and Corporate Change 18: 835-868.

Robertson, T.S. and Gatignon, H., 1998. Technology development mode: a transaction cost conceptualization. Strategic Management Journal 19: 515-531.

Royer, A., Ménard, C. and Gouin, D.M., 2016. Reassessing marketing boards as hybrid arrangements: evidence from Canadian experiences. Agricultural Economics 47(1): 105-116.

Saccomandi, V., 1998. Agricultural market economics: a neo-institutional analysis of the exchange, circulation and distribution of agricultural products (European perspectives on rural development). Royal Van Gorcum, Assen, the Netherlands.

Sauveé, L., 1998. Toward an institutional analysis of vertical coordination in agribusiness. FAO, Rome, Italy.

Teece, D.J., 1996. Firm organization, industrial structure, and technological innovation. Journal of Economic Behavior and Organization 31(2): 193-224.

Walker, G. and Weber, D., 1984. A transaction cost approach to make-or-buy decisions. Administrative Science Quarterly 29: 373-391.

Williamson, O.E., 1985. The economic institutions of capitalism. Firms, markets, relational contracts. The Free Press, New York, NY, USA.

Williamson, O.E., 1991. Comparative economic organization: the analysis of discrete structural alternatives. Administrative Science Quarterly 36: 269-296.

Williamson, O.E., 1996. The mechanisms of governance. Oxford University Press, Oxford, UK.

6. Traceability standards and vertical coordination

S. Stranieri[*] and A. Banterle

Department of Environmental Science and Policy, Università degli Studi di Milano, Via Celora 2, 20133 Milan, Italy; stefanella.stranieri@unimi.it

Abstract

Traceability plays a strategic role in assuring the safety and quality characteristics of food products, in managing supply chain risks, preventing unfair practices, and in providing an efficient distribution of liabilities among the agents of food supply chains. In recent years, an increasing number of traceability standards within the food system is observed, especially in relation to the sustainability of food supply chains. The aim of this chapter is to analyse the different existing traceability standards within the EU, and the effects that they imply on the dyadic relationships and the organization of food supply chains. In specific, the present chapter introduces, firstly, the concept of traceability and the costs and benefits associated to the implementation of this standard. Secondly, the legal framework on mandatory and voluntary traceability standards is presented. Third, an evaluation of the effects on vertical coordination due to the introduction of different traceability standards is presented. In the last section an attempt to classify different existing standards on the basis of their organizational consequences is

Keywords: traceability, supply chain relationships, transaction cost economics

6.1 Traceability standards

6.1.1 Definitions

The definition of food traceability is not univocal (Table 6.1). At international level it is not possible to find a common and shared definition. The International Standard organization (ISO) in 1994 defines traceability as the '...ability to trace the history, application or location of an entity by means of recorded identifications' (ISO, 1994). Moreover, the ISO-22005:2007 defines traceability as 'ability to follow the movement of a feed or food through specified stage(s) of production, processing and distribution'. The EU General Food Law defines traceability as 'the ability to trace and follow a food, feed, food-producing animal or substance intended to be, or expected to be, incorporated into a food or feed, through all stages of production, processing and distribution' (Reg. 178/02, 2002).

Two important aspects come out from this definition. First, the traceability refers to an integrated approach which considers all the stages and operators of food supply chain. This leads to consider traceability standard within a supply chain perspective. Second,

Table 6.1. Some recent definitions of traceability.

Source	Definition of traceability	Keyword for traceability description
Institutions		
ISO, 1994	ability to trace the history, application or location of an entity by means of recorded identifications	product history
ISO 22005, 2007	ability to follow the movement of a feed or food through specified stage(s) of production, processing and distribution	supply chain movement of product
Reg. 178/02, 2002	ability to trace and follow a food through all stages of production, processing and distribution	product history and movement
Literature		
Cheng and Simmons, 1994	ability to retrace steps and verify that certain events have taken place	verification of product properties
Moe, 1998	ability to trace in one of the steps in the chain	firm internal transparency
Opara and Mazaud, 2001	collection, documentation, maintenance and application of information related to all processes in the supply chain in a manner that provides a guarantee to the consumer on the origin and life history of a product	collection of information, product guarantee
Bollen *et al.*, 2006	the means by which the information is provided	information
Dupuy *et al.*, 2005	ability, in every point of the supply chain, to find origin and characteristics of a product from one or several given criteria	origin
Schwagele, 2005	ability to identify the origin of an item or group of items, through records, upstream in the supply chain	information record
Karlsen *et al.*, 2013	ability to trace the movement of an entity, and the identification of an entity	product identification

traceability implies the implementation of system aimed at managing information related to products flows within the different stages of production processes.

The implementation of different systems of traceability can lead to a reorganization of vertical relationships within the food supply chains. The effects on vertical coordination will depend on the type of standard and system applied, leading to a set of different situations from a low to a strong coordination (Canavari, 2010; Stranieri and Banterle, 2015). Without going into further details, the definition provided by the European law identifies the subjects who have to adopt traceability and the information which has to be traced, but this information is not connected to single products and it does not make any reference on the type of systems to adopt. This means that the agents of the supply chain are free to choose among the available traceability systems those considered more appropriate in relation to firms' strategic objectives.

Several definitions of traceability exist also in the literature because of the complexity of the field (Olsson and Skjöldebrand, 2008). According to Cheng and Simmons (1994) traceability is 'the ability to retrace steps and verify that certain events have taken place'. Opara and Mazaud (2001) describe it as 'the collection, documentation, maintenance and application of information related to all processes in the supply chain in a manner that provides a guarantee to the consumer on the origin and life history of a product'. Moreover, Bollen *et al.* (2006) define traceability differently as 'the means by which the information is provided', and García *et al.* (2008) refer to such standard as the ability 'to trace all the elements that can be considered relevant enough for the organization within a particular project or software product'.

Some authors also distinguish between 'tracing' and 'tracking'. For Dupuy *et al.* (2005), tracing is the 'the ability, in every point of the supply chain, to find origin and characteristics of a product from one or several given criteria', and tracking is the 'the ability, in every point of the supply chain, to find the localization of products from one or several given criteria'. Schwagele (2005) refers to tracking as the 'the ability to follow the path of an item as it moves downstream through the supply chain from the beginning to the end', and tracing as the 'the ability to identify the origin of an item or group of items, through records, upstream in the supply chain'.

Karlsen *et al.* (2013) identify common characteristics associated to the existing concepts of traceability, i.e. the ability to trace the movement of an entity, and the identification of an entity. This leads to the consideration that the definition of traceability can be referred to all the systems able to record specific information related to an entity within the food supply chain, which can be a subject, a batch, a substance, a food or specific food characteristics. With regard to this perspective it is possible to consider traceability standards all the certifications which imply the adoption of a system able to trace one or more information related to a product along the supply chain. In the present analysis we refer to this last concept of traceability.

6.1.2 Types of traceability standards

With regard to the concept of traceability introduced in the Section 6.1.1, several voluntary traceability standards are present within the food supply chains. Such standards can be classified in different ways, mainly depending on the object of the analysis, and can differ in terms of:

- target of agents involved;
- types of product involved;
- schemes' owners;
- legal validity;
- level of traceability complexity.

Voluntary traceability standards cover a wide range of different initiatives that function at different stages of the food supply chain. They can operate at the business-to-business (B2B) level or at the business-to-consumer (B2C) level. B2C standards usually make use of logos, whereas at the B2B-level may do not.

Moreover, traceability standards can relate to certifications which refer to all food products characterized by precise production rules (like ISO 22005:2007, Organic production, PDO/PGI, BRC, etc.) or to specific products (like Fair Trade for coffee, Friend of the Sea for fish, etc.). Existing voluntary traceability standards can be classified also on the basis of the schemes' owners. It is possible to distinguish among standards belonging to a single economic agent (producers, farmers, retailers), to NGOs (ISO 22005/2008, Fair Trade, etc.), groups of firms of the same sector (BRC, SQF, IFS) and public authorities (PDO/PGI, umbrella schemes, etc.).

Then, if we refer to the legal validity of traceability standards, it is also possible to distinguish between national and international voluntary traceability standards, which comprehend both European and global standards. A study conducted by European Commission in 2010 has revealed the presence of 145 food traceability standards with a national validity. The main countries represented are Spain, Italy and Germany. In Italy, such rules are mostly implemented by retailers for the management of private label supply chains and by institutions for the development of local economies through the implementation of systems aimed at differentiating traditional food products (Banterle and Stranieri, 2013).

With regard to the level of traceability complexity, standards can differ also in term of the degree of batch dimension of the system implemented (degree of accuracy of product identification), the amount of information to be recorded for the standard compliance, and the sectors traced (Golan *et al.*, 2004). The batch dimension used deals with the breadth of the system. The amount of information exchanged deals with the precision of traceability, and the sectors of the supply chain involved are connected to the depth of the traceability system implemented.

As for the depth of traceability, different kind of standards can be distinguished on the basis of the entities involved by the system. Moe (1998) distinguishes two kinds of traceability standards: supply chain and internal traceability. The first one refers to the 'ability to track a product batch and its history through the whole, or part, of a production chain from harvest through transport, storage, processing, distribution and sales'. Thus, it deals with all the sectors involved in a certain food supply chain. Internal traceability is the 'ability to trace in one of the steps in the chain'. It deals with the transparency of information inside a firm.

With regard to the degree of batch dimension of the system implemented, it is possible to distinguish between two systems: (1) supply chain traceability; and (2) supply chain and product traceability (Figure 6.1).

The first one refers to the so called 'one step backward, one step forward' traceability. Such system aims at identifying all the agents of the supply chain. Indeed, all suppliers and customers along the supply chain are identified through a specific document. In this case, the degree of product identification accuracy is low and it is not possible to reconstruct the history of the products. Therefore, the system does not allow attributing specific liabilities among the agents of the supply chain.

Supply chain and product traceability system deals with the management of flows for separated batches. This system is able to trace information flow at level of agents of the

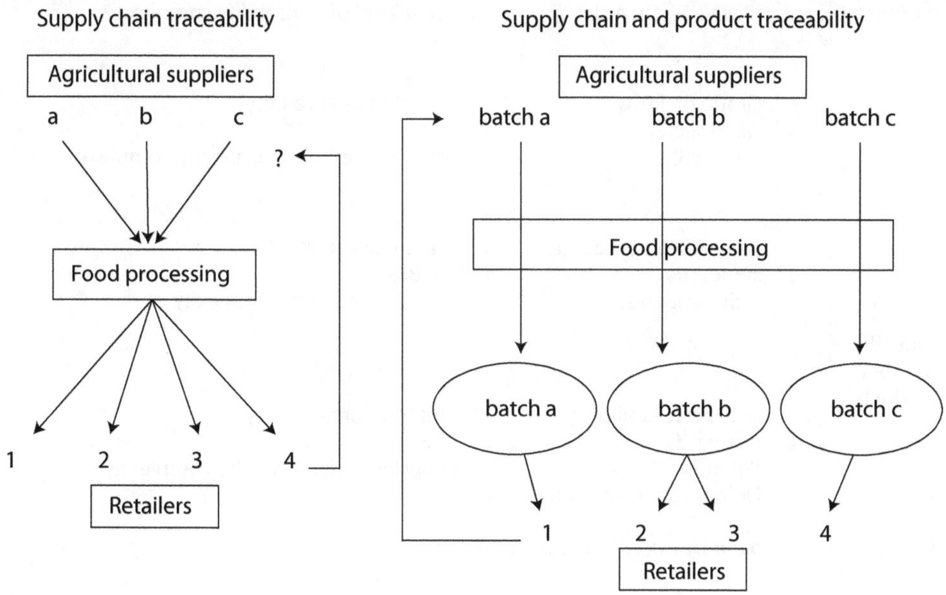

Figure 6.1. Supply chain and supply chain and product traceability system.

supply chain (supply chain traceability) and within a single firm by means of internal procedures (product traceability). Through this kind of traceability it is possible to trace back the product and to have the possibility to completely reconstruct the history of a certain product up to the identification of agricultural raw materials used in the production process. With such system it is possible to identify specific liabilities among the agents of the supply chain.

On the basis of the elements described above it is possible to draw a synthesis of existing traceability voluntary standards (Figure 6.2).

6.1.3 Costs and incentives for traceability standards

When firms implement traceability standards incur in several costs, which can be subdivided into (Banterle and Stranieri, 2005):

- costs for system implementation;
- costs for the system management.

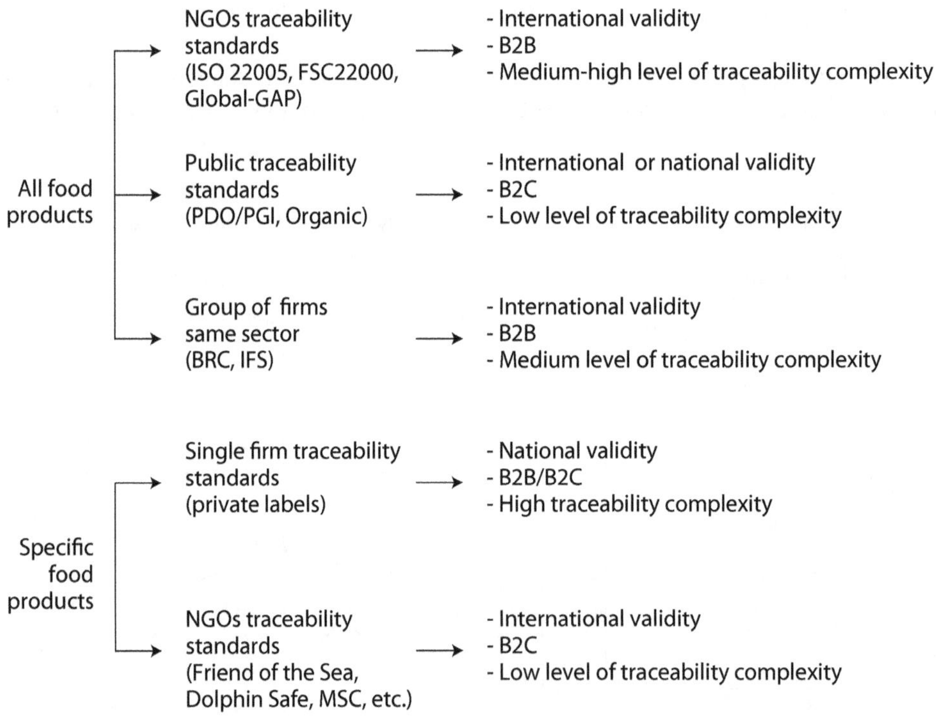

Figure 6.2. Existing voluntary traceability standards.

The first category of costs deals with all the *ex-ante* costs related to the traceability adoption. These costs relate to the investments firms occur when firms want to adopt a voluntary traceability standard. More precisely, they refer to fix costs, like for example:

- the costs for the system planning;
- the costs related to the implementation of the information software to trace information flows along the supply chain;
- costs of certifications;
- costs for the training of human resources.

The second category of costs refers to the management of traceability standards. Such costs arise *ex-post* the adoption of the standard and are necessary to apply and maintain the standard. They are variable and relate to:

- costs of personnel connected to data input;
- monitoring costs of the system implemented;
- marketing costs related to the promotion of the standard (advertisements, pricing strategy activities, etc.).

The type and quantity of costs which firms face for the implementation of traceability standards depend on the type of the standard adopted. Many authors stressed that the more complex the standard is, the higher the costs faced by food firms will be (Asioli *et al.*, 2014).

The presence of different kind of traceability standards has led different scientific contributions to explore the incentives (benefits perceived) connected to firms decisions on the type of standard to adopt. Based on existing literature it is possible to identify different kind of determinants related to traceability adoption. We can group them into (Stranieri *et al.*, 2016):

- market incentives;
- supply chain incentives;
- regulatory incentives.

The first group is related to those drivers derived from market stakeholders, namely consumers, food firms, and retailers. Marketing studies have concentrated the attention on the determinants that influence consumers attitudes towards traced products in terms of information labelled. Different are the quality and safety cues investigated, like for example product origin, environmentally-friendly, ethical, nutritional and healthy product characteristics (Heyder *et al.*, 2012; Loureiro and Umberger, 2007; Mai *et al.*, 2010; Manos and Manikas, 2010).

Several studies have investigated incentives for traceability from firm points of view. These include: increase of sales, the improvement of competitive advantage, the production procedures enhancement, the maintenance or upgrading of firm reputation, and the pressure exerted by retailers to implement food safety control systems (Brofman and Martinez, 2014; Fritz and Schiefer, 2009; Ruitz-Garcia *et al.*, 2010; Wilson *et al.*, 2008).

Supply chain incentives entail those dealing with an improvement of supply chain efficiency. The efficient management of supply chain complexities in terms of product identification (logistic management), the increase of information along the supply chain (information management), and the implementation and management of food safety and quality requirements (quality management) have been identified to influence firms behaviour towards the adoption of traceability standards (Ringsberg, 2014). Another important aspect related to supply chain incentives regards the efficient management of vertical exchanges, in terms of better liabilities distribution among the operators of food supply chains.

Finally, regulatory incentives refer to those incentives driven by regulatory requirements within the national or international markets (Liao *et al.*, 2011; Senneset *et al.*, 2007; Smith *et al.*, 2005).

According to the literature, the more complex the traceability standard implemented is, the higher the costs and benefits associated to the standard implementation will be. Moreover, when benefits perceived about a traceability standard outweigh the costs perceived by food firms, the probability to implement an accurate traceability standards increase (Holleran *et al.*, 1999).

6.2 Legal framework of traceability in the EU

6.2.1 General EU traceability requirements for food safety

Mandatory traceability standards are set up by governmental institutions, introducing rules which must be implemented by all the agents of the supply chain involved. In the global context, European legislation on food traceability is one of the most complete normative frameworks (Charlebois *et al.*, 2014). The EU Regulation 178/2002 (General Food Law; GFL) introduced a mandatory traceability system for food, feed, food-producing animals, and any other substance incorporated into a food at all stages of production, processing and distribution. The EU Regulation obliges food firms to be able to identify from whom a food product is supplied and to whom it is sold. This legal requirement relies on the 'one step back'–'one step forward' approach (Karlsen *et al.*, 2011). GFL does not specify the system that operators have to implement but it must keep the following information, at a minimum: name, address of supplier, and identification of products supplied; name, address of customer, and identification of products delivered; date and, where necessary, time of transaction/delivery; volume, where appropriate, or quantity delivered.

Very briefly the traceability standard introduced by Reg.178/2002 implies:
- the implementation of a traceability system established at all stages of production, processing and distribution;
- it obliges business operators to be able to identify any person from whom they have been supplied with any substance that has been incorporated into a food product;

- business operators must have in place systems and procedures to identify customers to whom their products have been supplied.

The traceability requirements introduced by the GFL specify the results that have to be achieved but it does not prescribe how such results can be reached. This allows certain flexibility for business operators in the implementation of these requirements. As GFL does not specify the specific system of traceability that firms have to adopt, it does not allow reconstructing the complete history of the product due to a missing link between information and product flows. In other words, GFL does not provide a mandatory scheme for supply chain and product traceability.

6.2.2 Specific requirements for meat products

The beef sector has been regulated in a stricter way through Regulation 1760/2000. Such Regulation obliges firms to implement a traceability system aimed at identifying and registering the product flows among the firms part of the supply chain and the product flows within the traced firms (supply chain and product traceability system) (Banterle *et al.*, 2006). Throughout such system operators are able to reconstruct the complete history of meat products, from the agricultural raw materials used for animals feeding, to all the practices used for animal breeding and for the processing phase. The traceability label has to report at least (Stranieri and Banterle, 2015):

- a reference number or code linking the meat to an animal or group of animals;
- the approval numbers of the slaughterhouse and cutting plant;
- the Member State or third country of birth;
- the Member State or third country of rearing;
- the Member State or third country of slaughter;
- the Member State or third country of cutting.

Moreover, Regulation 653/2014 has amended Reg.1760/2000 concerning the electronic identification of bovine animals and labelling of beef. Such Regulation has also simplified the management of the traceability system related to voluntary labelled information by reducing the costs of adoption and control of voluntary traceability system.

Recently, the traceability scheme has been extended to most of meat supply chains by Regulation 1337/2013. On the basis of this new requirement food business must apply a system able to ensure:

- the link between the meat and the animal or group of animals, from which they were obtained;
- the transmission of the traced information to operators in subsequent stages of production and distribution.

Every food business operator is responsible for the application of the system of identification and registration in the pre-production and distribution in which it operates. As for bovine meat, this new regulation has highlighted the important role of fine traceability system in assuring specific safety and quality attributes connected to the supply chain. Such Regulation has been applied to fresh, chilled and frozen meat of swine, sheep, goats and poultry.

6.3 A transaction cost perspective to traceability standards

6.3.1 Transaction cost economics and food supply chains

Coase (1937) identified different sunk costs associated to transactions: search, negotiating and monitoring costs. Search costs refer to costs that arise *ex-ante*. Negotiating costs take place during transactions, whereas monitoring costs occur after the exchange has been negotiated (*ex-post* to a transaction). Transaction cost economics argues that the governance structure of transactions depends on some transaction characteristics (Williamson, 1985, 1996):
- the degree of asset specificity;
- the degree of uncertainty;
- the degree of frequency.

Asset specificity relates to the bilateral dependency of economic agents due to investments which occur to perform a specific transaction. We can distinguish among different kinds of asset specificity depending on the nature of the investment. It is possible to differentiate among human asset, site asset, and physical or dedicated asset specificity.

The second factor depends on information asymmetry, bounded rationality of agents, and the random nature of some events. It is connected principally with the probability that subjects will respect contractual obligations. Transaction costs increase when the degree of uncertainty arises. This aspect of the transaction has become very important due to new uncertainties in the food market, like recent food crises and the increasing internationalization of markets (Banterle *et al.*, 2006). The third factor is connected to the time lapse between one transaction and another.

Transaction costs increase when there is an augmentation in the levels of transaction uncertainty and asset specificity of the transactions. In such situations the need to safeguard transactions from opportunistic behaviour increases.

According to Williamson (1985) it is possible to distinguish among three different forms of transaction governance:
- markets;
- hybrids;
- hierarchies.

Markets are short-term arrangements between transacting parties. They mostly refer to the organization of those transactions with a low level of idiosyncratic exchanges and with a high level of frequency. In the food sector examples of different forms of market are: market for perishable agricultural products (like fruits and vegetables) and market for non-perishable agricultural products, like cereals (Board of Trade).

Hybrids are forms of long-terms arrangements which mostly refer to contracts. These last are used when transactions are characterized by a certain degree of asset specificity and the presence of transaction uncertainty. Contracts usually entail the maintenance of transaction parties' independence and the provision of safeguard conditions in order to reduce the risks of arrangement infringements.

Finally, hierarchies comprehend all those forms of vertical integrations where two or more activities belonging to different steps of the supply chain are placed under the same ownership structure. Tirole defines vertical integrations as 'the vertical dimension that reflects the extent to which goods and services that can be purchased from outsiders are produced in house' (Tirole, 1991).

6.3.2 Traceability standards and vertical coordination

According to North (1991) we consider traceability standards as institutions, based on a set of rules, which have an influence on the management of economic relationships. Indeed, traceability standards play an important role in affecting economic relations among the agents of the supply chain. More precisely, such standards lead to an increase of information exchanged, an enhancement of supply chain transparency, and an increase of bilateral dependency due to specific investments in traceability systems.

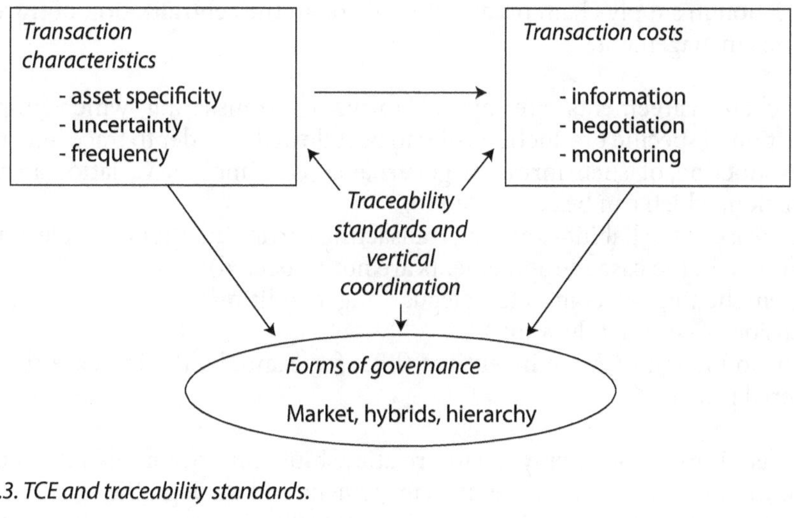

Figure 6.3. TCE and traceability standards.

Such rules can imply also a variation of transaction costs in dyadic relationships. More precisely, the introduction of traceability standards leads to (Figure 6.3):

- an increase in transaction costs due to a growth of physical asset specificity related to the investments necessary to implement the standards;
- an increase of transaction costs related to the improvement of human asset specificities;
- a reduction of transaction costs related to the decrease of the degree of transaction uncertainty due to a higher level of supply chain transparency.

More precisely, the variation of physical asset specificities depends on the investment related to the system implementation and the variation of processing facilities.

Human asset specificity variation deals with the costs related to the technical specialization of human resources for the application of voluntary traceability and of the costs associated to the selection of suppliers which can become part of the traced supply chain.

The reduction of transaction uncertainty relates with an increase in supply chain transparency and efficiency connected to the quantity of exchanged information, the timing of exchanged information variation and the precision of exchanged information variation.

With regard to the reorganization of vertical exchanges, the introduction of voluntary standards and the variations in transaction costs can lead to the necessity of (Banterle and Stranieri, 2008a,b):

- a variation in the organization of dyadic relationships through the introduction of supply chain agreements;
- a variation in supply chain organization through the centralization of traceability systems management.

Supply chain agreements are hybrid forms of transacting which imply the introduction of specific production rules to be followed in order to trace information. The introduction of such forms of governance can imply a variation in vertical coordination, which can be connected to:

- an increase of the liabilities among transacting parties (like, for example, sanctions applicable in the case the agreements are not respected);
- a strengthening of contracts safeguarding conditions (like, for example, the extension of contract duration);
- the introduction of price incentives (like, for example, the increase in the raw material price).

The reorganization of supply chain relationships can occur also through the centralization of traceability systems management and through the standardization of information flows through the chain. The centralization relates to the fact that the firm which decides to promote the standard, than, acts as the leader within the supply chain. This firm coordinates the activities of the other agents and it is in charge of the certification of the system. The supply chain leader can be a food processor, a retailer, an agricultural cooperative or, in general, any entity interested in building up a good market reputation. The creation of a guidance related to the implementation of the traceability standards leads to an incentive to create trust among the operators of the supply chains and to stimulate the correct executions of procedures.

The variation in transaction and supply chain organization would depend not only on the variation in transaction characteristics and costs associated to the introduction of the standard, but also on other factors, like the structure of transaction governance existing at the moment of the introduction of traceability standards and the drivers which lead firms to implement them. More precisely, the higher is the level of vertical coordination, the less would be the variation in transaction organization consequently the introduction of voluntary standards. Moreover, depending on the drivers leading firms to implement traceability standards the effects on supply chain and transaction organization would be different. If firms are driven by supply chain incentives, the level of transaction and supply chain reorganization will be more consistent than the reorganization associated to market or regulatory incentives (Stranieri *et al.*, 2016).

6.3.3 A focus on the effects of sustainability standards on vertical coordination

On the basis of the considerations reported above, we examined the effects of traceability standards on vertical coordination. In specific, we decided to focus the attention on sustainability standards because of their fast proliferation within the food sector in the last years. To reach our aim we followed different steps. First, we identified the existing sustainability standards present in Italy. Second, we constructed the sample of food processing firms applying the identified standards. Third, we conducted a survey in order to analyse the effect of such standards on the characteristics of transactions, on dyadic relationships and supply chain organization. The synthesis of the empirical investigation is reported below.

To detect the sustainability standards present in Italy we referred to the online platforms supported by International Trade Center (ITC). Such platform provides comprehensive, verified and transparent information on voluntary standards covering issues such as food sustainability, quality and safety. Such initiative provides information on two hundred traceability standards addressing environmental, social, economic feasibility, ethical and quality management aspects of agricultural and food products. The inclusion of the standards to such initiative is subject to the existence of the following criteria: a published set of requirements, an existing operational system, and the reference to principles that address at least one pillar of sustainable development (social, environment, economic, ethics or quality management).

For our analysis we considered all the traceability standards applied on Italian food products which were included in the platform described above in 2015. The lists of food firms certified was obtained by consulting the lists available of operators published on the web site of the selected standards. The standards which were considered are: FSSC 22000, BRC, Fair Trade, Rainforest Alliance, Roundtable on Sustainable Palm Oil certification (RSPO), Friend of the Sea, Dolphin Safe, and Marine Stewardship Council (MSC). In addition, we considered also private label standards.

On the basis of the survey conducted, it is possible to underline some common aspects among the standards analysed in terms of transaction characteristics and costs (Table 6.2). First, all the certifications analysed reveal an increase in transaction asset specificity. In some cases the increase relates only to the certification fee and the costs for auditing, whereas for other standards such augmentation is connected to the investment in a supply traceability system. This implies costs related to the technology investments, to the system implementation, and to the training of human resources used to manage such system. Second, transaction uncertainty generally decreases with the introduction of the standards thanks to an increase of transparency in vertical exchanges. The decrease of transaction uncertainty is expected because it is one of the aims of traceability standards. Indeed we refer to traceability standards as all those standards which imply an increase in information transparency within food supply chain. However, depending on the standard analysed differences in the supply chain transparency can be highlighted. For some standards the higher transparency is mainly connected to some sustainable procedures of the production processes, like for example for the standards Friend of the Sea, Dolphin Safe, and Marine Stewardship Council (MSC). For other standards, like BRC standard and FSSC 22000, the implementation of traceability imply transparency in agricultural procedures adopted and transparency in the food-processing, like quality, safety, operational criteria followed in the food processing.

With regard to the effect of such standards on the vertical coordination of food supply chain, it is possible to underline some factors, which could be considered as important in order to distinguish and classify sustainability standards. These elements mainly relate to:
- type of certification (B2B/B2C);
- incentives for the standard adoption;
- type of standard audit.

As for the type of certification, it is possible to highlight a difference between B2B and B2C standards. B2B are associated to a higher level of vertical coordination variation compared to that of B2C standards. In the two types of B2B standards analysed, BRC and FSSC 22000, a variation in dyadic vertical relationships is revealed. In the case of FSSC 22000 also a change of supply chain organization is observed.

The organizational difference between the two standards are related to their different characteristics. Indeed, BRC standard implies a variation in dyadic relationships, which is mainly connected to a better specification of contractual conditions and it often implies the possibility to build long-terms contracts with retailers. FSSC 22000 considers supply chain aspects important determinants of the implementation

Table 6.2. Organizational changes of sustainability standards.

Standard audit	Product traced	Type	Drivers	Physical asset specificity	Human asset specificity	Transaction uncertainty	Dyadic vertical coordination variation	Variation in supply chain organization
Third-party certification	food products	B2B	supply chain incentives and non-monetary incentives	x	x	x	x	x
Third-party certification	processed products and agricultural packaged products	B2B	non-monetary market incentives	x	x	x	x	
Third-party certification	product from sustainable fisheries and aquaculture	B2C	market incentives and non-market incentives (reputation)	x		x		
Third-party certification	wild capture fishery	B2C	market incentives and non-market incentives (reputation)	x		x		
Second-party certification (FLOCERT)	agricultural and processed food	B2C	non-monetary market incentives	x		x	x	
Third-party certification	tuna	B2C	market incentives	x		x		
Second-party certification (Rainforest Alliance's auditing division)	coffee,tea, cocoa	B2C	non-monetary market incentives	x		x	x	
Third party certification (RSPO approved certification bodies)	processed products with palm oil	B2C	market incentives	x		x		
First-party certification	processed products and agricultural packaged products	B2C	supply chain and market incentives	x	x	x	x	x

of the standards. The attribution of specific responsibilities in terms of safety and quality characteristics among the agents of the food supply chain and the possibility to improve a system aimed at recognizing specific liabilities of economic agents lead to the necessity to identify a leader within the supply chain (centralization) in charge for the management and monitoring of the supply chain.

With regard to B2C certifications, there exists different situations in relation to the consequences on vertical coordination. More precisely, there are standards which do not imply a variation in vertical coordination, other standards which lead to a variation in dyadic relationships and others characterized both by a variation in dyadic transaction organization and on supply chain organization.

The analysis suggests that one possible explanation of different variation in vertical coordination could be related to the standard incentives. More precisely, standard applied for a differentiation of product quality characteristics that does not require a redesign of vertical organization, like Friend of the Sea, Dolphin Safe, MSC and RSPO, will not reveal a variation in vertical coordination. On the other side, those standards whose incentives are based both on quality differentiation and on a reinforcement of supply chain relationships, in terms of sustainable economic relations (Fair Trade, Rainforest Alliance, and private labels), show a variation in the organization of vertical relationships.

Moreover, also the type of audit chosen for B2C standards seems to affect the effects of sustainability standards on vertical coordination. In the food sector, it is possible to distinguish between first, second and third party certifications for product quality standards. First-party certification relates to an individual firm, which is responsible for the truthfulness of the information certified by the standard. Generally, the higher the reputation of the firm is, and the closer is the relationship between consumers and sellers, the higher the probability to implement such certification. Second-party certification refers to associations that provide the assurance and the monitoring of a certification. Often the member of these organisms are also the subjects asking for a quality certification. Third-party certification is considered among the most appropriate way to assure a certain certification because it involves an independent entity that specified requirements and monitor for the correct executions of the rules.

Certifications based on first-party-certifications (private labels) seem to have important variations both in dyadic vertical relationships (the increase of contract last, the increase of raw material price) and on supply chain organization (centralization of information management, on traceability system management and on monitoring activities). Standards based on second-party certification reveal a variation in dyadic relationships. Standards based on third party certifications do not seem to have an influence on the reorganization of vertical exchanged and supply chain organization.

Summarizing, the results of this preliminary insights highlight the possibility to classify traceability standards on the basis of their effect on the reorganization of vertical relationships:

- standards with no variation in vertical coordination;
- standards with variations in dyadic vertical relationships;
- standards with variations in supply chain organization.

The first category comprehends those standards, which are mainly adopted for market incentives. More precisely, quality differentiation of products is one of the main determinants for the implementation of such traceability standards. From a transaction costs perspective, this quality differentiation implies a little change of transaction characteristics and costs and, thus, does not lead to consistent variations in vertical coordination. Moreover, the increase in transaction asset specificity and related costs due to the investments related to the traceability system implementation can be compensated by a decrease of transaction uncertainty and costs.

The second category refers to standards implemented mainly to strengthen firm reputation towards suppliers and consumers (non-monetary market incentives). The adoption of such standards implies an increase in transaction asset specificity due to the implementation of a traceability standard and the training of human resource. In addition, a reduction in transaction uncertainty is observed due to the higher level of information, which characterizes transactions. In this case the high level of investments leads to the necessity to modify vertical modes of transacting, by adding stricter safeguard conditions to preserve vertical relationships or price incentives to counterbalance the cost of traceability standards implementation.

The third type of traceability standards relates to supply chain incentives. Firms adopting such standards look for a more efficient functioning of the supply chain management, by introducing a system, which leads to a strengthening of vertical relationships and the identification of precise responsibilities among the agents of the supply chain. The adoption of these standards lead to an increase in transaction asset specificity and a decrease in transaction uncertainty. Moreover, such standards lead to a variation in the organization of both dyadic relationships and supply chain. With regard to dyadic relationships, the adoption of safeguard conditions is observed. With regard to the supply chain organization, the agent who promotes the standard becomes the leader of the supply chain. He is in charge with the management of information flows within the chain, with the management of product quality characteristics, and with the monitoring of the compliance with supply chain activities to the standard applied.

Limitations of the study relate both to the method used in the analysis and the specificity of the standards chosen. Future research will concentrate on finding appropriate statistical methods in order to study the determinants affecting the variation in vertical coordination due to the introduction of traceability standards. A further understanding of the variables that impact mostly on firm strategic decisions about the traceability standards to adopt need a deeper investigation. Moreover, the present analysis takes into consideration only some of the existing traceability standards. Further investigation could be address toward a broader understanding of the influence of different kinds of certifications on the reorganization of vertical relationships within food supply chains.

References

Asioli, D., Boecker, A. and Canavari, M., 2014. On the linkages between traceability levels and expected and actual traceability costs and benefits in the Italian fishery supply chain. Food Control 46: 10-17.

Banterle, A. and Stranieri, S., 2005. Coordinamento verticale e tracciabilità. Un'analisi della filiera lattiero-casearia. Aracne editrice, Roma, Italy, 112 pp.

Banterle, A. and Stranieri, S., 2008a. Information, labelling, and vertical coordination: an analysis of the Italian meat supply networks. Agribusiness 24(3): 320-331.

Banterle, A. and Stranieri, S., 2008b. The consequences of voluntary traceability system for supply chain relationships: an application of transaction cost economics. Food Policy 33: 560-569.

Banterle, A. and Stranieri, S., 2013. Sustainability standards and the reorganization of private label supply chains: a transaction cost perspective. Sustainability 5(12): 5272-5288.

Banterle, A., Stranieri, S. and Baldi, L., 2006. Traceability and vertical co-ordination in the Italian dairy chain: a transaction cost approach. Journal on Chain and Network Science 6: 69-78.

Bollen, F.P., Riden, C.P. and Opara, L.U., 2006. Traceability in postharvest quality management. International Journal of Postharvest Technology and Innovation 1(1): 93-105.

Brofman, E.F.M. and Martinez, M.G., 2014. The technological evolution of food traceability systems and their impact on firm sustainable performance: a RBV approach. International Journal of Production Economics 150: 215-224.

Canavari, M., Centonze, R., Hingley, M. and Spadoni, R., 2010. Traceability as part of competitive strategy in the fruit supply chain. British Food Journal 112(2): 171-184.

Charlebois, S., Sterling, B., Haratifar, S. and Naing, S.K., 2014. Comparison of global food traceability regulations and requirements. Comprehensive Reviews in Food Science and Food Safety 13: 1104-1123.

Cheng, M.J. and Simmons, J.E.L., 1994. Traceability in manufacturing systems. International Journal of Operations and Production Management 14(10): 4-16.

Coase, R.H., 1937. The nature of the firm. Economica 4(16): 386-405.

Dupuy, C., Botta-Genoulaz, V. and Guinet, A., 2005. Batch dispersion model to optimise traceability in food industry. Journal of Food Engineering 70(3): 333-339.

Fritz, M. and Schiefer, G., 2009. Tracking, tracing, and business process interests in food commodities: a multi-level decision complexity. International Journal of Production Economics 117: 317-329.

García, H., Santos, E. and Windels, B., 2008. Traceability management architectures supporting total traceability in the context of software engineering. In: Sixth International Conference in Information Research and Application i-Tech. Varna, Bulgaria, pp. 17-23.

Golan, E., Krissoff, B., Kuchler, F., Calvin, L., Nelson, K. and Price, G., 2004. Traceability in the US food supply: economic theory and industries studies. Agricultural Economic Report Number 830. USDA, Washington, DC, USA.

Heyder, M., Theuvsen, L. and Hollmann-Hespos, T., 2012. Investments in tracking and tracing systems in the food industry: a PLS analysis. Food Policy 37: 102-113.

Holleran, E., Bredahl, M.E. and Zaibet, L., 1999. Private incentives for adopting food safety and quality assurance. Food Policy 24(6): 669-683.

International Organization for Standardization (ISO), 1994. Quality management systems – fundamentals and vocabulary. International Organization for Standardization, Geneva, Switzerland.

Karlsen, K.M., Dreyer, B., Olsen, P. and Elvevoll, E.O., 2013. Literature review: does a common theoretical framework to implement food traceability exist? Food Control 32: 409-417.

Karlsen, K.M., Sørensen, C.F., Forås, F. and Olsen, P., 2011. Critical criteria when implementing electronic chain traceability in a fish supply chain. Food Control 22(8): 1339-1347.

Liao, P.A., Chang, H.H. and Chang, C.Y., 2011. Why is the food traceability system unsuccessful in Taiwan? Empirical evidence from a national survey of fruit and vegetable farmers. Food Policy 36: 686-693.

Loureiro, M.L. and Umberger, W.J., 2007. A choice experiment model for beef: what US consumer responses tell us about relative preferences for food safety, country-of-origin labeling and traceability. Food Policy 32: 496-514.

Mai, N., Bogason, S.G., Arason, S., Arnason, S.V. and Matthiasson, T.G., 2010. Benefits of traceability in fish supply chains e case studies. British Food Journal 112(8-9): 976-1002.

Manos, B. and Manikas, I., 2010. Traceability in the Greek fresh produce sector: drivers and constraints. British Food Journal 112(6): 640-652.

Moe, T., 1998. Perspectives on traceability in food manufacture. Trends in Food Science and Technology 9(5): 211-214.

North, D.C., 1991. Institutions. Journal of Economic Perspectives 5(1): 97-112.

Olsson, A. and Skjöldebrand, C., 2008. Risk management and quality assurance through the food supply chain e case studies in the Swedish food industry. Open Food Science Journal 2: 49-56.

Opara, L.U. and Mazaud, F., 2001. Food traceability from field to plate. Outlook on Agriculture 30(4): 239-247.

Ringsberg, H., 2014. Perspectives on food traceability: a systematic literature review. Supply Chain Management 19(5-6): 558-576.

Ruiz-Garcia, L, Steinberger, G. and Rothmund, M., 2010. A model and prototype implementation for tracking and tracing agricultural batch products along the e food chain. Food Control 21: 112-121.

Schwagele, F., 2005. Traceability from a European perspective. Meat Science 71(1): 164-173.

Senneset, G., Forås, E. and Fremme, K., 2007. Challenges regarding implementation of electronic chain traceability. British Food Journal 109: 805-818.

Smith, G.C., Tatum, J.D., Belk, K.E., Scanga, J.A., Grandin, T. and Sofos, J.N., 2005. Traceability from a US perspective. Meat Science 71: 174-193.

Stranieri, S. and Banterle, A., 2015. Consumer interest in meat labelled attributes: who cares? International Food and Agribusiness Management Review 18(4): 21-38.

Stranieri, S., Cavaliere, A. and Banterle, A., 2016. Voluntary traceability standards and the role of economic incentives. British Food Journal 118(5): 1025-1040.

Tirole, J., 1991. The theory of industrial organization. MIT Press, London, UK.
Williamson, O.E., 1985. The economics of capitalism. Free Press, New York, NY, USA.
Williamson, O.E., 1996. The mechanism of governance. Oxford University Press, New York, NY, USA.
Wilson, W.W., Henry, X. and Dahl, B.L., 2008. Costs and risks of conforming to EU traceability requirements: the case of hard red spring wheat. Agribusiness 24: 85-101.

7. Transactions in coffee and milk agri-food value chains: why measure?

J.P. de Souza and S.M.S. Bánkuti*

Department of Administration, State University of Maringá, Av. Colombo 5790, Maringá, PR 87020-900, Brazil; jpsouza@uem.br

Abstract

This study analyses agri-food systems considering the new institutional economics approach (NIE). In Brazilian agribusiness, NIE's constructs are especially suitable. In an environment of high uncertainty, the system relies on institutions capable of generating rapid response to structural and systemic determinants. The primary focus in this chapter is to understand the importance of measurement for transactions and relationships, and its consequences in the design of governance structures between producers and buyers. From the perspective of transaction cost economics and measurement cost economics governance structures' responsiveness to efficiency and performance is addressed in the light of property rights protection mechanisms. Empirical studies in milk and coffee sectors in Paraná state (Brazil) show the influence of mechanisms to protect property rights on measurable attributes. We discuss the importance of formal rules in defining the parameters for raw materials' quality evaluation. It shows that governance structures, even though not perfectly aligned with traded attributes, create sufficient coordination to support transactions. It does not mean that the purpose of efficiency is being met. Results showed that agents adopt mostly incomplete contracts in the case of coffee, and informal agreements, in the case of milk, for defining dimensions and the execution of the transaction, which leaves room for rent capture. The poor information sharing increases uncertainty and strengthens the possibility of rent capture from adverse selection, keeping agents under the condition of moral hazard even when long-term relations take place.

Keywords: measurement costs, transaction costs, property rights

DOI 10.3920/978-90-8686-844-5_7,

7.1 Introduction

In the context of the new institutional economics (NIE), focus on efficiency explains and justifies different governance structures, driven by transaction cost economizing. Several studies focused on the influence of coordination and governance costs in organizations' competitive performance (Alchian and Demsetz, 1972; Barzel, 2001, 2003, 2005; Coase, 1937; Farina, 1999; Hart, 1997; Klein *et al.*, 1978; Knight, 1964; Langlois and Foss, 1999; Ménard, 2004; North, 1990; Saes, 2009; Williamson, 1985, 1996; Zylbersztajn, 2005a). This perspective also has particular relevance when considering agri-food systems (AGS), notably in the study of governance structures and relations of coordination.

This chapter's proposal is added to those studies involving contractual relations under the theoretical assumptions involving property rights. From that perspective, it considers the measurement as the unit of analysis, in shaping governance structures and its consequences, as well as its influence on efficiency. It makes use of transaction cost economics (TCE) as the basis and the measurement cost economics (MCE) in a complementary basis, although it does not ignore that other perspectives (such as industrial organization, resource based view and incomplete contracts) could also contribute to understand firms in terms of their existence, size, scope and organization.

So, considering that AGS are organized into complex forms of interaction, formal and informal contractual mechanisms define governance structures that enable coordination and the implementation of strategic actions. The limits or deviations in that dynamic, however, characterize losses that can impact directly on the maintenance of agents competitive conditions in each segment, directly impacting on strategic choice. We intend to understand the dynamics and how to consider the distribution of guarantees related to performance and contractual agreements in agri-food transactions.

7.2 Applications of micro institutions study in agri-food systems

7.2.1 New institutional economics in agri-food systems

The consideration of NIE to discuss AGS is not new. Studies involving AGS have been improved since the pioneering work of Davis and Goldberg (1957), indicating the interdependency of chain's segments. Morvan's (1985) work inserted the concept of filière and the consideration of production chain and its systemic character. In Brazil, studies by Farina, Azevedo and Saes (1997), Zylbersztajn and Farina (1999), Azevedo (2000) and Zylbersztajn (2005a) initiated and inspired a series of studies and targeted research to understand AGS dynamics under the support of NIE, in its macro and micro analytical branches.

When considering the micro-institutional branch, Langlois (1992) noted that, since Coase (1937), studies identified patterns of ownership and contracts and its ability to minimize production and transaction costs. The author states that two traditions in transaction costs analysis can be observed. One focused on the management, administration, negotiation and monitoring of joint production at the firm level, which he calls 'asset-specificity view'. The other, which is called 'measurement-cost view', emphasizes quality assurance and performance of contractual agreements. In the former, the emphasis is mainly on Klein *et al.* (1978) and Williamson (1985); in the second, we highlight Barzel's work (1997, 2001, 2003, 2005).

Given these settings, the problem in the study of contractual form, and its focus on agri-food system is consolidated here, initially, concerning the role of institutions or institutional environment, following North (1990), Williamson (2000) and Barzel (1997). In Brazil, Azevedo (2000, p. 34, authors' translation) points out that: 'Rights of ownership of land, minimum pricing policies, agrarian reform, as well as food security policies – in its double meaning of access to food (food security) and minimum quality assurance (food safety) – are elements of the institutional environment (macro-institution) that have important effects on the actions of those who comprise the agro-industrial systems'. At a second stage, it involves theoretical stream of measurement costs (MCE), analysing the formation of governance structures, following the guidelines by Barzel (1982, 1997, 2001, 2003, 2005). It should be noted that the institutions in the MCE, as Zylbersztajn (2005b, p. 7) points out, 'take as given [...] matter, since legal ownership is based on the state's capacity to offer guarantees to the transaction based on an existing contract'. So, while the first considers the rules and forms of monitoring, appropriate to maintain the competitive environment, and offer guarantees and stable relations, the latter deals with guarantees of property rights on the dimensions of traded assets and its influence on formation of governance structures.

It recovers here the observation of Langlois (1992) that in the MCE approach, the assumption is that it is often costly to measure quality and even the amount results at a production stage, which influences the organizational form. In this sense, the more difficult to measure a contributing factor to the value in relation to another factor it is, the greater the tendency for the owner of the factor assume the residual claimant position will be. This is strengthened when it comes to AGS. Exposures to diversity and environmental uncertainties (Williamson, 1985) demand appropriate forms of coordination in order to obtain regular qualitative and quantitative results in the inputs and outputs with competitive performance.

Hence, we consider that the pursuit of better coordination, in horizontal and vertical levels, may indicate strategic behaviour to seek a better competitive performance. This behaviour, however, can be prevented by horizontal or vertical bottlenecks, which justifies the search for coordination and cooperation, as an alternative to vertical integration (Farina, 1999). In addition, strategic moves, in turn, may be limited when considering the mechanisms to guarantee property rights, relating to transferred assets and their potential uncertainties. These can reduce the strategic options and the expected results, since the losses related to the absence of guarantee and protection mechanisms can be real. In this respect, transaction costs arise because of the search for guarantees and protection against possible variations and value capture (Barzel, 2005).

According to this statement, analysis of the competitive strategies sustained in contractual relations, common to AGS, allow at least two considerations: even facing high levels of asset specificity, the adoption of contracts as a governance structure alternative to vertical integration (Barzel, 2005) may provide a reduction of transaction costs (Williamson, 1985) or enables competitive strategies (Farina, 1999); these relationships, however, can limit the expected positive effects when the appropriate mechanisms to guarantee property rights are not clearly defined, which makes the enforcement mechanisms limited. Barzel (2005, p. 361) states: 'to be effective, then, contract stipulations must be objectively measurable'.

This indicates that the rationale present in the TCE should be complemented with the assumptions inherent in measurement branch, justifying the contractual relationship even in potential conditions of quasi-rent appropriation. In this respect, the perception of efficiency, which may have conditioned their effectiveness to existing contractual commitments, supports itself with appropriate measurement mechanisms, influencing strategic decisions. In AGS, the contractual relationships clearly stand out and the deployment of this rationale seems more evident. The resulting assumption is that the governance structure, as such, is explained as a result of the search for alignment of the attributes present in the transaction, and the possibility of measuring attributes, to ensure property rights (Barzel, 2002, 2005; Zylbersztajn, 2005a), supplementing the rationale proposed by Williamson (1985, 1991).

MCE involving contractual relations, considers at least two propositions. First, the conduct targeted by trying to appropriate income in formal or informal agreements can join the search for a favourable position in the relevant market or competition, notably in specific assets, with an impact on production costs or product availability. Conversely, the guarantees associated with appropriate mechanisms for measurement and control (information) allow for adjustments between the parties and maintenance of contractual relations, given that it minimizes uncertainty in the transaction. As a consequence, the possibilities to sustain performance become effective as suitable distribution mechanisms can be structured.

7.2.2 TCE and MCE: complementary theoretical approaches

In the context of the NIE, concerning institutions, two levels of analysis can be identified, the macro and micro institutional ones (Azevedo, 2000; Williamson, 2000; Zylbersztajn, 2005a). The macro-institution, or institutional environment, is composed of restrictions designed by implemented political structure, economic and also social interaction (North, 1991). Williamson (2000) argues that much of the economics of property rights relates to that level, whose enforcement is not easy, something that leads to the need for consideration of micro analytical level or governance.

At the micro analytical level, empirical work is traditionally related to governance structures (Williamson, 1985, 1996), taking as a guide the model entitled TCE. From the propositions of Coase (1937), Williamson (1985) formalizes its analytical model oriented to minimize transaction costs, condition obtained by the alignment of transaction attributes (asset specificity, frequency and uncertainty) and governance structures. In that case, the information is handled in a context of asymmetry, a characteristic that justifies the uncertainty and the possibility of opportunistic behaviour among agents in the presence of limited rational capacity. The component called MCE is presented in Barzel's studies (1982, 2002, 2003, 2005), by inserting a new dimension, measurement, for empirical analysis, in addition to the model of TCE, proposed by Williamson. Note that this strand (MCE and TCE) is suitable for treating field research in the agro-industrial systems, given that bilateral relationships and interdependencies context associated with the exchange of attributes and dimensions configure its operational dynamic.

Williamson's work (1985), involving the economic institutions of capitalism, finds special supplement and deepening in Barzel's work (1982, 2001, 2003, 2005), which deals with the measurement and its relationship with property right and the consequent decision about firm boundaries. When discussing the relationship of TCE, Williamson (1985) recognizes the measurement as a branch of Transaction Cost Economics, pointing reference to Barzel's work (1982). Barzel (2003), in turn, notes that the analysis of transaction costs was not about rights. This would be an important aspect, according to him, since the use of an asset typically requires the owner cooperation of other assets, determining a dependency and the involvement of property rights in the transaction. This justifies the property rights theory in the discussion of the boundaries of the firm.

When Williamson (1985) defines a rationale for conducting this choice, it gives propositions of Coase (1937) effective conditions for the empirical work on the purpose of efficiency, defining how the choice of governance structures is suitable for conducting the transaction. As a strategic guidance to agents, this rationale states that: in the presence of high asset specificity, recurrent frequency and uncertainty of future events, taking for granted bounded rationality and opportunistic behaviour, the transactions tends to take place within the firm (vertical integration). This choice would allow the reduction of transaction costs, since it provides protection against possible attempts of quasi-rent appropriation. The absence of those attributes, in turn, would indicate the market as the best option. The hybrid form, or contracting, appears alternatively as an intermediate, especially in cases of medium specific assets, where prohibitive costs do not occur.

The works by Barzel (1997, 2001, 2003, 2005), which are associated with the so-called Old Property Right Approach (OPRA), highlight the importance of the definition of institutional mechanisms to guarantee property rights. They highlight the right at the economic and legal levels, focused on the contractual relationship and coordination. In this respect, information plays a key role to efficiently carry out a transaction, enabling the exchange market (Barzel, 2005). However, for the author, the information is not only about production costs, as it presents difficulties in transmission, generating value appropriation problems, or to guarantee property rights, and consequently transaction costs.

Barzel (2005) offers a complementary look at the process of choosing the governance structure, and property rights guarantee, which the author calls 'forms of enforcement'. Focusing on information (measurement), the author notes that the exchanges require mechanisms for governance, given the characteristics of products variation and their sale, as well as upon the necessary inputs. Thus, the enforcement structures, when considering the perception of the author, and the proposal of Williamson (1985), should align a set of traded attributes, and the availability of related information, given the possibility of opportunistic behaviour, bounded rationality and value appropriation.

The consideration of measurement in Barzel's proposition (1982, 2003, 2005) defines an alternative way to explain vertical integration. For the author, high measurement costs may indicate that the market may be the most expensive way to govern the transaction (Barzel, 2003). In this case, the necessary performance measurement is configured in transaction costs and vertical integration is an alternative to reduce these costs. Vertical integration would be appropriate when the cost of measuring intermediate commodities were high (Barzel, 2005). The scheme suggested in Figure 7.1 by Souza and Bánkuti (2012), developed from Williamson (1985), illustrates that rationale.

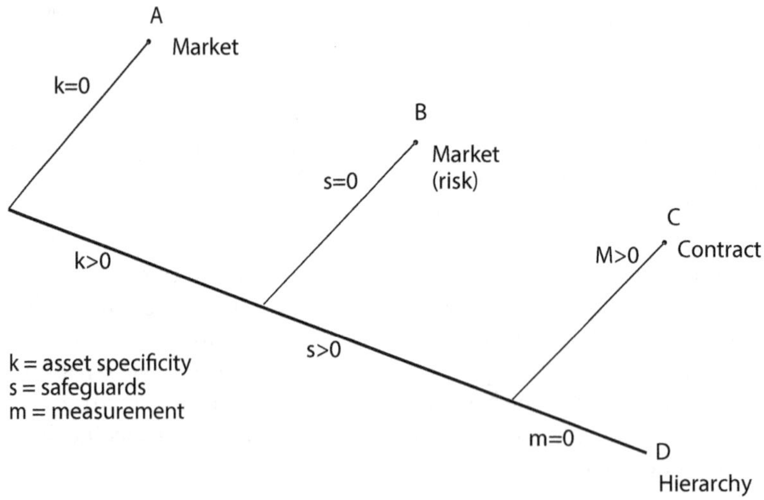

Figure 7.1. Contractual scheme in the context of transaction costs economics and measurement costs economics (Souza and Bánkuti, 2012).

For Barzel (2005), the measurement is a 'particular form of information'. In this case, the author notes that certain measurements are made during the process and others, which are necessary for the production, are only obtained during consumption. The former is easy to obtain and could be contracted. Those obtained in the consumption tend to be subjective, and may be self-controlled (self-enforced). The author concludes that the exchanges may be governed by different enforcement mechanisms, defining an additional rationale for the governance structures choice. For him this guarantee of rights can be obtained by different governance structures. These enforcement structures can be contracts, long-term relationships and those carried out within organizations (vertical integration), or even involving auctions, when the information is obtained *ex-post* (up front).

The implementation of the measure directs this analysis to the field of property rights. In this case, measurement difficulties generate transaction costs *ex-ante*, to guarantee the economic rights to the assets, tending to maintain the ownership of the asset within the firm. The ease of measurement, in turn, would enable contracting as an alternative, since legal and economic rights could be maintained. Notice that, for Barzel (2005), contract could be used together with long-term relationship enforcement to protect valuable promised difficult-to-measure attributes. The existence of the State and the rule of law are thus required.

It defines the rational choice of the most appropriate governance structure to address the measurable dimensions of the asset. Zylbersztajn (2005b) proposes understanding Barzel's rationale as follows:

> The easy-to-measure dimensions are contracted, whilst the difficult-to-measure attributes make room for capture of value and are expected to remain within the firm.

The legal rights and economic rights on assets come to play an important role in the decision to make or buy, i.e. property rights matter.

Figure 7.2 summarizes institutional arrangements under transaction cost and measurement cost theories. In this case, for TCE, high specific assets (denoted +) require vertical integration to protect against the possibility of rent appropriation. When considering Barzel's proposal, even under those conditions, the possibility and ease of performing measurement (denoted +) allows contracting. In this case, the search for safeguarding property rights by making use of the legal system, negotiation, or even residual claim, will be defined by the enforcement system adopted, the institutional framework and the structure of the relationship. The difficulty of measuring dimensions (denoted -) justifies vertical integration as an appropriate mechanism for property rights protection.

Barzel (2003: 57) notes, 'Any given transaction may have multiple attributes subject to variability in outcome, but only those elements guaranteed by the firm's equity capital

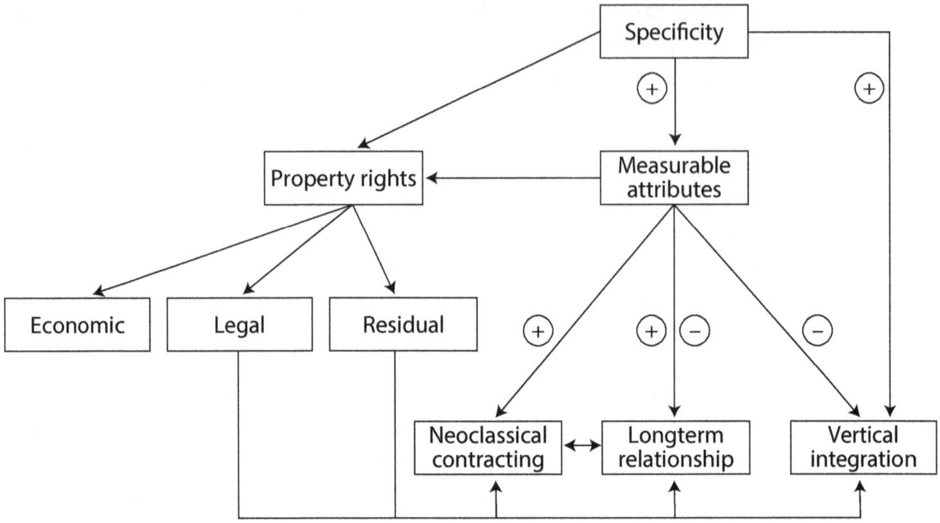

Figure 7.2. Measurement, property rights and governance structures scheme.

are included within the firm'. As a result, one that offers guarantees has the right to integrate. Foss and Foss (2001) state that the great contribution of Barzel was the indication of the need, in the economic sense, to pay attention to the ownership of the attributes of assets, not on assets. That's because many assets have several attributes, and some hard specification, justifying the categorization between legal and economic ownership.

Zylbersztajn (2005a) states that MCE takes for granted the implicit opportunism, bounded rationality and contractual incompleteness, as perfect measurement is impossible. As rationale, it considers that the hierarchical structure results from the value maximization perspective, and such decisions are made at any time.

Thus, Barzel's (2003, 2005) and Williamson's (1985, 1991) propositions are complementary, identifying that property rights in bilateral transactions are subject to the forces of the present governance structure, the presence of the institutional environment and availability of information. At the same time, measurement provides information for assistance, or mere presence of the state to guarantee property rights. It also defines the mechanisms so that the uncertainty related to the transferred assets is not set on the possibility of opportunistic behaviour. It is noteworthy that the governance defined by Williamson (2000) (as a way to mitigate conflict and realize mutual gains), have in measuring the means to operationally stimulate the process of aligning mutual, legal and economic interests, and generate performance.

These strands guided the discussion of two important AGS in the state of Paraná, Brazil: milk and coffee. The discussion of governance structures and its effect on the guarantee of property rights in transactions between producers and buyers took into account Williamson's rationale (TCE) and Barzel's perspective (MCE). In the analysis of the existing governance structure, aspects associated with explicit opportunistic capture and search for the correct distribution of value were then considered.

7.3 Milk and coffee systems in the state of Paraná, Brazil: empirical analysis

Although Brazil is one of the world's largest producers in dairy sector, low productivity per animal stands out, mainly explained by the extensive and grass-fed production systems (Mezzadri, 2012). Paraná has a remarkable position in Brazilian dairy sector, with an increasing production in recent years. The state shifted from 1.7 billion litres of milk in 2000, to 4.3 billion litres of in 2013, according to Department of Rural Economy (DERAL, 2016). Paraná is one of the major milk producing states, ranking the third position in Brazil. Mezzadri (2012) noted that, although productivity features relative improvement, regional discrepancies arise in the state, especially in the technological aspect, producer's expertise and climatic conditions. In the processing segment, the predominance of small farms with low production, and concentration of production in medium and large processors (according to SEAB (2010) 5% of large processors accounted for 65.9% of the state production), showed difficulties, especially for the performance of micro and small-sized agents. The observed concentration and organizational and productive specificities may require appropriate mechanisms for coordination and legal guarantees in order to ensure AGS performance, since this condition can affect the income distribution process.

In the case of coffee, Brazil stands out as the world's largest producer and exporter, accounting for 35% of the world production (ICO, 2013). Paraná has low participation in this result, accounting for about 2% of total Arabica coffee production (ABIC, 2016). The low participation of the state in domestic production can be explained by factors such as the existence of old coffee plantations, low productivity and low product quality. In addition, there is the rent appropriation problem (Revista Cafeicultura, 2013). By studying the Northern region, Bronzeri (2010) casts aspects that indicated possibilities for income appropriation as: vertical integration of different coffee producers; concentration in the processor segment; information asymmetry in transactions between producers and exporters, involving coffee brokers. However, efforts to improve quality and productivity are observed in the North and Northwest regions, enhancing studies of institutional arrangements in this AGS.

These two systems consist of small farmers exposed to difficulties to compete and protect property rights, requiring appropriate coordination mechanisms. It may be noted that for these systems at least two ideas are evident: (1) the interdependency and the need for coordination, in order to generate growth in relations involving the producer and the processor segment; (2) dependency on a suitable system of property rights on specialized assets involved, with effect on investment decisions, particularly in the segments upstream. These aspects are added to threats that destabilize the contractual relationships in these AGS affecting their performance, which are: the economic and environmental uncertainties; rising product standards and process demanding investments, especially in the producer segment; high inter-segment dependence (on one hand the processors need appropriate raw materials, on the other hand, producers require purchase warranty and return on assets with high sunk cost); the use of measurement as a transaction instrument suitable for selection and setting value; limited enforcement mechanisms, which are often sustained in reputational mechanisms (credible commitment).

Considering those statements, a descriptive qualitative research was conducted to have a better comprehension of the dynamic involving governance and measurement in these two AGS, and the effects over property rights. Under TCE and MCE lenses, producers and buyers in the state of Paraná were investigated under the coordination of the Study Group of Coordinated Structures (GECOR). In order to collect data it was used semi-structured interviews and such information was categorized and analysed according to the theory.

In coffee AGS the developed work covered the North and Northwest regions of the State. The interviews were conducted in 2014 and 2015, involving 24 producers, six processors and three brokers. For milk AGS, the agents were from Northwest, West, Mid-East and Mid-South of the state. In this case, 30 producers (small-size, medium-size and large-size) and five processors (cooperatives and non-cooperatives) were interviewed. The results are shown in the following sections.

7.4 Governance structures and property rights in coffee agri-food systems in Paraná state

The coffee production in Brazil can be divided into three major operating systems, according to the size of the enterprise: small-sized farmers, mostly family-based, which is grown to approximately ten hectares of coffee plantations by farm; medium-sized farmers, owning between ten and fifty hectares of land; and large-sized farmers, cultivating more than fifty hectares of coffee (MAPA, 2009). According to that source, large growers are located in Bahia and Minas Gerais. The medium-sized coffee growers are concentrated in the south of Minas Gerais and São Paulo, and small farmers are located in Zona da Mata of Minas Gerais, Paraná, Espírito Santo and Rondônia.

Paraná, which once stood out in the Brazilian production in the decade of 60/70, nowadays has low participation in the production, of about 2% (ABIC, 2016). In Paraná, there is arabica coffee production, and also the production of specialty coffee is being developed. Regarding governance structures, Ferreira *et al.* (2009) stated that transactions in coffee AGS are made mostly through the market due to the fact that it does not involve high asset specificity, and green coffee is treated as a commodity.

The authors stated that as coffee becomes differentiated, companies adopt vertical governance structure, enabling greater control over the steps until the product reaches the consumer.

Nevertheless, contracting has been more and more used to govern rights and duties among agents, in order to ensure product quality. In transactions involving producers and processors, Demoner (2016), in a study concluded in 2003, found that 40% of coffee is sold directly by producers, without processing the product, which could add between 10 and 15% of value through stripping, grading and packing. The authors also claim that the buyers have some advantage over producers in the negotiations because they have more information concerning the product's attributes. In addition, Bronzeri (2010) identifies that brokers are the ones who take coffee to buyers for evaluation without the presence of the producer. Thus, the existence of intermediaries can raise doubts in the process and limits the guarantee of property rights. Bronzeri (2010) also identified the existence of vertical integration by specialty coffee producers that began to process the product, in the Northern of Paraná.

The study conducted in 2014 and 2015 found that the conditions indicated by Demoner (2016) and Bronzeri (2010) remain. In this case it was found that the producers sell their product directly to buyers or use brokers to carry out the transaction. This is possible due to the consideration of: formal rules for classification, defined by the Ministry of Agriculture, Livestock and Food Supply (MAPA), a public entity responsible for formal rules concerning food safety and agricultural and livestock productive and sanitary aspects; the marketing rules to the final product, as humidity and impurities defined by the Brazilian Coffee Roasters Association (ABIC), a private entity which certifies the product in the domestic market; the action of the National Company of Food and Supply (CONAB), which influences the price by providing information on crops and industry statistics.

Regarding governance structures, two of the three ways proposed by Williamson (1985, 1991) are presented: spot market and vertical integration. It can be inferred that this is due to the uncertainty (climate and price) which is independent of the actions of the chain actors, and thus affect both producers and the processors in a similar way. Therefore, agents adopt governance structures that suit their specific contexts, so as to minimize the consequences of uncertainties, mainly in prices and the accomplishment of short-term contract (neo-classic contracts).

From the interviews, it was found that hybrid form is rarely found empirically. Contracts are not used as a form of guarantee of rights, and agents prefer to run the risks of the market. In the case of contracts made through brokers, the relationship between producers and processors is similar to the trust format. As brokers observed, the relationship is based on reputation, which is developed between agents, with cases in which no contracts are signed at all, but are fulfilled in a verbally agreed way.

In the case of green coffee, measurement of physical and organoleptic dimensions depends on an expert performing the product tasting. It should not be so difficult to measure, although performing it is not trivial. The low cost associated with the measure directs the governance structure for the market contract, as stated by Barzel (2001, 2005). Producers can compare information related to dimensions with third party evaluators and avoid possible appropriation of value in the transaction, which they often do.

Vertically integrated producers process only their own coffee, producing roast & ground coffee for the market. For them, vertical integration is justified to ensure product quality, and to capture rents. When considering TCE, that choice could be justified, if one considers that special coffee could be classified as a physical specificity asset. However, for the MCE, according to Barzel (2005), the existence of specific assets is not sufficient for vertical integration, since the definition of rights becomes easier when the measurement is also easy and not expensive, not encouraging integration. Although within-organization quality and traceability can be assured, avoiding the capture of property rights, these producers integrate to get some value adding.

It is noticed that the information, which is related to property rights, according to Barzel (2005), is essential for green coffee producers transact their product. Coffee producers submit their samples for evaluation for more than one processor / buyer. In this case, information gathering generates transaction costs, particularly in the identification of the attribute values in order to ensure property rights. Here, rent appropriation is feasible, wherein the various measurements performed are characterized as a form of protection against opportunistic attitudes therefore costly.

Mechanisms of control are mainly established in a market relationship, using written contract, as pointed out by Barzel (2005). The measurement of attributes is made at the time of purchase by both buyers and sellers. The buyer performs the measurement of the attributes at the time of purchase through experts in sensory coffee analysis, and producers get their measurement to give their sample for at least two or three different buyers. These transactions are supported by one-time contracts, and it is possible to guarantee legal right, with the price being determined according to market price. However, the economic right is also displayed at the time of product evaluation by coffee evaluators, through discounts or performance results, allowing a variation in price set according to the market price (Figure 7.3).

Despite the presence of one-time contracts, transactions between buyers and suppliers tend to be recurrent, which enhances trust and strengthens the enforcement mechanisms of the MAPA and ABIC. This practice is confirmed by reducing the likelihood of moral hazard, since the transaction parameters are established and ratified for each transaction. Reputation can be established by the recurrence of measurement (every transaction requires measurement), as indicated by Barzel (2005). Note that the cases of vertical integrated producers can be explained by the

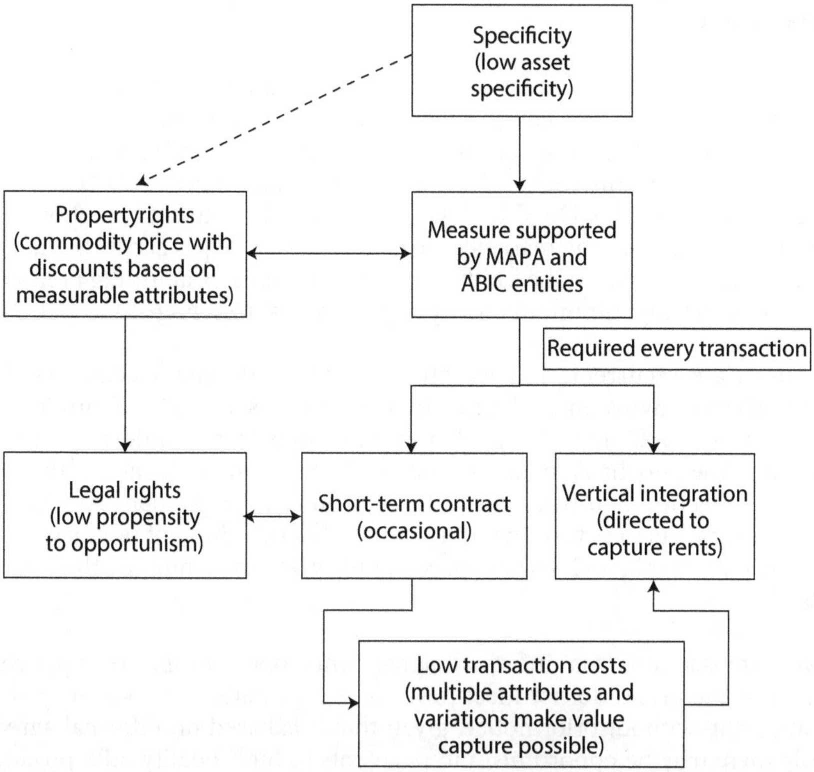

Figure 7.3. Governance structures and property rights in coffee agri-food systems in Paraná.

search for a small part of producers to produce differentiated coffees mainly to retain added-value.

It was observed that the market structure allows the appropriation at the time dimensions are measured, through economic rights (Barzel, 2001, 2005). In that context, the product price is set according to the market price, guaranteed by legal right, but the discounts and the classification can allow variations in that price, enabling evaluators to appropriate rents. Since product can generate different quality levels, it favours different levels of value adding or discounting (e.g. buyers can buy green coffee at a lower price, as a low-quality product and sell it at higher prices, as a better quality product, even to external market, because of classification and segregation).

7.5 Governance structures and property rights in milk agri-food systems in Paraná state

In the discussion involving property rights, the understanding of governance structures present in transactions between producers and processors becomes relevant. Two important factors guide transactions between farmers and processors in Paraná's dairy system: (1) Normative Instruction 62 (NI 61/2011), from MAPA; (2) Dairy Industry / Farmer Parity Council (CONSELEITE-PARANÁ), a private entity that operates for determining the price to the marketed product. Conseleite primarily operates in the definition of the reference value for the marketing of milk, establishing premium and discounts considering minimum quality requirements from NI 62.

Transactions are recurrent, and we observed site and time specificities, besides behavioural and environmental uncertainties. In this case, both producers and processors seek to maintain stable relationships, supported by informal agreements or contracts. The perishable nature of the product and the influence of the location affect milk collection and price and enforce producers' dependency allowing even informal agreements to govern the transactions. The definition of sanitary and safety parameters for raw milk in NI 62 impacts on milk price and simplifies the contractual process.

However, transactions via informal agreements open room for opportunistic behaviour. If the prices do not meet producer's expectations, he/she may abandon the relationship without prior notice, given that it is based on informal agreement. Similarly there may be opportunism in payments to high-quality milk producers in poor milk collection routes. Moreover, since recurrence does not necessarily guarantee the continuity of transactions, some farmers only keep transaction based on price. It should be considered, in turn, that this kind of attitude, according to respondents, is isolated, since most producers reported to trust their buyers. Even if transactions are not supported by a contract, processors, which are interested in maintaining their suppliers, define criteria for the payment of milk and, in some cases, individually negotiate with each producer. For transactions via formal contract, recurrence occurs in the terms established between the segments. The observed recurrence condition confirms theoretical directions.

According to Williamson (1985), the higher the frequency, the lower the costs concerning information gathering, mainly related to the reputation of a particular agent. In that case, recurrence enables the development of reputation and, therefore, opportunistic actions are minimized (Azevedo, 2000), since the values and behaviours are already known by both parties. It means that it reduces *ex-post* behavioural uncertainty (moral hazard), and creates credible commitment between them.

When considering the rationale suggested in TCE (Williamson, 1985), it was identified that transactions involve recurrent frequency (collecting every two days) and a monthly payment (cooperatives may pay twice a month); site and time specificities (milk perishability and farm's location matter) and physical assets are of general investments; the main uncertainty is related to price and on the continuity of transactions. The presence of uncertainties exists for both producers and processors. There is the possibility of opportunistic behaviour and the presence of bounded rationality for both parties. According to interviews, currently in Paraná milk AGS, governance structures have been identified in hybrid form (informal and formal contracts) and vertical integration, supported by guidelines of the NI 62 and Conseleite. Trust, vertical integration and formal contracts are found in cooperatives. This finding shows that transaction costs, for informal contracts, should emerge as there is no alignment, under TCE perspective. Yet, the presence of adverse selection due to non-distribution information indicates that the limited rationality claims uncertainty, increasing transaction costs. Formal contracts seem to be the suitable governance structure. This governance structure was found only in transactions involving large-size producers and cooperatives, mainly to retain producers, according to interviewees.

Analysis through MCE shows that the measurement of milk characteristics has become one of the most important steps of the transaction for both segments, which are aware of the importance of that procedure. The result of measurement defines additional amounts received on the base price, and allows processors to direct milk to a specific production line, according to its characteristics and quality attributes, defining transactions terms. These procedures can be identified for any kind of governance structures (Vertical integration, formal and informal contracts). Despite long term enforcement, transactions involve measurement, contrary to Barzel (2005).

Although measurement is widespread in milk AGS, with the definition of reference standards set by NI 62, it is clear that, in general, there are differences regarding payment criteria and results from analysis. This is because the measurement of milk attributes demands complex physical, chemical and microbiological processes, and it depends not only on a qualified person, but also on a number of equipment and chemical reagents. Thus, measurement costs are high, supported in Barzel (2005), which limits this ability only to processors and to some high-volume capitalized farmers. That does not mean, according to those interviewed processors, that producers cannot perform such tests or request a new analysis by an independent party. Because of the possibility of errors, both of interpretation or analysis, this test is not ruled out. Besides that, according to the interviewees it ratifies the reputation of the dairies for not engaging in quasi rent capture.

According to producers, the possibility of reaching into better terms of agreement is linked to the volume being traded. Thus, it is clear that small producers are likely to continue transacting through informal agreements, and as a result, use informal enforcement (economic right) for settling disputes. The presence of NI 62, in turn, allows some protection to producers, given that it ensures the minimum conditions for transactions.

In that context and considering MCE, transactions should be governed via contract (formal agreement) or by integration, as stated by Barzel (2005). In this case, the control and the producer's residual right of control tend be more effective. Thus, even if the producer segment has reported that the measurement on the property is impractical, all participants reported the possibility of carrying out analyzes of their production for laboratories that do not belong to the dairy companies which they transact, confirming the contractual relationship. However, contrary to the theory, to the relationship between milk producers and processors, the hybrid governance structure through informal agreements is the predominant model. In the few cases of producers using formal contract, the criteria and access to bonus percentages are clearer than those established between producers who make transactions via informal agreements (Figure 7.4). This allows better protection of rights, since under the protection established in the designed contract the producer can trade in legal terms or, at least, rely on court enforcement (Barzel, 2005).

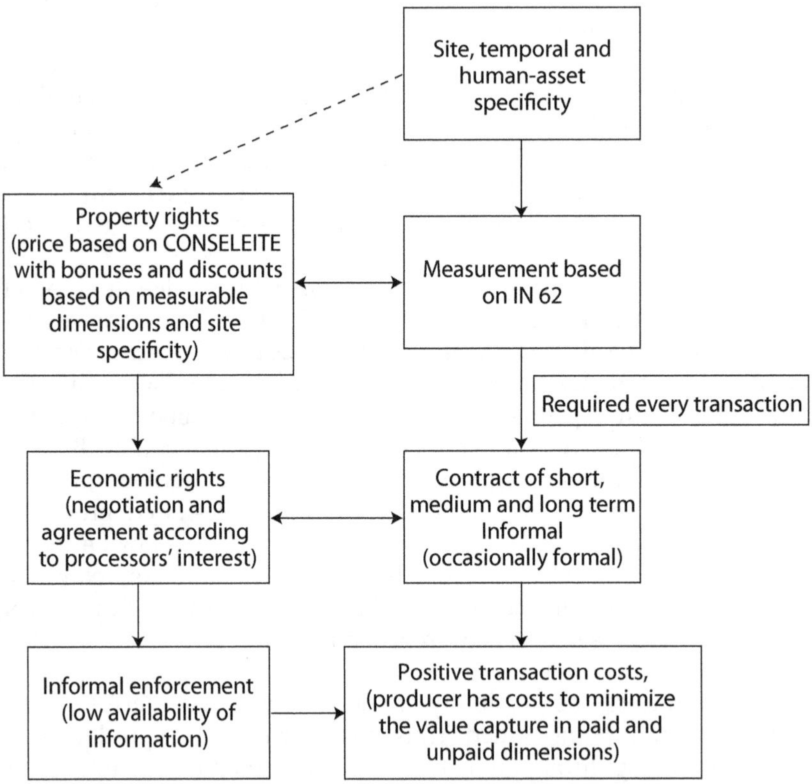

Figure 7.4. Governance structures and property rights in milk agri-food systems in Paraná.

However, it is clear that both producers with formal and informal contract are in the same context of information asymmetry, since they do not follow the whole process. In addition, the processor segment may hide information on attributes, supported by contractual and residual rights, which might not be communicated to producers. Thus, although contracts assure the right of control, and ownership and the informal agreement protects the control residual right (New Property Rights Approach; NPRA), it is not exercised by producers. The possibility of appropriation focuses on price and additional amounts to be obtained by the producer, or discounts, which are under control of the processor. In that case, for the agreements to be effective, product dimensions should be measurable and verifiable.

Ex-post, the problem is associated with information asymmetry, since the measurement process is in charge of processors, the information to producers are always late, resulting in the late search for economic rights. This condition can facilitate the capture of economic rights of each product attribute (Barzel, 2005), generating transaction costs for both parties. A not clear criterion in payment and low level of information symmetry generate costly enforcement (sometimes duplicating measurement). Transaction costs ratifies Barzel's definition, indicating that those transactions spend resources against capture of rents in transfer and protection of raw material. This context is observed mainly among the producers who transact via informal agreements, given the lack of accuracy to the criteria for calculating additional payments and discounts, according to product quality (Figure 7.5).

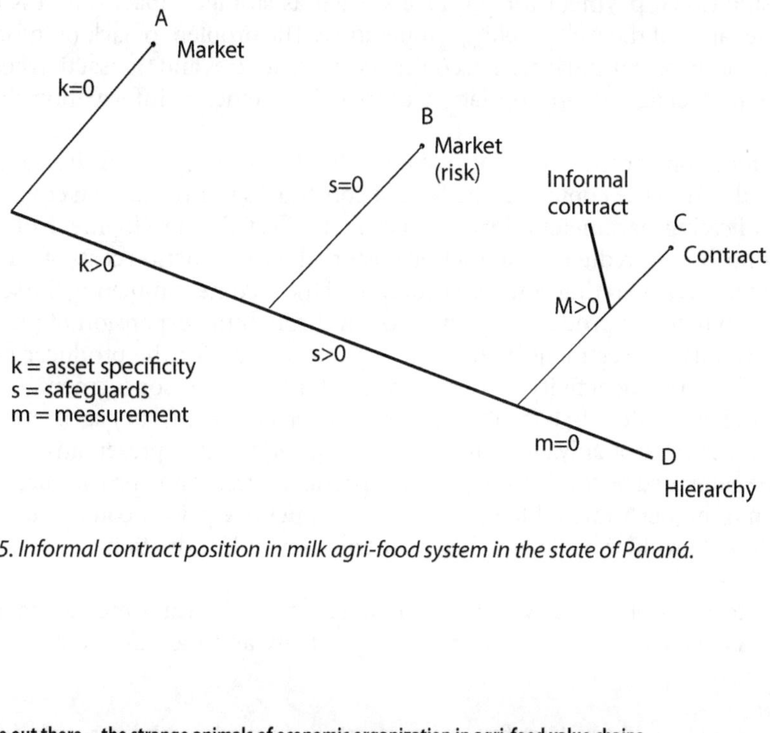

Figure 7.5. Informal contract position in milk agri-food system in the state of Paraná.

Thus, due to limitations in defining property rights on milk attributes and informal contractual relationship, the application of legal right is reduced, and safeguards disappear in the event of disputes (Barzel, 1997). Thus, economic right takes effect on agreements between parties, allowing greater chances of rent appropriation by processors. To be effective, the terms of agreement also need to be measured, but considering interviews, the information is not distributed efficiently, implying value appropriation indications or questions about the results. In this case, again, long-term relations seem unable to sidestep rent capture, as Barzel (2005) predicted.

For producers who transact via formal contract, they can also suffer from some kind of information asymmetry, due to the processor dependency on measurement process. That asymmetry, however, is limited since with the contract and the definition of dimensions in the pricing process, the possibility of income allocations is restricted. That's because, according to Barzel (1997), with well defined property rights, there is greater protection of legal rights. This condition ensures producers on their rights in the transactions and it motivates investments improving quality standardizing. As rights are established and more explicitly defined, it economizes on measurement (Barzel, 1997).

Finally, based on producers' statements, most processors offer a bonus system for just a few attributes established by NI 62, which are more valuable, such as the protein and fat levels, total bacterial count (TBC) and somatic cell count (SCC). The list of items that can generate discounts is wider, but it does not mean that if the product does not present problems, the producer will get a premium. Still, it is worth remembering that other measurable milk dimensions, beyond physical and microbiological aspects, could generate additional payment for producers, such as storage capacity, access to farm and maintenance of the milk cooling temperature. The problem of lack of information remains for both formal and informal contracts. Inefficiencies and transaction costs arise in order to protect against appropriation, as there is no efficient information sharing.

Thus, the adoption of contractual form to formalize transactions could be an alternative to be considered in the context of the formal constraints in institutional environment. Besides indicating parameters, formal rules can: affect the development of product quality; contribute to reducing transaction costs and the production costs of the activity as a consequence; generating better distribution of the income stimulating the search for innovation to increase productivity and growth through the expansion of production capacity; stimulate investment by minimizing uncertainty for the producer segment, related to the return of activity; control capacity for the processor segment, on supply and quality of raw materials; contribute to insertion of farmers in dairy agri-food system, to make the activity not only alternative for diversification but representative activity in the income generated by the farm; expand the product offering and provide access to the population of higher value-added products at competitive prices; contribute to better organization of producers allowing better distribution of information.

The characteristics of property rights and governance structures are summarized in Table 7.1. As one can see, the conditions of specificity and measurement are present,

Table 7.1. Coffee and milk agri-food systems contrasts: property rights and governance structures.

	Coffee	Milk
Asset specificity	• low specificity	• site • time • human
Measurement	• supported by MAPA's and ABIC's institutions	• based on in 62 (quality attributes)
Property rights	• commodity price • discounts based on measurable attributes for each transaction	• bonuses and discounts over basis price, considering measurable dimensions and site specificity
Legal rights	• occasional short term contracts • low opportunism	• occasionally formal contracts • informal enforcement
Economic rights	• multiple attributes and quality variations make value capture possible	• information asymmetries • negotiation and agreement according processors' interest
Governance	• market • vertical integration directed to value adding	• informal contracts

supported by formal mechanisms (public or private), and the misalignment with governance structures evoke transactions costs, mainly to protect against rent capture.

7.6 Conclusions

Enforcement mechanisms used in the coffee AGS, are given by the use of short-term contracts and vertical integration. In milk AGS, informal agreements are predominant, close to the relational, with some cases of vertical integration and formal contract, the latter involving cooperatives. In both cases, the presence of strong public rules, notably related to the traded product parameters, but enforcement driven by economic right (informal enforcement), brings condition to ensure basic property rights in transactions.

In the milk AGS, these parameters allow to guarantee minimum dimensions, at least for basic price payment, favouring processors and maintaining a minimum return to producers, although the amount of appropriation situations by not paying attributes are reported. In coffee system, the diversity of parameters allows the occurrence of implicit opportunistic behaviour, especially the exploitation of elements not considered, constituting rent appropriation. In both cases, processors' residual right to perform the measurement and set the price to be paid provide processors better conditions for conducting the transaction and the power to distribute value. In addition, uncertainty regarding the maintenance of the measured parameters is all for the producer, while recognizing that processors face market risk.

The presence of formal measurement parameters to guide transactions in both AGS limit the uncertainties about product and stimulate investments in production and better coordination. The establishment of rules, defined for both sides, gives parameters for payment and management of disputes, reducing transaction costs and favouring efficiency. Information asymmetries reduce incentives to invest due to the real possibility of rent appropriation.

On the other hand, the absence of public enforcement mechanisms aimed at regulating agreements and enforcing information sharing in both sectors may limit investments, generating transaction costs, which confirms the importance of the state. In addition, the lack of rules provides opportunity for informal agreements and rent appropriation, indicating moral hazard. The search for better regulation of the relationship, using the parameters already in place, can help to reduce the movement of farmers exclusion in the AGS studied, and of the rural activity.

The diversity of dimensions that can be decomposed and create value in coffee AGS demand a systematic aggregation or composition dimensions that must be considered in the payment system, and deserve further studies. In the milk AGS, a system of incentives to obtain better rates in value dimensions, should be adopted in order to drive for a higher value. For the processor segment, it means replacing the reduction of the discount costs for the gain in productivity. Initially a better information sharing seems to be a primary direction to be adopted. In addition, an establishment of rules aimed at generating long-term relationships, may be the way to reduce uncertainty and induce investments in the producers' side.

Acknowledgements

This research was financially supported by the Brazilian National Council for Scientific and Technological Development (CNPq).

References

Alchian, A.A. and Demsetz, H., 1972. Production, information costs and economic organization. American Economic Review 62(5): 777-795.

Azevedo, P.F., 2000. Nova Economia Institucional: referencial geral e aplicações para a agricultura. Agricultura em São Paulo 47(1): 33-52.

Barzel, Y., 1982. Measurement cost and the organization of markets. Journal of Law and Economics 25(1): 27-48.

Barzel, Y., 1997. Economic analysis of property rights. Cambridge University Press, Cambridge, UK.

Barzel, Y., 2003. Property rights in the firm. In: Anderson, T.L. and McChesney, F.S. (eds.) Property rights: cooperation, conflict, and law. Princeton University Press, Princeton, NJ, USA.

Barzel, Y., 2005. Organizational forms and measurement costs. Journal of Institutional and Theorical Economics 161(3): 357-373.

Bronzeri, M.S., 2010. Estratégia na cadeia produtiva do café: uma análise de empresas e produtores no norte pioneiro do Paraná. In: XIII SIMPOI – Simpósio de Administração da Produção, Logística e Operações Internacionais. São Paulo, Brazil.

Coase, R., 1937. The nature of the firm. Economica 4(16): 386-405.

Coffee Roasters Association (ABIC), 2016. Estatísticas. Available at: http://tinyurl.com/ycpo27re.

Davis, J.H. and Goldberg, R., 1957. A concept of agribusiness. Harvard University, Boston, MA, USA.

Demoner, C.A., 2016. Estudo técnico-econômico da cultura do café. Available at: http://tinyurl.com/y7x2vja2.

Department of Rural Economy (DERAL), 2016. Leite – análise da conjuntura agropecuária: ano 2015/16. Available at: http://tinyurl.com/ybz9tcoo.

Farina, E.M.M.Q., 1999. Competitividade e coordenação dos sistemas agroindustriais: a base conceitual. In: Jank, M.S. (ed.) O Agribusiness do leite no Brasil. Editora Milkbizz por PENSA, São Paulo, Brazil, pp. 20-31.

Farina, E.M.M.Q., Azevedo, P.F. and Saes, M.S.M., 1997. Competitividade: Mercado, Estado e Organizações. Singular, São Paulo, Brazil.

Ferreira, G.M., Waquil, P.D. and Gonçalves, W.M., 2009. Governança na cadeia do café: um caso no sul de Minas Gerais. Revista Custos e Agronegócios 5(3).

Foss, K. and Foss, N.J., 2001. Assets, attributes and ownership. International Journal of the Economics of Business 8(1): 19-37.

Hart, O., 1997. Firms, contracts and financial structure. Claredon Press, Oxford, UK.

International Coffee Organization (ICO), 2013. Trade statistics. Available at: http://www.ico.org.

Klein, B., Crawford, R.G. and Alchian, A.A., 1978. Vertical integration, appropriable rents, and the competitive contracting process. Journal of Law and Economics 21(2): 297-326.

Knight, F.H., 1964. Risk, uncertainty and profit. Reprint of Economic Classics. Augustus M. Kelley, Bookseller. Sentry Press, New York, NY, USA.

Langlois, R.N., 1992. Transaction-cost economics in real time. Oxford University Press, Oxford, UK.

Langlois, R.N. and Foss, N.J., 1999. Capabilities and governance: the rebirth of production in the theory of economic organization. Kyklos 52(2): 201-218.

Ménard, C., 2004. The economics of hybrid organizations. Journal of Institutional and Theoretical Economics 160(3): 345-376.

Mezzadri, F.P., 2012. Análise da Conjuntura Agropecuária Ano 2011/2012. Estado do Paraná – Secretaria da Agricultura e do Abastecimento. Departamento de Economia Rural, São Paulo, Brazil.

Ministry of Agriculture, Livestock and Food Supply (MAPA), 2009. Análise estrutural da cafeicultura brasileira. Ministério da Agricultura, Pecuária e Abastecimento, Secretaria de Produção e Agroenergia, São Paulo, Brazil.

Morvan, Y., 1985. Filière de production: fondementes d'economie industrielle. Economica, Paris, France.

North, D.C., 1990. Understanding the process of economic change. Princeton University Press, Princeton, NJ, USA.

North, D.C., 1991. Institutions. Journal of Economic Perspective 5(1): 97-112.

Revista Cafeicultura, 2013. Programa Cadeias Produtivas – O Café No Paraná. Available at: http://tinyurl.com/yapnazst.

Saes, M.S.M., 2009. Estratégias de diferenciação e apropriação da quase-renda na agricultura: a produção de pequena escala. Annablume, Papesp, São Paulo, Brazil.

Secretary of Agriculture and Food Supply of Paraná (SEAB), 2010. Secretaria de Estado da Agricultura e do Abastecimento. Caracterização da Indústria de Processamento e Transformação do Leite no Paraná. Curitiba, Brazil.

Souza, J.P. and Bánkuti, S.M.S., 2012. Uma análise dos contratos no sistema agroindustrial suinícola no Oeste paranaense pela ótica da mensuração e da transação. Gepec 16(2): 1-19.

Williamson, O.E., 1985. The economic institutions of capitalism: firms, markets, relational contracting. New York Press, New York, NY, USA.

Williamson, O.E., 1991. Comparative economic organization: the analysis of discrete structural alternatives. Administrative Science Quarterly 36(2): 269-296.

Williamson, O.E., 1996. The mechanisms of governance. Oxford University Press, New York, NY, USA.

Williamson, O.E., 2000. The new institutional economics: taking stock, looking ahead. Journal of Economic Literature 38: 595-613.

Zylbersztajn, D., 2005a. Papel dos contratos na coordenação agro-industrial: um olhar além dos mercados. In: Congresso da Sociedade Brasileira de Economia e Sociologia Rural, XLIII, 2005. Ribeirão Preto, Brazil.

Zylbersztajn, D., 2005b. Measurement costs and governance: bridging perspectives of transaction cost economics. In: International Society for the New Institutional Economics. Barcelona, Spain.

Zylbersztajn, D. and Farina, E.M.M.Q., 1999. Strictly coordinated food-systems: exploring the limits of the Coasian firm. International Food and Agribusiness Management Review 2: 249-265.

8. The apparent paradox of unadvertised private food safety standards[1]

C. Russo[1], M.A. Perito[2,3] and A. Di Fonzo[1]*

[1]University of Cassino and Lazio Meridionale, Via S. Angelo loc. Folcara, 03043 Cassino, Italy; [2]University of Teramo, Via R. Balzarini 1, 64100 Teramo, Italy; 3ALISS-INRA, 65 Boulevard de Brandebourg, 94200 Ivry-sur-Seine, France; carlo.russo@unicas.it

Abstract

Large retail chains have spent considerable resources to promote production protocols and traceability across the supply chain, aiming at increasing food safety. Yet, the majority of consumers are unaware of these private food safety standards (PFSS) and retailers are not informing them. This behaviour denotes a pooling paradox: supermarkets spend a large amount of money for food safety and yet do not inform consumers. The result is a pooling equilibrium where consumers cannot discriminate among high quality and low quality products and supermarkets give up a potential price premium. This paper provides an economic explanation for the paradox using a contract-theory model supporting two major conclusions. First, we found that PFSS implementation may be rational even if consumers have no willingness to pay for safety, because the standard can be used as a tool to solve asymmetric information along the supply chain. Using the PFSS, supermarkets can achieve a separating equilibrium where opportunistic suppliers have no incentive to accept the contract. Secondly, even if consumers exhibit a limited (but strictly positive) willingness to pay for safety, advertising may be profit-reducing. If the expected price margin is high enough, supermarkets have incentive to supply both certified and uncertified products. In this case, we show that, if consumers perceive undifferentiated products as 'reasonably safe', supermarkets may maximize profits by pooling the goods and selling them as undifferentiated. This result is not driven by advertising costs, as we derive it assuming free advertising.

Keywords: advertising, food safety, private standard, consumer information

[1] The chaper is the result of the joint effort of the authors. However, Maria Angela Perito is responsible for Section 8.1, 8.2 and 8.6, Carlo Russo for for Section 8.3 and 8.4, Antonella Di Fonzo for Section 8.5.

DOI 10.3920/978-90-8686-844-5_8,

8.1 Introduction

In the last decade, large retailers have spent considerable effort in establishing private standards to ensure the safety of their products (Henson and Reardon, 2005; Reardon *et al.*, 2001). Yet, despite of the large investments, consumers seem to be largely unaware of the very existence of these private food safety standards (PFSS) and supermarkets are not putting any significant effort to advertise and promote them (Fulponi, 2006; Grunert, 2005; Hammoudi *et al.*, 2009; Rozan *et al.*, 2004). Several PFSS limit communication to the B2B environment and often labelling or advertising to final consumers is restricted (Theuvsen *et al.*, 2007).

This behaviour denotes an apparent 'pooling paradox'. Supermarkets invest a non-negligible amount of money to supply safer products but then they do not pursue a differentiation strategy as they do not advertise the standard nor inform consumers. Consumers' lack of information determines a sort of pooling equilibrium where certified and uncertified products are sold for the same (or very similar) price and supermarkets give up the potential price premium.

The economic literature addressed the pooling paradox. Rozan, Stenger and Willinger (2004) conclude that consumers perceive food safety to be a basic characteristic that they expect from all products and hence they may not willing to pay a premium for it (see also Giraud-Heraud *et al.*, 2009; Roosen, 2003). These explanations postulate that PFSS are adopted for other reasons in addition to food safety, as there is no economic rationale in incurring in the cost of providing attributes that consumers do not want to pay for. Economics literature suggests PFSS might serve other purposes than the mere food safety. In this regard, many authors provide an institutional analysis of PFSS and conclude that they can be used as a tool for governing the food system (Fulponi, 2006; Gereffi *et al.*, 2005; Henson and Humphrey, 2010; Humphrey and Schmitz, 2001; Kariuki *et al.*, 2012; OECD, 2006).

Our paper contributes to this literature providing a theoretical model explaining the paradox. We reached two main results. Firstly, we find that PFSS implementation may be rational even if the consumers' willingness to pay for safety is equal to zero, because the standard may be a tool to solve asymmetric information problems along the supermarket's supply chain. Secondly, we find that advertising may reduce expected profits if the supermarket optimal strategy consists in offering consumers a mix of certified and non-certified products, even if consumers exhibit a limited (but strictly positive) willingness to pay for safety. This result is not driven by advertising costs, as we derive it assuming free advertising, and requires that the retailers' price margin is large and that consumers perceive undifferentiated products as 'reasonably safe'.

Our findings provide an explanation to the apparent pooling paradox. PFSS are implemented because they are valuable tools to solve asymmetric information, regardless of consumers' willingness to pay for safety. If the willingness to pay is strictly positive, a supermarket may still decide to give up the price premium on the standard products in order to avoid losses on the non-standard goods. In this framework, the decision to advertise (or not) depends on case-specific cost-benefit evaluation.

8.2 Private standards and food supply chain

Over the past few years, food safety concerns have become extremely dynamic and a significant driver of agri-food systems. The risk of food crises has brought the consumers to demand more food safety and, at same time, retailer and food industry have developed a plethora of private food safety and quality management standards as complements of public regulations.

While much of the focus of the economics literature has been on the role of public food safety and quality standards both as policy instruments to influence public regulations about food safety (Lutz *et al.*, 2000; McCluskey and Winfree, 2009) and as non-tariff barriers to trade (Balsevich *et al.*, 2003; Boselie *et al.*, 2003; García Martinez and Poole, 2004; Hobbs, 2010), it is evident that private standards are playing an increasing role in the governance of agricultural and food supply chain (Grazia *et al.*, 2014; Russo *et al.*, 2014).

In the context of global sourcing, the food system has an incentive to supply safe food, reducing the risks related to microbial contamination and pesticide residues, with the implementation of food safety standards. The private standards are often more restrictive than public regulations and their development can increase costs for the farms. Private initiatives of standardization are mostly undertaken by retailers, with high market power and a competitive strategy based on private standard (especially business-to-business standards as: GlobalGap, BRC, SSQF, GFSI, etc.), to bind the company's reputation on quality products.

Humphrey (2011) distinguishes between premium and baseline private standards. Premium standards attempt to distinguish specific quality attributes (e.g. organic, fair trade, origin, animal welfare, etc.) and are generally communicated to consumers with a specific logo or symbol. Instead, baseline standards are mainly business-to-business (e.g. GlobalGap, IFS and BRC) and they give access in global supply chains.

Baseline standards can be pure private standards and collective standards. The first typology concerns firm-specific private standards that are defined, controlled and used by an individual retailer (e.g. Tesco Nature's Choice and Filières Qualité Carrefour) (Berdegué *et al.*, 2005; Codron *et al.*, 2005; Giraud-Héraud *et al.*, 2006). Collective standards are supported by the big retailers and food industry with the vertical coordination realized through intermediate markets and third-party certification (e.g. the BRC, IFS and Globalgap) (Henson and Reardon, 2005).

The efforts to implement collective standards are not communicated to consumers and the specific logo can only be placed on the pallets that will not be displayed at the point of sale. In particular, retailers ask their suppliers for these certifications and the compliance with private standards becomes a condition for the market access. The costs to verify the compliance to the private standards are transferred along the supply chain from retailers to the supplier (Fischer *et al.*, 2009; Graffham *et al.*, 2007; Hatanaka *et al.*, 2005). The supplier is responsible for product safety and retailers do not offer a premium price because the certification is considered a prerequisite to enter the transaction (Mènard and Valceschini, 2005). The collective private standard can be considered as an instrument for the coordination of supply chain through the standardization of product requirements to the suppliers. A number of studies suggest that expected competitive advantages are important reasons for firms to embrace private standards, also if they operate on a business-to-business basis (Henson and Caswell, 1999). Retailers ask their suppliers for these certifications, and compliance with private standards becomes an addition to the list of services which retails are asked to suppliers as a condition of sustained market access. The control costs necessary to private standards are transferred the grower and the exporter upstream the supply chain by retailers (Fischer *et al.*, 2009; Graffham *et al.*, 2007; Hatanaka *et al.*, 2005).

Finally, all of these standardization efforts not necessarily imply lacks of sanitary food risks along supply chain. In fact, it is possible that certified and uncertified products are sold together to the final consumer and also at the same (or very similar) price. From this point of view, as it is above-mentioned, private collective standards may be mainly a tool to solve asymmetric information along the supermarket's supply chain.

8.3 The model

In this section we develop a simple theoretical model to explain why a profit-maximizing supermarket may decide to implement a PFSS without extensive advertising to its customers. Our analysis concludes that PFSS can be used as a supply management tool, ensuring that suppliers comply with an incomplete contract with the buyer in the presence of asymmetric information. This result is independent of any impact that PFSS may have on consumers. Furthermore, the theoretical model shows that a buyer facing stochastic prices, under specific conditions, maximizes profits by mixing products complying with the PFSS with other products during peak seasons. This strategy requires that consumers are not able to tell the difference between certified and uncertified products, otherwise the supermarket could incur in a price penalty on the potentially unsafe goods. Hence the advertising of the PFSS may not be efficient, since it makes consumer suspicious of uncertified, and potentially unsafe, food.

We apply a standard principal-agent setting to model the PFSS. The principal is a supermarket buying a product from farmers and selling it to the final consumers for an exogenous price PR. Similarly to Russo *et al.* (2014), we postulate that the profit of the supermarket depends on a 'critical attribute': if the attribute is missing the supermarket incurs in a profit loss D. Examples of such critical attribute are the compliance with the logistic requirement of the supermarket's supply chain (timeliness of the delivery, reliability, etc.) or the respect of public safety standards as imposed by the current regulation. A key feature of our model is that the critical attribute may or may not concern the quality or safety of the product. It is sufficient that: (1) the lack of the attribute determines a profit loss; and (2) the attribute is not freely observable by the supermarket.

The farmers act as the agents of the model. We assume that there are M farmers in the market, producing one unit of the product each. The farmers are homogenous except for the parameter c_i, representing their cost efficiency in delivering the critical attribute. The parameter is not observable by the supermarket and is uniformly distributed as $c_i \sim U[0, c_H]$. The distribution of the parameter is public knowledge. Each farmer can sell the unit of product to the spot market for an exogenous price P_M or deliver it to the supermarket for a price P_S. If the farmer decides to deliver to the supermarket, he/she can deliver the hidden attribute or behave opportunistically.

We assume the existence of a spot market where an infinite number of small independent retailers compete with the supermarket. These independent retailers are passive players, buying one unit of the product from farmers for the price P_M and selling it to consumers for the price P_R.[2]

Prices P_R and P_M are assumed to be exogenously determined and stochastic. They follow a joint distribution such that the price margin $PM=P_R-P_M$ is distributed as a random variable with mean $E(PM)$ and variance σ^2_{PM}. We assume that, in the absence of advertising, the PFSS do not affect the consumers' willingness to pay so that the independent retailers and the supermarket are exposed to the same retail price. The price distributions are public knowledge and all players are risk-neutral.

We represent the supply contract using a simple principal-agent model. At time t_0 the supermarket offers n_C contracts to randomly selected farmers. The supermarket offers a price $P_M + \Delta$, where Δ is a price premium and requires the delivery of products with the *critical attribute* at time t_1. The n_C farmers under contract must decide if they will deliver the attribute or if they will behave opportunistically. To the purpose of our analysis, the opportunistic behaviour is defined as accepting the contract, asking for the price premium Δ without delivering the critical attribute.

[2] We assume that M is large enough, so that n^* is always lower than M, where n^* is the profit-maximizing quantity traded by the supermarket. The assumption greatly simplifies the presentation of the results and does not affect the key findings. Furthermore, the absence of a monopolistic/monopsonistic retailer is consistent with observed reality.

The expected payoff from opportunism is $P_M + p\cdot\Delta$, where $p \in (0, 1)$.[3] Summarizing the strategy sets of the supermarket and the ith farmers are, respectively, $S_S[n_C, n_M, \Delta]$ and $S_i[AR, OPP]$, where n_M is the quantity of product that the supermarkets buys on the spot market (i.e. without the critical attribute), *AR* and *OPP* are binary variables with *AR*=1 if the farmer accepts the contract and zero otherwise and *OPP*=1 if the farmer behaves opportunistically and zero otherwise.

At time t_1 production is revealed. The *n* farmers under contract will deliver the products to the supermarkets and the other *M-n* farmers will sell their product to the independent retailers.

Assuming that the cost function is separable in the price of the product, the buyer expected payoff is:

$$E(\pi_S) = n_C[E(PM) - \Delta - z\cdot D] + n_M[E(PM) - D] - C_S(n_C+n_M) \qquad (8.1)$$

where *n* is the number of contracts offered by the buyer, which is equal to the quantity produced since each farmer produces one unit, *z* is the probability that a farmer behaves opportunistically. *C*(*n*) is the buyer's marketing cost function, which is assumed to be $C(n_C+n_M)=a\cdot(n_C+n_M)^2$. Under the specified cost function, the marginal marketing cost for certified and uncertified products is the same.

Farmer *i*'s expected payoff is:

$$E(\pi_i) = \begin{cases} E(P_M) \text{ if sells to an independent retailer} \\ E(P_M) + \Delta - c_i \text{ if delivers the attribute} \\ E(P_M) + p\cdot\Delta \text{ if behaves opportunistically} \end{cases} \qquad (8.2)$$

From Equation 8.2, the farmer's individual rationality (IR) constraint requires $\Delta \geq 0$ and the incentive compatibility (IC) constraint is satisfied if $c_i < (1 - p)\Delta = \bar{c}$. Since the cost parameter c_i is uniformly distributed, the probability of contracting an opportunistic farmer (i.e. with $c_i > \bar{c}$) is:

$$z = \max\left[1 - \frac{(1-p)\Delta}{c_H}, 0\right]$$

[3] The parameter *p* determines the expected benefit from opportunism. It may summarize several factors such as unrecoverable down-payments made by the supermarket, the probability that the opportunistic behavior goes undetected (for example if the supermarket can observe only the aggregate profit loss), or the probability that the supermarket cannot reject the delivery (for example because it cannot find another supplier on a short notice). We restricted the parameter to the interval (0,1) because for *p*=0 there is no incentive for opportunistic behavior. For *p*≥1 the model is undetermined, as there is no incentive for delivering the attribute. In this case, the key findings of this paper can still be derived using an infinitely repeated game, under the assumption that the supermarket can punish the opportunistic farmers by refusing to offer them contracts forever.

and the problem of the buyer is:

$$\max_{\Delta,n} E(\pi_S) = n_C\left[E(PM) - \Delta - \frac{c_H - (1-p)\Delta}{c_H}D\right] + n_M\left[E(PM) - D\right] - a\cdot(n_C + n_M)^2$$

subject to:

$(a)\ \Delta \geq 0$

$(b)\ \Delta \leq \dfrac{c_H}{1-p}$

$(c)\ n_C \geq 0, n_M \geq 0$

where inequality (a) is the farmer's IR constraint, inequality (b) states the buyer is rational and does increase the price premium beyond the level corresponding to z=1, inequalities (c) are the usual non-negativity constraints on quantities. The farmer's IC constraint has been substituted into the objective function.

The supermarket's optimal strategy depends on the sign of the derivative of the objective function with respect to Δ:

$$\frac{\delta E(\pi_s)}{\delta\Delta} = n_C\left[\frac{(1-p)D}{c_H} - 1\right] = n_C\left(\frac{D}{\overline{\Delta}} - 1\right)$$

where:

$\overline{\Delta} = \dfrac{c_H}{1-p}$, i.e. the incentive price for the most inefficient farmers.

If D is larger than $\overline{\Delta}$, the derivative is always positive and the marginal benefit of reducing the probability of opportunistic behaviour is greater than the marginal cost of increasing the price premium. In this case, constraint (b) is binding, the buyer offers the optimal price premium $\Delta^* = \overline{\Delta}$ and no farmer has incentive to behave opportunistically. The buyer has no incentive to use the spot market, because the per-unit margin on the certified goods is higher than the one on the uncertified products and the marginal cost is the same.

If D is smaller than $\overline{\Delta}$, the derivative is negative and the buyer maximizes profits by setting the price premium to zero and incurring in the loss of D for each unit of product (i.e. acting as an independent retailer).

The strategy of the buyer is driven by the relative magnitude of the expected profit loss from the opportunism and the price premium. If the expected loss is large, the buyer uses contracts (and incentive pricing) to organize the supply chain. If the expected loss is small, the buyer organizes the transaction using the spot market. If the price margin is negative the supermarket exits the market. Table 8.1 summarizes the

Table 8.1. Supermarket's optimal strategy in the absence of PFSS as a function of D (the profit loss from opportunistic behaviour and $\bar{\Delta}$ (the incentive price for the least efficient farmers).

	$D<\bar{\Delta}$	$D\geq\bar{\Delta}$
Per-unit margin is negative	exit	exit
Per unit margin is non-negative	pure spot market strategy (spot market only, no contracts) act as independent retailer	pure contract strategy (contracts only, no spot market) coordinated supply-chain

strategic options of the supermarket. Simple algebra shows that profits for a spot market only strategy are:

$(4a)^{-1}\left[E(PM)-D\right]^2$,

profits for a contract only strategy are:

$(4a)^{-1}\left[E(PM)-\bar{\Delta}\right]^2$, and profits from the exit strategy are equal to zero.

8.4 The PFSS as supply management tools

In this section we show that the buyer can improve profits by imposing a PFSS to farmers, even if the standard itself does not affect consumer demand. In the typical literature about PFSS, the buyer benefits from the adoption of the standard because of consumers' willingness-to-pay for quality and safety. In our model, we show that the PFSS can be used as a device ensuring that opportunistic farmers have no incentive to subscribe the contract with the buyer.

We assume the adoption of the standard requires that farmers bear an observable implementation cost *K*, which is equal for all farmers. The payment of the cost *K* is a pre-requisite for being offered a contract. In this setting the expected payoff for the *i*-th farmer is:

$$E(\pi_i)=\begin{cases} E(P_M) \text{ if sells to the spot market} \\ E(P_M)+\Delta-c_i-K \text{ if delivers the attribute} \\ E(P_M)+p\cdot\Delta-K \text{ if behaves opportunistically} \end{cases} \quad (8.3)$$

Following Russo *et al.* (2014), we assume also that the function *K*=*f*(PFSS), linking the implementation cost to the technical provisions of the PFSS, is public knowledge. Banterle and Stranieri (2008) suggested that *K* can be considered as a specific investment. The supermarket can set the implementation cost *K* strategically, in order to create a separating equilibrium where the inefficient farmers opt out of the contract. Therefore if a PFSS is adopted the strategy of the supermarket is defined as the set $S_B\{n_C, n_M, \Delta, K\}$.

The separating equilibrium requires that the expected payoff is negative if the farmer behaves opportunistically and positive otherwise, i.e. $\Delta - c_i - K \geq 0$ and $p{\cdot}\Delta - K \leq 0$. These conditions are satisfied for any pair:

$$\bar{\Delta} \geq c_i(1-p)^{-1} \text{ and } \bar{K} \in [p{\cdot}\bar{\Delta}, \bar{\Delta} - c_i] \tag{8.4}$$

If the pair $[\Delta, K]$ satisfies conditions (Equation 8.4), the probability z of incurring in opportunistic behavior is zero, because the inefficient farmers have no incentive to join the contract. The supermarket expected payoff adopting the PFSS is:

$$E(\pi_s \mid PFSS) = \sum_{i=C,M} n_i\left[E(PM) - W_i\right] - a\left(\sum_{i=C,M} n_i\right)^2 \tag{8.5}$$

with $W_C = \Delta$ and $W_M = D$

and the problem of the principal is:

$$\max_{\Delta, n_C, n_M, K} E(\pi_S \mid PFSS)$$

subject to:

$$(a)\ n_C \leq \frac{(1-p)\Delta}{c_H} M$$

$$(b)\ \Delta \geq 0,\ n_C \geq 0, n_M \geq 0$$

where

$$\frac{(1-p)\Delta}{c_H} = \frac{\Delta}{\bar{\Delta}}$$

is the share of producers that are willing to accept the contract, since farmers with $c_i > (1-p)\Delta$ have no incentive to join. Constraint (a) states the price premium Δ must be high enough to elicit supply. Because the supermarket's only incentive to pay a price premium is eliciting supply, at the optimum the constraint is satisfied with equality (assuming that K* belongs to the interval K). The non-negativity constraint on n_C is satisfied as long as Δ is non-negative.

Solving the constrained maximization program and checking the values of the Lagrange multipliers, we found that the non-negative constraint on Δ is not binding and the non-negative constraint on n_M is binding if:

$$D > \frac{\bar{\Delta}}{\bar{\Delta} + a \cdot M} E(PM)$$

For notational convenience we define:

$$\lambda = \frac{\bar{\Delta}}{\bar{\Delta} + a \cdot M}$$

This result implies that there are two possible optimal strategies, depending on the value of the parameter *D*. If *D* is large enough, the supermarket organizes the supply using contracts only (pure contract strategy). If *D* is small, the supermarket uses a mix of contracts and spot-market purchases (mixing strategy). In this latter case, the supermarket sets a small price premium, elicits a small supply for contract production and uses the spot market up to the optimal production level. Noticeably, under a PFSS, the mixing strategy strictly dominates the pure spot market and the exit strategies. Once the information asymmetry is solved, contracting always offer profit opportunity.

The optimal strategic set $S_B\{n_C, n_M, \Delta, K\}$ for a pure contracting strategy is:

$$S^*_{B|PCS}\left\{\frac{\lambda \cdot E(PM)}{2\bar{\Delta}} \cdot M, 0, \frac{\lambda}{2} \cdot E(PM), \tilde{K}\right\}$$

And the corresponding profit level is:

$$\pi_{S|PCS} = \frac{\lambda}{4\bar{\Delta}} E(PM)^2 M$$

If the non-negative constraint on n_M is not binding, the supermarket adopts a mixing strategy and the optimal strategic set is:

$$S^*_{B|MS}\left\{\frac{D}{2\bar{\Delta}} M, \frac{E(PM) - D}{2a} - \frac{D \cdot M}{2\bar{\Delta}}, \frac{D}{2}, \tilde{K}\right\}$$

and the corresponding profit level is:

$$\frac{\left[E(PM) - D\right]^2}{4a} + \frac{D^2}{4\bar{\Delta}} M$$

Simple (yet tedious) algebra shows that the PFSS strategies always dominate the no-PFSS strategies. The intuition behind this result is straightforward as the optimal strategy without PFSS lies in the set of the admissible solutions for a supermarket using PFSS (i.e. setting K=0). Therefore, the level of expected of the profits from PFSS strategies is at least the same as the no-PFSS strategies. Using PFSS solves the information asymmetry at no cost for the principal, because the adoption cost is paid by the farmers.

Compared to the no-PFSS strategies, the supermarket must pay a smaller information rent to the farmers and gains higher profits. This result was obtained under general assumptions: the existence of a critical attribute, asymmetric information regarding farmers' efficiency and the delivery of the attribute, the possibility of imposing *ex-ante* an observable cost strategically. The PFSS increases expected profits even if consumers have no willingness to pay for the increased quality and safety attributes of the certified products.

This conclusion provides a possible explanation for the pooling paradox. If the supermarket cannot charge a price premium, there is no incentive to advertise. Yet, we can still observe PFSS, as they are a device solving information asymmetry along the supply chain.

8.5 To advertise or not to advertise

In this section we relax the restrictive assumption that consumers are not willing to pay for safety. Therefore, we introduce a premium for certified safe products and a penalty for uncertified goods. Differentiation strategies are now feasible, if consumers were informed about the PFSS. Yet, we argue that advertising the PFSS may not be rational under specific circumstances, because it may reduce the profits from Mixing strategies. If the supermarket plans to supply a mix of certified and uncertified goods, it might be rational to give up the premium on the certified products in order to save on the penalty on the uncertified goods. The optimal advertising decision depends on the relative magnitudes of these two components and it is an empirical question. Our model provides a theoretical rationale for the apparent pooling paradox.

We developed a simple model showing that advertising does not necessarily increase expected profits even if consumers exhibit strictly positive willingness to pay for quality and safety and advertising cost is zero. The intuition behind this conclusion is straightforward: if the supermarket's profit-maximizing strategy is mixing certified and uncertified products, under specific conditions (that will be derived later on in the paper) advertising might reduce the expected profits on the uncertified products more than it increases the returns on the certified products. The results are consistent with previous findings that supermarkets may avoid advertising in order to prevent consumer wariness of uncertified products (e.g. Giraud-Heraud *et al.*, 2009).

Assume that in a differentiated market, where consumers are informed about PFSS, the expected price margin of the uncertified product is $E(PMU)=E(PM)-Z$ and the expected price margin of the certified products is $E(PMC)=E(PM)+Z\cdot\beta$, with $Z>0$ and $\beta\geq0$. In our modelling, the parameter Z represents the 'safety concern effect' and it is the discount that consumers require to buy products that are advertised as 'less safe'. The product $Z\beta$ is the 'value-added effect' and denotes the consumers' willingness to pay for the improved food safety and certification.

The parameter β may be interpreted as the consumers' perception of food safety in the absence of advertising. $\beta=0$ implies that, in the absence of advertising, the consumers consider all products as safe and they take safety as granted. In this case, the only effect of the advertising is to make them aware of the existence of unsafe products. High values of β, instead, imply that the consumers are aware that products might be unsafe and the PFSS can increase the value of the product. The parameter Z can be interpreted as a measure of consumers' concern for safety and quality. The higher the value of Z, the more valuable the certified products are compared to the uncertified products. The expected price $E(PM)$ can be interpreted as the emerging pooled price when certified and uncertified products are undistinguishable.

The supermarket's strategy set when advertising is possible and consumers are willing to pay for quality can be defined as:

$$S_A\{n_C, n_M, \Delta, K, A\}$$

where A is a binary variable being equal to 1 if the supermarket advertises the standard or 0 otherwise.

We want to show that it exists a strictly positive value β^* such that for any $\beta \in (0,\beta^*]$ setting A=1 is not a dominant strategy, even if advertising is free. This result implies that it exists a set of market conditions such that free advertising is profit-reducing even if consumers have a strictly positive willingness to pay for safety.

To show this result we find the supermarket's optimal strategy, conditioned to advertising, and then we compared the expected profits with the no-advertising case from Section 8.2. Our goal is to show that it exists a non-empty set of parameter values such that

$$E(\pi_S|ADV) \leq E(\pi_S|\text{PFSS}),$$

where the right-hand side of the inequality represents the expected profits without advertising and the left-hand side is the expected profits with advertising.

The advertising supermarket profit function is defined as:

$$E(\pi_S|ADV) = n_C[E(PM)+Z\beta-\Delta] + n_M[E(PM)-Z-D] - a(n_C+n_M)^2$$

Similarly to the no-advertising case, the advertising supermarket must maximize the objective function under the constraints:

$$(a)\ n_C \leq \frac{(1-p)\Delta}{c_H} M$$

$$(b)\ \Delta \geq 0,\ n_C \geq 0, n_M \geq 0$$

where constraint (a) is assumed to be satisfied with equality and the non-negativity constraint on n_C is satisfied for any non-negative Δ.

The non-negativity constraint on Δ proved to be non-binding. The constraint on n_M is binding for:

$$D > \lambda \cdot E(PM) - Z\left(1 + \beta \frac{a \cdot M}{\bar{\Delta} + a \cdot M}\right)$$

As expected, the economic interpretation of the condition is similar to the non-advertising case. If *D* is large, the supermarket will use a pure contract strategy (with advertising). If *D* is small enough, the supermarket will adopt a mixing strategy.

The optimal strategic set for a pure contract strategy is:

$$S^*_{A|PCS}\left\{\frac{\lambda\left[E(PM)+Z\beta\right]}{2\bar{\Delta}}M, 0, \frac{\lambda}{2}\left[E(PM)+Z\beta\right], \tilde{K}, 1\right\}$$

and the resulting profits are:

$$\pi_{A|PCS} = \frac{\lambda}{4\bar{\Delta}}\left[E(PM)+Z\beta\right]^2 M$$

The optimal strategic set for a mixing strategy is:

$$S^*_{A|MS}\left\{\frac{D+Z+Z\beta}{2\bar{\Delta}}M, \frac{E(PM)-D-Z}{2a} - \frac{D+Z+Z\beta}{2\bar{\Delta}}M, \frac{D+Z+Z\beta}{2}, \tilde{K}, 1\right\}$$

and the resulting profits are:

$$\pi_{A|MS} = \frac{\left[E(PM)-D-Z\right]^2}{4a} + \frac{(D+Z+Z\beta)^2}{4\bar{\Delta}}M$$

The comparison with results from Section 8.2 shows that the advertising/pure contract strategy weakly dominates the no advertising/pure contract strategy in the admissible parameter range. If the supermarket offers only certified products, free advertising increases profits unless *Z* and/or *β* are equal to zero.

The strategy advertising/mixing weakly dominates the strategy no advertising/mixing only if *β* is large enough compared to *E*(*PM*). In fact, the difference in the expected profits is:

$$\pi_{A|MS} - \pi_{S|MS} = Z \cdot \frac{M\left[Z(1+\beta)^2 + 2D(1+\beta)\right]}{4\bar{\Delta}} - Z \cdot \frac{\left[2E(PM)-2D-Z\right]}{4a}$$

The difference is negative for any:

$$E(PM) > \frac{2D\left[\bar{\Delta} + a \cdot M(1+\beta)\right] + a \cdot M(1+\beta)^2 Z + Z \cdot \bar{\Delta}}{2\bar{\Delta}} \quad (8.6)$$

which means that for high values of the expected price margin, the profit loss from the reduction in the sale price of the uncertified products more than offsets the profit gain from the increase in the sale price of the certified goods.

Basic algebra shows that the a similar result holds for the range:

$$E(PM) \in \left[\frac{D(\bar{\Delta} + a \cdot M)}{\bar{\Delta}}, \frac{(D+Z)(\bar{\Delta} + a \cdot M) + MaZ\beta}{\bar{\Delta}} \right]$$

where the optimal strategies are no advertising/mixing and advertising/pure contracting.

Our model shows that product differentiation may allow the supermarket to extract profit from consumers' willingness to pay for safety, but may reduce profits on the sales of the uncertified products, if consumers demand a discount on the potentially unsafe goods. If the latter effect prevails on the former, advertising reduces expected profits. The model shows that advertising is not rational if the expected price margin is high and the consumers' consider the undifferentiated products 'reasonably safe' (i.e. β is small). This conclusion offers an explanation of the pooling paradox: if the above mentioned conditions are met, we can observe PFSS's, because of their value of supply management tools, but we do not observe advertising, as the supermarkets have no incentive to promote product differentiation.

8.6 Conclusions

Supermarkets are actively promoting their PFSS among producers, yet they do not put any significant effort to bring the improved food safety to consumers' attention. Our paper presented a theoretical model providing possible explanations to this apparent paradox. We found that the supermarket behaviour can be rational if expected price margin for the pooled goods is high compared to the consumers' concerns for safety, measured in monetary terms.

Our theoretical model provided two major conclusions. Firstly, we showed that PFSS can be used to improve coordination along the supply chain. Consequently, their scope goes beyond exploiting consumers' demand for safety, as they can be used as institutional arrangement to organize the supply chain. In this context, the adoption of PFSS is rational even if consumers had no willingness to pay for safety.

Secondly, we showed that advertising the PFSS (and the higher quality of the certified products) might reduce supermarkets' expected profits, under specific circumstances. If the expected price margin is high enough, supermarkets have incentive to mix certified and uncertified products. In such instances, supermarkets may benefit from pooling the goods and selling them as undifferentiated. In this way they avoid the losses from selling uncertified (and potentially unsafe) products.

These two results provide a possible economic justification to the existence of unadvertised PFSS and they suggest that the adoption of such standards is not driven by the concern for consumer safety only. Efficient management of the supply chain is a key issue too.

References

Balsevich, J., Berdegué, A., Flores, L., Mainville, D. and Reardon, T., 2003. Supermarkets and produce quality and safety standards in Latin America. American Journal of Agricultural Economics 85: 1147-1154.

Banterle, A. and Stranieri, S., 2008. The consequences of voluntary traceability system for supply chain relationships. An application of transaction cost economics. Food Policy 33(6): 560-569.

Berdegué, J.A., Balsevich, F., Flores, L. and Reardon, T., 2005. Central American supermarkets' private standards of quality and safety in procurement of fresh fruits and vegetables. Food Policy 30: 254-269.

Boselie, D., Henson, S. and Weatherspoon, D., 2003. Supermarket procurement practices in developing countries: redefining the roles of the public and private sectors. American Journal of Agricultural Economics 85: 1155-1161.

Codron, J., Giraud-Heraud, E. and Soler, L., 2005. Minimum quality standards, premium private labels, and European meat and fresh produce retailing. Food Policy 30: 270-283.

Fischer, C., Hartmann, M., Reynolds, N., Leat, P., Revoredo-Giha, C., Henchion, M., Albisu, L.M. and Gracia, A., 2009. Factors influencing contractual choice and sustainable relationships in European agri-food supply chain. European Review of Agricultural Economics 36: 541-569.

Fulponi, L., 2006. Private voluntary standards in the food system: the perspective of major food retailers in OECD countries. Food Policy 31: 1-13.

García Martinez, M. and Poole, N., 2004. The development of private fresh produce safety standards: implications for developing mediterranean exporting countries. Food Policy 29: 229-255.

Gereffi, G., Humphrey, J. and Sturgeon, T., 2005. The governance of global value chains. Review of International Political Economy 12: 78-104.

Giraud-Heraud, E., Grazia, C. and Hammoudi, A., 2009. Agrifood safety standards, market power and consumer misperceptions. Journal of Food Products Marketing 16(1): 92-128.

Giraud-Héraud, E., Hammoudi, H. and Soler, L.G., 2006. Food safety, liability and collective norms. Cahiers du Laboratoire d'Econométrie de l'Ecole Polytechnique. Available at: http://tinyurl.com/y6vomrlc.

Graffham, A., Karehu, E. and MacGregor, J., 2007. Fresh insights 6: impact of EurepGAP on smallscale vegetable growers in Kenya. IIED, DFID, London, UK.

Grazia, C., Hammoudi, A. and Hamza, O., 2014. Quelle légitimité à la réglementation SPS Européenne: une approche d'économie industrielle. In: Hammoudi, A., Grazia, C. and Surry, Y. (eds.) Sécurité sanitaire des aliments. Régulation, analyses économiques et detours d'expérience. Hermès-Lavoisier, Paris, France.

Grunert, K.G., 2005. Food quality and safety: consumer perception and demand. European Review of Agricultural Economics 32: 369-391.

Hammoudi, A., Hoffmann, R. and Surry, Y., 2009. Food safety standards and agri-food supply chains: an introductory overview. European Review of Agricultural Economics 36: 469-478.

Hatanaka, M., Bain, C. and Busch, L., 2005. Third-party certification in the global agrifood system. Food Policy 30: 354-369.
Henson, S. and Caswell, J., 1999. Food safety regulation: an overview of contemporary issues. Food Policy 24: 589-603.
Henson, S. and Humphrey, J., 2010. Understanding the complexities of private standards in global agri-food chains as they impact developing countries. Journal of Development Studies 46: 628-1646.
Henson, S. and Reardon, T., 2005. Private agri-food standards: implications for food policy and the agri-food system. Food Policy 30: 241-253.
Hobbs, J.E., 2010. Public and private standards for food safety and quality: international trade implications. Estey Centre Journal of International Law and Trade Policy 11: 136-152.
Humphrey, J. and Schmitz, H., 2001. Governance in global value chains. IDS Bulletin 32(3): 19-29.
Humphrey, J., 2011. Standards, food safety and trade: the European Union and the United States. October, 2011, NTM-IMPACT, Working Paper 11/14, Brussels, Belgium, 67 pp.
Kariuki, I.M., Loy, J.-P. and Herzfeld, T., 2012. Farmgate private standards and price premium: evidence from the GlobalGAP scheme in Kenya's French beans marketing. Agribusiness 28: 42-53.
Lutz, S., Lyon, T.P. and Maxwell, J., 2000. Quality leadership when regulatory standards are forthcoming. Journal of Industrial Economics 48: 331-348.
McCluskey, J.J. and Winfree J.A., 2009. Pre-empting public regulation with private food quality standards. European Review of Agricultural Economics 36: 525-539.
Ménard, C. and Valceschini, E., 2005. New institutions for governing the agri-food industry. EuropeanReview of Agricultural Economics 32: 421-440.
Organisation for Economic Co-operation and Development (OECD), 2006. Final report on private standards and the shaping of the agro-food system. Directorate for food, agriculture and fisheries, committee for agriculture. Available at: http://tinyurl.com/yaj3ydhg.
Reardon, T., Codron, J.-M., Busch, L., Bingen, J. and Harris, C., 2001. Global change in agri-food grades and standards: agribusiness strategic responses in developing countries. International Food and Agribusiness Management Review 2: 421-435.
Roosen, J., 2003. Marketing of safe food through labelling. Journal of Food Distribution Research 34: 77-82.
Rozan, A., Stenger, A. and Willinger, M., 2004. Willingness-to-pay for food safety: an experimental investigation of quality certification on bidding behavior. European Review of Agricultural Economics 31: 409-425.
Russo, C., Perito, M.A., Di Fonzo, A., 2014. Using private food safety standards to manage complexity: a moral hazard perspective. Agricultural Economics Review 15: 113-127.
Theuvsen, L., Plumeyer, C.H. and Gawron, J.C., 2007. Certification system in the meat industry: overview and consequences for Chian-wide communication. Polish Journal of Food and Nutrition Science 57: 563-569.

9. Contract farming, a strange type of contract: more than obligation 'to give'

K. Watanabe[1], A.E. Lourenzani[2] and S.M.S. Bankuti[3]*

[1]Center of Biological, Environmental and Agrarian Science, Federal University of Reconcavo da Bahia, Rua Rui Barbosa 710 Centro, Cruz das Almas, BA 44380-000, Brazil; [2]Department of Administration, Sao Paulo State University, Av. Domingos da Costa Lopes 780, Jd. Itaipu, Tupa, SP 17602-496, Brazil; [3]Department of Administration, State University of Maringa, Av. Colombo, 5790, Jd. Universitário, Maringa, PR 87020-9000, Brazil; wtkassia@ufrb.edu.br

Abstract

Contract farming is based on agreements that are made before production decisions, providing a production coordination and uncertainty minimization for both farmer and agro-industry/distributor, unlike the simple purchase and sale operations through the open market view, when the product is delivered immediately to a price. The activities might be interdependent and the transactions are performed in a 'strange' form of governance. These transactions are carried out through contracts between farmers and agroindustry/distributor. Contract farming is used in agribusiness supply chains to ensure access to agricultural products with certain specifications, such as quality, quantity, and origin, among others. In Brazil, for instance, agricultural activities are widely performed under contract farming; however, there is no specific law for typifying it. The lack of a legal framework for this 'strange' type of contract leads to misunderstanding because it might not fit properly with existing legal types. This chapter aims at discussing the 'strange' contract named contract farming. We assume that, contracts and agreements are typically related to hybrid structure of governance. Therefore, more than formal rules, informal institutions are taken into account. The analysis of contract farming is based on new institutional economics focusing on transaction costs economics.

Keywords: new institutional economics, transaction costs economics, hybrid form, institutions

9.1 Introduction

Food markets have evolved in the last centuries. In the beginning of the 19th century, the farmers produced their own food and the surplus was sold to the local market. Vertical integration was common for agricultural activities since the decision of food consumption and production process was concentrated in the same hands (Penn, 1958).

Over time, with the growing population, urbanisation process and rising demand for food, specialisation processes have emerged in several activities including agriculture. The process of specialisation is a gradual disintegration of functions in agriculture to form the agricultural production system (Davis and Goldberg, 1957). The farmers have specialised their agricultural activities and the production was oriented to satisfy the consumers. The farmers began to devote themselves to agricultural production; companies for agricultural services have arisen and farmers started to subcontract services, such as planting, harvesting, processing and transporting agricultural products.

However, as different agents are involved in the agricultural system, a governance challenge arises. Despite the specialisation, all activities are interdependent to each other; so, besides the decision of making or buying, 'strange' forms of governance arise (Ménard, 1996). The interdependence of sectors involved in the agricultural production system lead to an institutional arrangement in order to incentivise the entire production to answer the particular characteristics of the product required. Therefore, the transactions are carried out through contracts between farmers and agro-industry/distributor and favour hybrid form of governance.

The issues about the institutional arrangements in the agro-food chain are widely discussed and it is not a new argument in Economics (Allen and Lueck, 2002; Azevedo, 2000; Ménard, 1996, 2005a,b; Ménard and Klein, 2004; Zylbersztajn, 1995; Zylbersztajn and Farina, 1999). In terms of a typical hybrid governance structure, there are contracts to coordinate the production (Ménard, 1996; Williamson and Masten, 1999). The hybrid governance structure consists, mainly, of long-term relations based on formal and informal institutions. The 'strange' type of contract in the hybrid governance mode arises from the fact that the farmers have obligations 'to do' in respect to the agro-industry/distributor, besides the obligations 'to give' (Benincasa, 1992; Jannarelli, 2012). Different institutions and the complexity of economic-legal relations compose a complex governance structure that increasingly manifests the interdependence between different sectors and the need for an appropriate legal framework (Jannarelli, 2000).

Contract farming is a type of relation between farmers and buyers organised in a hybrid governance structure (Jannarelli, 2012; Jia and Bijman, 2012; Pultrone, 2012; Silva, 2005). Contract farming is used in the agribusiness supply chain to ensure access to agricultural products with certain specifications, such as quality, quantity, and origin, among others. For this type of contract, in general, farmers provide land and human capital, grow and supply the raw agricultural commodity to the buyer, who monitors the farmer production and coordinates the transaction for transaction costs reduction (Pultrone, 2012; Silva, 2005).

The agricultural activities performed under contract farming have become a common arrangement in both developed and developing countries (Jia and Bijman, 2013). In Brazil, for instance, the significant growth and importance of poultry and hog supply chains are due to the adoption of contract farming (Bánkuti and Souza, 2014; Souza and Bánkuti, 2012; Souza and Zylbersztajn, 2011). Moreover, besides poultry and hog sectors, with the need of controlling milk production, especially concerning quality standard, contract farming has been performed in dairy sector in Brazil. Besides food sector, contract farming is used in biodiesel production system with the intervention of State, which requires the purchase of feedstock from family farmers (Paiva, 2009; Silva Jr. *et al.*, 2013). Although contract farming is widely performed in Brazil, as in many countries, there is no specific law to regulate it.[1]

The lack of a legal framework for this 'strange' type of contract, which might not fit properly with the existing legal types, leads to misunderstanding, such as labour contract, future selling contract, agrarian partnership, amongst others. Moreover, the interdependence between farmer and agro-industry/distributor characterised by investment dependence may bring abuse in the relation of contract farming (Bánkuti and Souza, 2014).

This chapter aims at discussing the 'strange' contract named contract farming. It is an attempt to analyse the contract farming based on two complementary points of the New Institutional Economics: (1) institutional arrangement; (2) institutional environment. This study is based on Transaction Costs Economics, which has the transaction as the unit of analysis and emphasises that institutions matter (Williamson, 1985). Considering two analytical levels of New Institutional Economics, this chapter has the following structure: in Section 9.2, a briefly context of hybrid governance structure is presented; Section 9.3 is focused on institutional environment and it is divided into two subsections (9.3.1: legal nature of contract farming; and 9.3.2: informal institution); Section 9.4 presents the analysis of dependence abuse in contract farming; Section 9.5 brings two examples of contract farming in Brazil, dairy and biodiesel sector; Section 9.6 draws some conclusions.

[1] Law n. 13,288, promulgated in May 2016 regulates contract farming in Brazil (Brazil, 2016). It is a recent Law, which will take time to be implemented by agents in agri-food chains. The present research was previously conducted, thus that Law was not considered in data collection, results and analysis.

9.2 Contract farming, a hybrid governance structure

The New Institutional Economics has addressed the institutions in two different levels: institutional environment and institutional arrangement. The former establishes the 'rules of game' that influence the agents' behaviour for governance structure choice. The latter is the micro analytical level that presents the transactions forming the governance structures. This section presents the institutional arrangement for contract farming.

Ménard (2004) refers institutional arrangement as a mode of governance or organisational mode that economic agents structure their production activities and exchange within a framework of rules (institutional environment). Therefore, governance structure (market, hybrid, hierarchy, vertical integration) exists within an institutional environment, which influences the trade-off of forms of production, depending on transaction attributes and economising transaction costs (Ménard, 2005a,b; Williamson, 1985, 1999).

The structure of agrarian institutions under the perspective of transaction cost economics can be observed within the spot market when the farmers, agro-industry/distributor engage in exchange and a previous planning is not necessary. On the opposite, vertical integration (hierarchy) is observed when one agent (farmer, agro-industry, or distributor) develops the full life cycle production of agricultural products that will be used in industrialization process or trade.

Hybrid governance structure consists of a long-term contractual relation that preserves autonomy and provides transactions specific safeguard (Jannarelli, 1997; Joskow, 1985). Hybrid structure of governance is understood as a mode for which an incentive compromise has been reached (Williamson, 1990). Moreover, the parts involved in transaction of hybrid governance are interdependent to coordinate the entire production system. The hybrid form is observed in the contract farming, which its parts are interdependent and legally independent (Albisinni, 2003; Jannarelli, 2002; Prévault, 1996; Traisci, 1992). In other words, the farmer is autonomous and the downstream agro-industry/distributor does not appropriate farm assets, but they specify the use of these assets (Jia and Bijman, 2013).

Within the hybrid structure of governance, different modes of contracts are carried out because of different levels of dependency between the parts involved in the transaction. Moreover, the temporal characteristic of these contracts might be long-term or short-term with the possibility of renewal. In terms of contract farming, different types of arrangements were observed.

Worley and McCluskey (2000) researched wheat industry and observed different types of contracts that involved vertical coordination. They classified the contracts according to the levels of control by the contractor as following: market contract, production management contract and production contract. The market contract has the lowest degree of control and the agreement is based on price and delivery of products. Moreover, the whole process of production and the risk belong to farmer. The production management contract has an intermediate level of control and the parts share liability since the contractor interferes at the producer's decisions in the production process. The production contract has the highest level of control by the contractor. The contractor provides all inputs to the agricultural production, besides the technical assistance. This type of contract is widely used in the poultry sector, in which the contractor supply the birds, feed and technical advice, while the farmers care for the chickens by providing the land, housing and labour.

Ménard (1996) analysed hybrid form in the French poultry sector with Label certification. The Label system products were created in 1960 by a group of producers who searched a quality certificate to differentiate their products. In this system, there are different forms of contractual arrangement. He observed the temporal characteristic is related to the degree of coordination, being the long-term contract more coordinated than the short-term contract. Moreover, the contracts are incomplete and price is mostly established through negotiation. Martinez (1999) focused his research on hog and poultry production system and named the contract as contractual agriculture. He observed the presence of interprofessional agreement that is performed by the farmer association and agro-industry in order to reach the equilibrium among the participants of the transaction. Jannarelli (2012) pointed three different types of contracts related to risk bearing upon the farmer that reflects to the decision making of the agricultural agents. He classified the contract as following: market specification contracts; resource providing contracts and management and income guaranteeing contract.

It is clear, thus, the complexity of contracts in agricultural systems, understood as hybrid forms of organisation. Moreover, besides the differences in taxonomies, authors highlight the degree of coordination, interdependence and obligations between parties, which brings to light such discussion in legal field.

9.3 The institutional environment of agricultural production system

This section presents the macro level that consists of the institutions facilitating and enforcing the transfer of goods or services. Institutions influence the trade-off among institutional arrangements since all transactions are embedded in an institutional environment. There are two classes of institutions: informal and formal institutions, which compose the 'rules of game' (North, 1990). According to North (1991), the informal institutions are sanctions, taboos, customs, tradition, and codes of conduct; formal institutions are constitutions, laws, and property rights.

The fundamental role of institutions on economic development is to reduce uncertainty and create a favourable environment for inducing investments decisions (North, 1990, 1991, 2000). Instead, the instability of institutions reduces their effect on human behaviour and enhances the degree of uncertainty in the economic environment, which in turn can reduce the level of investment.

Institutions can motivate, positively or negatively, individual parties to carry out the agreements and contracts; courts are a mode of contract enforcement; and market motivates contractual relationship to trade goods and services (Arruñada, 2001). Contract farming has the role of establishing links between farmers and agro-industry/distributor. From agro-industry/distributor side, contract helps to plan the production and ensure the achievement of standard agricultural product with suitable price. From the other side, contract allows farmers receive technologies and guarantee of selling their product, what might encourage farmers' participation in agricultural production system.

Contract farming is a hybrid form of governance, in general relying on long-term contract and relational exchange. Relational exchanges are based on longer-term interactions that involve repeated transactions without a formal contract. According to Macneil (1978), relational contract sustains ongoing relations in long and complex contracts by adjustment processes of a more thoroughly transaction-specific. Moreover, the long-term contract is incomplete by reason of bounded rationality (Williamson, 1990). The analysis of incomplete contract focuses on the debate related to the rules for filling gaps, either recommended by the State or by private institutions.

The State intervention in the economy of contract aims at transaction costs reduction. Since it is not possible to establish a relationship that accounts for all contingencies, renegotiation might occur and informal institutions are considered. In other words, in order to face problems that arise from contract farming, parts of this transaction seek solutions through legal, private, or complementary (legal and private) resources. Therefore, the trade-off among alternative institutions is at stake.

In agricultural transactions, informal constrains are present in many different ways (Azevedo, 2000). Even though informal rules exist, typifying the contract through a law might reduce uncertainty for agents. Indeed, the classification of contracts that have achieved an important social typology is a tendency in the countries of civil law system to conduct more legal certainty for the relationship.

The formal and informal institutions allow exchange of agricultural goods and services. In terms of formal institutions, the laws aim facilitating trade and the conflicts solutions, which are relied on centralised third part, such as arbitrations and judges. In terms of informal institutions, the conflict solutions are carried out through private parts in a decentralised mode (Arruñada, 2001).

The decentralized system is practiced on agricultural system and institutions have the task to ensure the ability to adapt to changes being more flexible and less predetermined to deal with globalization (Campbell and Harris, 2005; Ferrarese, 2002). The privatization of standards is quite committed to the operation of the market. It can be said that the market presents, besides the traditional legal measures, the production of new forms, which are not pre-defined character, but are arrangements adapted following the markets, according to their needs (Ferrarese, 2002; Jannarelli, 2000).

More than contracts, informal institutions interfere in the coordination process of agricultural production; reduce uncertainties in trade of agricultural goods and services; and guarantee for contract performance through self-enforcement (North, 1991, 2000; Sztajn *et al.*, 2005). Despite the informal institutions being important to the market by providing quick solutions to the functioning of the agricultural production system, they may be insufficient to sustain economic growth, so the importance of state interference with formal institutions to carry out exchanges goods and more impersonal services (Keefer and Shirley, 2000). Institutions governing the behaviour of participants in the agricultural production system might allow for more accurate coordination of their actions insofar as they are well defined. Therefore, institutions are taken into account to ensure the efficiency of the whole agricultural production system (Albisini, 2003; Fanfani and Montresor, 1998; Janarelli, 2000).

The contractual relations considered in this work transcend the instant exchange observed in the spot market organisation and consider contracts and agreements typically related to the hybrid structure of governance. Therefore, more than formal rules, organisational solutions and informal institutions should be taken into account.

The comparison of the institutional alternatives serves to establish the benefits of institutions at stake: formal or informal institutions. A formal rule would supply gaps and offer legal certainty to address key elements of contract farming. As a result, investments might increase. However, with the long-term contract, usually performed by contract farming, the parts aim at utility maximization directly through long-term co-operative behaviour based on trust (Campbell and Harris, 2005). Besides the view of individual institutions, both formal and informal institutions can coexist and they are complementary instruments (Klein, 1992).

9.3.1 Contract farming, a strange legal type of contract

The transaction between farmer and agro-industry/distributor, which are interdependent and autonomous, is known as integration contract (Lorvellec and Danet, 2004; Prevault, 1996; Traisci, 1992), but it has other names, such as agricultural contract, agro-industrial contract, contract farming, among others. For this work we use the term contract farming. The contract farming is characterised by the presence of various obligations to do and its legal type can be easily misunderstood (Benincasa, 1992; Prévault, 1996). The characteristic of contract farming is the economic dependence between farmer and agro-industry/distributor, although they are autonomous. Transaction Costs Economics (TCE) explains the economic dependence. When farmers make specific investments to attend the demands of agroindustry/distributor; thus, the parts develop safeguards through long-terms contracts to prevent opportunistic behaviour, instead of instantaneous transaction in spot market.

The study of contract farming legal framework is important since it eliminates the costs of searching legal type of contract that might adjust the contract farming and reduces opportunistic behaviour based on legal uncertainty. In *latu sensus*, the contract farming is an agrarian contract for agrarian firm transactions (Galloni, 1978).

The transaction in contract farming is characterised with the intervention of agro-industry/distributor in the process of agricultural production and the interdependence among the parts of contract. Therefore, contract farming is neither a simple contract of selling and buying, nor a future sale contract since the obligation to do is observed, besides the obligation to give (Casadei, 1984; Confortini and Zimatore, 1983). For the classification of contract farming, discussions among the jurists arose. Moruzzi (1962) defended the thesis of future sale contract and compared the contract farming to contract with price reference (*contratto con prezzo di riferimento*). Despite the similarities between these contracts, the difference is the fact that in the contract farming the price may be fixed *ex-ante*. Moreover, for contract farming, the integrator agro-industry/distributor interferes in the agricultural production process and the farmer has obligation to do.

Other thesis for contract farming legal type was the work contract (*contratto di appalto*) defended by Bivona (1979). His thesis was supported on the fact that farmer signing a contract farming assumes the obligation to give the agricultural product with the quality pre-established. In terms of work contract, the risk would belong only to farmers. However, for contract farming, the risk might belong to both farmer and the integrator agroindustry/distributor since farmers must follow the orientation of integrator (Pavone la Rosa, 1990).

The submission of the farmer to a number of obligations required from integrator agro-industry/distributor could resemble the subordinate employment contracts. However, the parts of contract farming are autonomous and they cooperate to each other (Benincasa, 1992). We could think of several types of contract to try to frame the contract farming. Although each type of contract can be observed in contract farming, as Pavone la Rosa (1990) argues, the contract farming is not included in unitary contract frame and a mix of contracts is not the solution. Because of the complexity of contract farming that presents a mixed of contracts types, *sui generis*, it is difficult to find a typical legal contract.

The conclusion about this 'strange' type of contract named contract farming is the fact that it is neither a contract of buying and selling nor a work contract. The farmers have a duty to do, i.e. obligations to do, besides the obligation to give. Moreover, the difficulty of fitting the contract farming in typical legal contract brings the need to typifying the contract farming, especially in countries in which legal systems rests on the rules of law as the main normative source (Civil Law system), such as Brazil. Some

European countries, such as France and Italy, have typified the contract farming. The reason for a legal framework for contract farming is to offer certainty to economic, more impersonal transaction, less *ex-ante* efforts to agents to perform the contract since there is the guarantee of State for fulfilling the contractual gaps.

The absence of a specific legal framework for contract farming might lead to misunderstand especially when the Judiciary is called upon to settle the disputes on contract farming. As Posner (1999, p. 5) wrote: '[…] courts, which are complete foreigners to the industry and to the relationship, are not likely to have, or be able to obtain, the information that is necessary for this understanding.' In Brazil, for example, where there is the lack of a specific legal framework for contract farming, labour issue is observed in sentences of the Brazilian Labour Justice.[2]

However, with the absence of a legal framework for contract farming, private rules prevail to farmers transact their agricultural goods and services. Without a legal type of contract farming, a better knowledge of the parts involved in the transaction is required, taking into account their reputation. The presence of legal and private rules might enhance the transaction of agricultural goods and services in a institutional environment.

9.3.2 Informal institutions

Transactions are faced with both formal and informal institutions. Thus, economic agents select the mode of coordination to deal with their transactions and draw contracts to reduce transaction costs. In the absence of a specific law for contractual relation between farmer and agro-industry/distributor, the parts involved in the transaction seek to have a balanced negotiation and develop private mechanisms in order to minimise opportunistic behaviour that may arise. In this case, reputation of the parts has a key aspect for the proper functioning of transaction. However, more *ex-ante* efforts are necessary, for example, knowing the parts involved in the transaction, a greater completeness in the preparation of the contract, and so on.

The lack or weakness of formal institutions motivates the emergence of informal institutions. The various institutions of transactions have their distinct advantages and disadvantages, and each complement or replace each other. For transactions based on bilateral reputation mechanisms, the economic agent is able to monitor the action of the other part, thus, the contingencies are more predictable and do not require large *ex-ante* efforts. In terms of public institutions, they are directed to the most complex and impersonal relationships and transactions are relatively large and very dynamic.

[2] Labour Court in Matão, 15th Region, State of São Paulo, in a lawsuit filed by the Labour Public Prosecutor against the Sucocítrico Cutrale Ltd., Louis Dreyfus Commodities Agroindustrial S/A, Citrovita Agro Industrial Ltd., and Fischer S/A, condemned them to abstain hiring farmers to perform activities, such as production and harvesting citrus fruits that are used in the industry activity (juice production) (Brazil, 2013).

Various institutional levels may exist. In this context, according to the work of Brosseau and Raynaud (2006), institutional levels are: 'general or public institutions'; 'private institutions'; and 'bilateral governance structures'. The 'general institutions' are the formal and informal institutions named 'rules of game' by North (1990). These institutions provide the agents general solutions that could be incomplete. Incompleteness of solution motivates the agents to build 'bilateral governance structures' for transaction performance. Therefore, 'bilateral governance structures' are contracts made between particular agents and their rules come voluntary. When several agents meet similar difficulties of coordination, they are motivated to act collectively for better coordination achievement and lower costs.

The private institutions arise to ensure the agreements performance within a weakness of State; i.e. when the State does not give guarantee for the fulfilment of agreements performed by economic agents. Thus Brousseau and Raynaud (2006, p. 5) posit, '[...] when State are unable to make laws and agreements enforceable, when trade takes place across boundaries where no trans-territorial order exists, (etc.) entrepreneurs and traders invent devices to enable them to perform trade.' However, private institutions can coexist with public institutions and various institutional levels can coexist within the same institutional environment, such as rules for certification, labels, and denomination of origin.

In terms of contract farming, farmers and agro-industry/distributor can create private institutions, even though there is a formal institution. Moreover, the farmers can create an organisation with power of representation to conduct negotiations with agro-industry/distributor. In this context, it is observed interprofessional agreements for agri-food chains for an efficient coordination. Interprofessional agreements minimise dispute between the parts and avoid the solution of conflicts by traditional mechanism of legal Courts (Sirsi, 1990). The economic agents involved in the interprofessional agreement may have informational advantage over the Judges. In this sense, the settlement of dispute is internal in the interprofessional agreement that represents the 'Court of first instance' (Williamson, 1996, 2005). State might incentive the achievement of interprofessional agreement for contract farming, such as in France and Italy. Therefore, private and public institutions coexist that is named 'quasi-private' institutions by Greif (2005).

The 'quasi-private institutions' are hybrid institutions that combine elements of public and private order. The complementarity of private and public institutions suggests that the use of both formal and informal arrangements provides more efficient outcomes than the use of either arrangement in isolation (Klein, 1992; Lazzarini *et al.*, 2004; Mazé, 2005; Watanabe and Zylbersztajn, 2014). The argument supporting complementary view is the idea that incomplete formal contracts facilitate the self-enforcement of informal agreement; i.e. self-enforcement of dimensions that are no contractible, according to Lazzarini *et al.* (2004).

9.4 Contract farming: abuse of power under the transaction costs approach

The transaction between farmer and agroindustry/distributor can be interdependent and, according to the theory of transaction costs, it would be a hybrid structure of governance. The bilateral dependence is when both farmer and agroindustry/distributor support specific investments, so one is linked to another. This dependence makes the companies interdependent and with reciprocal obligations.

One of the features of contract farming is the reciprocity of obligations. By analysing the legislation in France that was the first country to typify legally the contract farming, the feature of reciprocity of obligations of this contract was interpreted literally by law n. 64-678 of July 6th, 1964. Economic dependence considered by French legislature is not characteristic of abuse of economic dependence since there is the reciprocity of obligation. However, the abuse can arise in contract farming, because, generally, imbalanced parts perform it.

The contractual imbalance of parts involved in contract farming might exist due to the fact that agroindustry/distributor are, generally, economically more empowered than farmers. Moreover, there is information asymmetry since agroindustry/distributor is better informed of the needs of the consumers compared to farmers. The better-informed part might behave opportunistically arising a contractual power abuse. These characteristics represent the market failures and formal or informal institutions reduce or correct such failures.

Although the contractual imbalance exists, the hybrid governance structure can reach efficiency when compared with other governance modes (Williamson, 1985). According to Natoli (2003), the abuse of economic dependence is related to the asymmetry of specific investment and different of the asymmetry of information.

In contract farming with specific investment involved, there is a risk of opportunistic behaviour of one part taking advantage of the other part who has the specific investment, what can cause a loss of efficiency (Klein, 1980; Klein *et al.*, 1975). The asymmetry of specific investment might determine the power and economic dependence. The law should intervene when abuse of power or market failures are observed, for instance monopoly or damages caused by information asymmetry. Moreover, the contract farming may contain abusive clauses that generate conflicts. Abusive clauses are the imbalance rights and obligations typically of a determined contract, or excessive unilateralism of clauses of contract (Marques, 1998).

Parts treat contract terms during the negotiation and many aspects might not be mentioned in the written contract since it is very costly to anticipate *ex-ante* contingencies. Indeed, the agents' rationality is limited; therefore, the contracts are incomplete. The contractual gaps are fulfilled by third party, such as judge, arbitrator, and among others (Arruñada, 2001). Many aspects of transaction in contract farming are handled internally, because it is expensive to use the Judiciary Court and difficult to specify and measure the contract (Klein, 1980). The self-enforcement is determined by the reputational capital of the economic agents involved in the contract farming. However, self-enforcement does not always work well to fulfill all gaps of incomplete contract. Therefore, third part is required and Barzel (2002, p. 34) posits, that third-party enforcement often works better, because they are able to provide the principles to an agreement an altered set of incentives such that their net gains from interacting will exceed those they could attain under self-enforcement. However, the author affirms that third-party enforcement is costly and cannot be used in all cases. Abuse may arise from farmer and agroindustry/distributor transaction under a contract farming relation, thus, in these cases; legal rules to curb it more effectively are necessary. The contractual power abuse is a welfare loss since it shows uncertainty to those who would make investments and perform contract farming.

The role of law is to prevent the incentives of abuses that arise from economic dependence (Longu, 2000). However, the excess of regulation to protect the dependent part results in incentives for dominant parts internalise all activities (vertical integration) and do not renovate the contract. Moreover, the effect of excessive legal protection impacts in the entry of new investments, besides the transformation of governance mode.

Some evidence of contract farming are observed in Brazil. The following cases illustrate the discussion.

9.5 Contract farming, two examples in Brazil: milk and biodiesel

9.5.1 Milk

Dairy sector in Brazil is known for its heterogeneity and has undergone restructuration process since the 90's decade. Although public institutions have been created to face sanitary problems in that productive chain (e.g. Normative Instruction 62), quality problems are still a weakness, sometimes leading to vertical coordination (Farina *et al.*, 2005). In this context, strange forms of organisation have emerged.

As empirical evidence, we discuss a medium sized dairy processor in Paraná state. Paraná is located in the south of Brazil, being the third largest producer in the country, with 13% of national production in 2013, which means 4.35 billion litres of milk IBGE, 2014). Milk production is accomplished mostly by family-based farmers in the state: official data states that more than 100 thousand dairy farmers in Paraná were family-based farmers ones in 2006, comprising 84% of total dairy farmers in the state (IBGE, 2006). Also, Fernandes (2013) states that 110.6 thousand dairy farmers in Paraná produces up to 100 litres of milk per day, which indicates that dairy production is structured in small-scale basis in Paraná. Transactions between dairy farmers and downstream agents in Brazil are mostly carried out through hybrid forms, mostly in informal contract basis. Amongst hybrids is an innovation form of organisation, the contract farming.

Transactions in this case involve plural forms, since six suppliers are under contract farming and other 92 farmers supply milk under long-term relationships, through verbal agreements. Contract farming between processor and farmers establishes that the former provides cows in milk, feed and technical assistance. Farmers are obliged to follow a protocol of best practices in pasture, nutrition, sanity and milking management, as well as provide appropriate infrastructure and corn silage. Quality standards of raw milk, as well as other terms related to obligations 'to do' are not settled in the contract, being verbally agreed between parties. Farmers are charged 12% of the value of milk in order to pay for the cow 'rent', as well as for feed. Technical assistance provided by processor, on one side, helps farmers to follow the protocol of best practices and, on the other side, supports contractor's monitoring and control over assets.

Since it involves high investments from both sides, contract farming is highly directed to obligations 'to do'. In addition, since controlling the process is important to warrantee quality in product in this case, contract farming emphasizes the protocol of best practices. In that contract, only volume is put as an obligations 'to give', since the processor has an expectation over cow productivity. In this sense, one can observe that milk daily productivity is higher; around 30 liters per lactating cow, then average production in conventional system (20 litres/cow in milk), and average in Brazil (4.5 litres per cow in milk) (USDA, 2015).

Thus obligations 'to do' comprise quality standards, being much more severe than those demanded from other suppliers. Even though these obligations are not written in the contract, the technical assistance works as a mechanism of control, allowing the processor to advise but also monitor suppliers' performance.

Contract farming in this case is motivated by, the need of high quality standards for raw milk, and processor's difficulty to obtain it from independent dairy farmers in Brazil. Thus, institutional arrangement is supported on private institutions. While public institutions (Normative Instruction 62 – NI62/2011, from the Ministry of Agriculture, Livestock and Food Supply) state lenient minimum quality requirements (somatic cell count: 600,000/ ml; total bacterial count: 600,000/ml; fat: 3.0%; protein: 2.9%), this processor's standards are more rigid, requiring from contracted farmers higher quality levels (somatic cell count: 200,000/ml; total bacterial count: 50,000/ml; fat: 3.2%; protein: 3.2%).

This innovative form of organisation in dairy chain brings some remarks on contract farming in Brazil. NI 62/2011 is part of a larger program in Brazil, named 'National Program for Milk Quality Improvement', comprising a set of actions to promote quality and sanity in dairy chain. It is important to mention that Brazilian dairy context is embedded in informal rules comprising consumer habits and production practices, indicating the importance of path dependence. In this sense, although formal institutions imply quite lenient quality standards, historical context in Brazilian dairy chain indicates such standards are still difficult to be reached by farmers. Pinto (2013), for instance, indicates that 27 and 40% of dairy farmers did not reach minimum SCC and TBC standards in Brazil in 2012, respectively. In the South, such percentages were, respectively, 34 and 52% (Pinto, 2013).

Once official standards are not severe, and formal/informal institutional background gives support to it, it might be difficult to find farmers willing and capable to adjust production system to higher quality standards. Efforts to achieve high quality standards might be unattractive for most farmers, since one can find buyers for raw milk produced according to formal minimum quality requirement (NI 62). Also, informal institutions related to dairy systems and milk quality may raise a barrier, once it demands deep changes in production systems, compared to traditional practices of handling, infrastructure, feed system, milking methods, among others. Indeed, farmers engaged in contract farming in this case are those beginning in the activity (less than four years in dairy activity), thus not embedded to old practices. Thus, contract farming in this case shows to be a way of supplying high quality milk, involving obligations 'to do', not only obligations 'to give'.

9.5.2 Biodiesel

The Brazilian National Program for Production and Use of Biodiesel (BNPB) was elaborated in December of 2004 to develop a sustainable chain of biodiesel production and use. Brazilian biodiesel was regulated through Federal Law no. 11097/2005, which introduced biodiesel into the Brazilian energy matrix.[3] One of the purposes of BNPB is the inclusion of family farmers in the biodiesel production system; however, the mechanisms alone is insufficient to promote their participation (Watanabe and Zylbersztajn, 2012; Watanabe *et al.*, 2012).

The inclusion of family farmers into the biodiesel production system does not happen spontaneously. Thus, Brazil's Ministry of Agrarian Development (MAD) has promoted the 'Social Fuel Label' to certify biodiesel refineries source from family farmers. The requirements for biodiesel plants to participate in the 'Social Fuel Label' program are: (1) to purchase a minimum percentage of feedstock from family farmers; (2) to enter into legally-binding contracts with family farmers under the supervision of workers' trade unions or institutions authorised by MAD; and (3) to provide technical assistance, seeds, input subsidies, and crop transportation to family farmers to facilitate their transaction with biodiesel refineries.

[3] The Law no. 1109/2005 authorized a voluntary 2% blend (B2) from 2005 to 2007. The letter 'B' means biodiesel and the number is the percentage biodiesel mixed to diesel oil. In 2008 this mix became mandatory, and by 2013 a 5% blend (B5) of was to be required. However, the National Energy Policy Council (CNPE), which supervises the gradual increase in the percentage of biodiesel, the time frame and determined that a 5% blend (B5) of biodiesel would be required from January 1st 2010.

The MAD established the percentage of feedstock to be acquired from family farmers according to the region where a biodiesel plant is located. The minimum feedstock percentage is calculated based on the investments allocated to family farmers. On the other hand, the Social Fuel Label certifies biodiesel refineries that process feedstock produced by family farmers. Then, the refineries with this label are entitled to tax benefits and favourable loans from the Brazilian Development Bank (BNDES). Moreover, these refineries with Social Fuel Label can participate in auctions organised by the Brazilian National Agency for the Petroleum, Natural Gas, and Biofuels (ANP, 2005, 2007). The government established that 80% of total sales in auction is reserved for refineries with Social Fuel Label.

The contract between family farmers and biodiesel refineries is controlled by workers' trade unions and their approval is required for signing contracts between family farmer and the biodiesel plant. The workers' trade union participates actively in the formulation and the execution of the contracts, besides defending the interests of family farmers (Abramovay and Magalhães, 2007).

In addition to worker's trade union, the inclusion of family farmers in the biodiesel production system has involved other agents that are part of the organisational environment: technicians, cooperatives, financial institutions and farmers' associations. These agents are important in facilitating transaction between family farmers and biodiesel plants. For instance, the technicians are the bridge between family farmers and the biodiesel program. Besides, assisting family farmers that is a requirement for Biodiesel Company to obtain the 'Social Fuel Label', the technicians organise them to participate in the BNPB.

The contracts performed in the BNPB are based on contract farming and their characteristics are: the reciprocity of obligation to do and to give; the supply of inputs, technical services to farmers; the guarantee of buying the feedstock produced by family farmers; cooperation from family farmers to supply feedstock to refineries. These contracts that involve family farmers in the biodiesel production system have the logic of hybrid structure of governance that occurs in the agro-food production system. Even though there is no legal framework for contract farming, indirectly the BNPB brought a model of contract farming with the intervention of State that regulates the transaction giving incentives to involve family farmers into the biodiesel production system. In other words, the BNPB uses the promotional function as normative technic. This positive technic encourages the economic agents to act through positive sanctions, also named premium sanctions (Paiva, 2009). These positive sanctions are: tax reduction for refineries with Social Fuel Label, which gives incentive to buy feedstock produced by family farmers. Because of State intervention, biodiesel industry buys feedstock from family farmers under hybrid structure of governance; otherwise, spot market would prevail.

9.6 Conclusions

Some transactions in agricultural production system are carried out through contract farming. The contract farming in the hybrid governance structure is a 'strange' type of contract because the farmers have obligations 'to do' in respect to the agro-industry/distributor, besides the obligation 'to give'. Because of imbalance contractual relation, contractual abuse may arise and the part with more power may behave opportunistically. In this sense, law is important to avoid opportunistic behaviour. Therefore, a formal institution can be complement to informal institution for a more balanced and transparent transaction. The lack of a specific law for contract farming implies to legal uncertainty and, consequently, the increase of transaction cost since the parts should draw a more complete contract and have more efforts to carry the transaction on. Many countries do not have a legal law typifying the contract farming, as Brazil for instance. The efforts of UNIDROIT, FAO and IFAD to build a Legal Guide on Contract Farming will help to fulfil the gap in these countries.

The attention in the contracts between farmer and agroindustry/distributor should be towards its operation in the market with the repression of opportunistic behaviour of parts of transaction. The incentives can be introduced through rules to encourage the balance between the parts of contract and the transparency of the agreements. It means that the law should ensure to the agents an efficient access to contract.

The contract farming is mainly long-term contract that allows the renegotiation and informal institutions are observed. The reputation of the parts is taken into account. However, private institutions might not be sufficient to perform contract farming more impersonally. The comparison of institutional alternatives serves to establish the benefits of rules that offer solutions out of legal apparatus. On the other side, a legal framework for contract farming would bring more certainty for the parts to perform this contract. Two cases were presented, milk and biodiesel production systems. For milk case, private institutions prevail; for biodiesel case, there is State intervention to enforce economic agents transact through contract farming. However, development of data is necessary for assessing the comparative efficiency for alternative institutions.

References

Abramovay, R. and Magalhães, R., 2007. The access of family farmers to biodiesel markets: partnerships between large companies and social movements. Available at: http://tinyurl.com/yctfh6wx.

Albisinni, F., 2003. Sistema agroalimentare. In: Digesto delle discipline privatistiche. Sezione Civile. Tomo Ii. Utet, Turin, Italy, pp. 1244-1261.

Allen, D.W. and Lueck, D., 2002. The nature of the farm. contracts, risk and organization agriculture. Massachusetts Institute of Technology, Cambridge, MA, USA.

Brazilian National Agency for Petroleum, Natural Gas and Biofuels (ANP), 2005. Resolution ANP No. 31. November 4, 2005. Regula Leilões Para Aquisição De Biodiesel. Available at: http://tinyurl.com/y8o65z2g.

Brazilian National Agency for Petroleum, Natural Gas and Biofuels (ANP), 2007. Resolution ANP No. 33. October 30, 2007.

Arruñada, B., 2001. the role of institution in the contractual process. In: Deffains, B. and Kirat, T. (eds.) Law and economics in civil law countries. Elsevier Science, Amsterdam, the Netherlands, pp. 149-167.

Azevedo, P.F., 2000. Nova economia institucional: referencial geral e aplicações para a agricultura. Agric. São Paulo, Sp, 47(1): 33-52.

Bankuti, S.M.S. and Souza, J.P., 2014. Arranjos contratuais e assimetria de poder no sag suinicola no Oeste Paranaense. Organizações Rurais e Agroindustriais (UFLA) 16: 92-107.

Barzel, Y., 2002. A theory of the state. economic rights, legal rights and the scope of the state. Cambridge University Press, Cambridge, UK.

Benincasa, M., 1992. Il problema della tipicità dei contratti agro-industriali. In: Il Diritto Dell'agricoltura 1.

Bivona, G., 1979. I contratti d'integrazione verticale in agricoltura. Giuffrè, Milan, Italy.

Brazil, 2013. Justiça do Trabalho 15a Região – Vara do Trabalho de Matão. Ação Civil Pública. Proc. 88.2010.5.15.0081. Ministério Público Do Trabalho. Réus: Sucocítrico Cutrale Ltda; Louis Dreyfus Commodities Agroindustrial S/A; Citrovita Agro Industrial Ltda; Fischer S/A – Comércio, Indústria E Agricultura. Relator: Flavio Allegretti De Campos Cooper. Sentença 13 Mar 2013. Available at: http://tinyurl.com/y86v3nlb.

Brazil, 2016. Law No. 13.288 of May 16, 2016. Diário oficial da república federativa do Brasil, Brasília, Df. Available at: http://tinyurl.com/ycn4ga68.

Brousseau, E. and Raynaud, E., 2006. The economics of private institutions: an introduction to dynamic of institutional framework and to the analysis of multilevel multi-type governance. Available at: http://dx.doi.org/10.2139/ssrn.920225.

Campbell, D. and Harris, D., 2005. Flexibility in long-term contractual relationships: the role of co-operation. Journal of Law and Society 20(2): 166-191.

Casadei, E., 1984. Prospettive di regolamentazione dei contratti agro-industriali nel diritto Italiano. Giurisprudenza agraria Italiana, pp.7-16.

Confortini, M. and Zimatore, A., 1983. Contratti agro-industriali. In: IRTI, Natalino (a cura di). Dizionari del Diritto Privato 4. Diritto Agrario, Giuffrè, Milan, Italy.

Davis, J.H. and Goldberg, R., 1957. A concept of agribusiness. Harvard University, Boston, MA, USA.

Fanfani, R. and Montresor, E., 1998. Instituzioni ed imprese nel percorso di sviluppo dei sistemi locali di produzione agroalimentare. La Questione Agraria 69: 87-108.

Farina, E.M.M.Q., Gutman, G.E., Lavarello, P.J., Nunes, J. and Reardon, T., 2005. Private and public milk standards in Argentina and Brazil. Food Policy 30(3): 302-315.

Fernandes, E.N., 2013. Leite 100. Programa de fortalecimento da atividade leiteira na agricultura familiar. Reunião da Comissão Nacional de Pecuária de Leite da CNA. Brasilia, Brazil.

Ferrarese, M.R., 2002. Il diritto al presente. Globalizzazione e tempo delle istituzioni. Il Mulino, Bologna, Italy.

Galloni, G., 1978. Nozioni e classificazione dei contratti agrari. In: IRTI, Natalino (org.). Manuale di Diritto Agrario Italiano. Turin, Italy.

Greif, A., 2005. Commitment, coercion, and markets: the nature and dynamics of institutions supporting exchange. In: Ménard, C. and Shirley, M. (eds.) Handbook of new institutional economics. Springer, New York, NY, USA, pp. 727-786.

Instituto Brasileiro de Geografia e Estatística (IBGE), 2006. Censo agropecuário. Available at: www.sidra.ibge.gov.br.

Jannarelli, A., 1997. La disciplina dell'atto e dell'attività: i contratti tra imprese e tra imprese e tra imprese e consumatori. In: NICOLÒ, Lipari. Diritto Privato Europeo 2: 489-552.

Jannarelli, A., 2000. Istituzione e mercato nel governo del «sistema agricoltura». In: Rivista di Diritto Agrario. Giuffrè, Milan, Italy, pp. 523-557.

Jannarelli, A., 2002. l'impresa agricola nel sistema agro-industriale. In: Diritto e giurisprudenza agraria e dell'ambiente. Tellus, Rome, Italy, pp. 213-219.

Jannarelli, A., 2012. Contractual frameworks and inter-firm co-operation in the agricultural sector. Uniform Law Review 1-2: 247-262.

Jia, X. and Bijman, J., 2013. Contract farming: synthetic themes for linking farmers to demanding markets. In: Silva, C.A. and Rankin, M. (ed.) Contract farming for inclusive market access. FAO, Rome, Italy, pp. 21-38.

Joskow, P.L., 1985. Vertical integration and long-term contracts: the case of coal-burning electric generating plants. Journal of Law, Economics and Organization 1: 33-80.

Keefer, P. and Shirley, M.M., 2000. Formal versus informal institutions in economic development. In: Ménard, C. (ed.) Institutions, contracts and organizations. Edward Elgar, Cheltenham, MA, USA, pp. 88-107.

Klein, B., 1980. Transaction cost determinants of 'unfair' contractual arrangements. American Economic Review 70(2): 356-362.

Klein, B., 1992. Contracts and incentives: the role of contract terms in assuring performance. In: Werin, L. and Wijkander, H. (eds.) Contract economics. Blackwell Publishing, Oxford, UK, pp. 149-180.

Klein, B., Crawford, R.G. and Alchian, A.A., 1975. Vertical integration, appropriable rents and the competitive contracting process. Journal of Law and Economics 21(2): 297-326.

Lazzarini, S.G., Miller, G.J. and Zenger, T.R., 2004. Order with some law: complementarity versus substitution of formal and informal arrangement. Journal of Law, Economics and Organization 20(2): 261-298.

Longu, T., 2000. Il divieto dell'abuso di dipendenza economica nei rapporti tra le imprese. Rivista di Diritto Civile 46: 345-387.

Lorvellec, L. and Danet, J., 2004. Production et marches. Contrats individuels d'integration. In: Juris-Clausseur (ed.) Rural.

Macneil, I.R., 1978. Contracts: adjustment of longterm economic relations under classical, neoclassical and relational contract law. Northwestern University Law Review 72: 854- 905.

Marques, C.L., 1998. Contratos no código de defesa do consumidor. O novo regime das relações contratuais, 3rd edition. Revistas dos Tribunais, São Paulo, Brazil.

Martinez, S.W., 1999. Vertical coordination in the pork and broiler industries: implications for pork and chicken products. Food and Rural Economics Division, Economic Research Service, Report No. 777, USDA, Washington, USA.

Mazé, A., 2005. Contract law and the range of self-enforcing contracts in agriculture: private institutions and multilateral-reputation mechanisms. Second French and Germany talks in law and economics. December 2-3, 2005. Sarebrucken, Germany.

Ménard, C., 1996. On clusters, hybrids, and other strange forms: the case of the French poultry industry. Journal of Institutional and Theoretical Economics 152: 154-183.
Ménard, C., 2004. L'économie des organisations. Repères, La Découverte, Paris, France.
Ménard, C., 2005a. A new institutional approach to organization. In: Ménard, C. and Shirley, M. (eds.) Handbook of new institutional economics. Springer, New York, NY, USA, pp. 281-318.
Ménard, C., 2005b. New institutions for governing the agri-food industry. European Review of Agricultural Economics 32(3): 421-440.
Ménard, C. and Klein, P.G., 2004. Organizational issues in the agrifood sector: toward a comparative approach. American Journal of Agricultural Economics 86(3): 746-751.
Moruzzi, L., 1962. Profili della vendita di prodotti agricoli com prezzo a riferimento e contratti di integrazione verticale. Giuffrè, Milan, Italy.
Natoli, R., 2003. Abuso di dipendenza economica. In: Digesto delle discipline privatistiche. Sezione commerciale, Utet, Turin, Italy, pp. 1-27.
North, D.C., 1990. Institutions, institutional change and economic performance. Cambridge University Press, Cambridge, UK.
North, D.C., 1991. Institutions. Journal of Economic Perspectives 5: 97-112.
North, D.C., 2000. A revolution in economics. In: Ménard, C. (ed.) Institutions, contracts and organizations. Edward Elgar, Cheltenham, MA, USA, pp. 37-41.
Paiva, N.S.V., 2009. Novos modelos contratuais para uma nova matriz energética. Aspectos jurídico-econômicos para produção de biocombustíveis no Brasil. Revista de Informação Legislativa, ano 46, No. 184.
Pavone la Rosa, A., 1990. Profili della nuova disciplina dei contratti agro-industriali. In: Massart, A. (Org.) Accordi interprofessionali e contratti agroindustriali. Problemi di inquadramento giuridico. Ets, Pisa, Italy, pp. 131-144.
Penn, R.J., 1958. Tenure innovations and tenure problems associated with vertical integration. Journal of Farm Economics 40(5): 1383-1390.
Pinto, M.S., 2013. Legislação sobre qualidade do leite no Brasil. Desafio de cumprimento dos índices de qualidade do leite impostos pela instrução normativa No. 62/2011. Painel Setorial Sobre Medição do Leite. Instituto Nacional de Metrologia, Qualidade e Tecnologia – Inmetr. Available at: http://tinyurl.com/y7l3rtae.
Posner, E.A., 1999. A theory of contract law under conditions of radical judicial error. Law and Economics Working Papers.
Prévault, J., 1996. Contrat d'intégration en agriculture. Encyclopédie Dalloz – Civil Iv.
Pultrone, C., 2012. An overview of contract farming: legal issues and challenges. Uniform Law Review 1-2: 263-289.
Silva Jr., A.G., Clemente, F. and Perez, R., 2013. Inclusion of small-scale farmers in the Brazilian biodiesel production chain. In: Silva, C.A.D and Rankin, M. (ed.) Contract farming for inclusive market access. FAO, Rome, Italy, pp. 183-196.
Silva, C.A.B., 2005. The growing role of contract farming in agri-food. FAO, Rome, Italy.
Sirsi, E., 1990. Profili metodologici dello studio sui contratti agro-industriali. In: Massart, A. (ed.) Accordi interprofessionali e contratti agroindustriali. Problemi di inquadramento giuridico. ETS, Pisa, Italy.
Souza, J.P. and Zylbersztajn, D., 2011. Poder de mercado e poder de contrato envolvendo integrados cooperados e não cooperados: percepções na cadeia de frango. Informações Econômicas 41: 2-13.

Souza, J.P. and Bankuti, S.M.S., 2012. Uma análise dos contratos no sistema agroindustrial suinícola no oeste paranaense pela ótica da mensuração e da transação. Informe Gepec 16: 1-15.
Stajn, R., Zylbersztajn, D. and De Azevedo, P.F., 2005. Economia dos contratos. In: Zylbersztajn, D. and Sztajn, R. (eds.) Direito and economia. Elsevier, Rio De Janeiro, Brazil, pp. 102-136.
Traisci, F.P., 1992. I contratti di integrazione verticale in agricoltura in Francia, Germania e Italia. In: Rivista di Diritto Agrario, Anno 71, pp. 551-594.
United State Department of Agriculture (USDA), 2015. Production, supply and distribution on line. Available at: http://apps.fas.usda.gov/psdonline.
Watanabe, K. and Zylbersztajn, D., 2012. Building supply systems from scratch: the case of the castor bean for biodiesel chain in Minas Gerais, Brazil. International Journal on Food System Dynamics 3(2): 185-198.
Watanabe, K. and Zylbersztajn, D., 2014. Contract farming in the Brazilian agribusiness system: private institution and state intervention. Uniform Law Review: 1-20.
Watanabe, K., Bijman, J. and Slingerland, M., 2012. Institutional arrangements in the emerging biodiesel industry: case studies from Minas Gerais – Brazil. Energy Policy 40: 381-389.
Williamson, O., 1985. The economic institutions of capitalism. Firms, Markets, Relational Contracting. Macmillan, New York, NY, USA.
Williamson, O., 1990. A comparison of alternative approaches to economic organization. Journal Of Institutional and Theoretical Economics 146: 61-71.
Williamson, O.E., 1996. The mechanism of governance. Free Press, New York, NY, USA.
Williamson, O.E., 1999. Comparative economic organization: the analysis of discrete structural alternatives. In: Williamson, O.E. and Masten, S.E. (eds.) The economics of transaction costs. Elgar Critical Writings Readers Series, pp. 101-128.
Williamson, O.E., 2005. Why law, economics and organization? Annual Review of Law and Social Science 1: 369-96.
Williamson, O.E. and Masten, S.E., 1999. The economics of transaction costs. Elgar Critical Writings Readers Series, pp. 101-128.
Worley, T. and Mccluskey, J.J., 2000. Production contracts as a means of vertical coordination with application to the wheat industry. Journal of Food Distribution Research 31(1): 215-224.
Zylbersztajn, D., 1995. Estruturas de Governança e Coordenação do Agribusiness. Uma Aplicação da 'Nova Economia das Instituições'. Tese de Livre-Docência Apresentada no Departamento de Administração da Faculdade de Economia, Adiministração e Contabilidade da Universidade de São Paulo, São Paulo, Brazil, 238 pp.
Zylbersztajn, D. and Farina, E.M.M.Q., 1999. Strictly coordinated food systems: exploring the limits of the Coasian firm. International Food and Agribusiness Management Review 2(2): 249-265.

10. Contractual scope in the agri-food market: the case of protection consortia in the wine sector[1]

L. Paoloni[*] and M. Gioia

Department of Law, University of Molise, Via Alessandro Manzoni 1, 86100 Campobasso, Italy; lorenza.paoloni@unimol.it

Abstract

This paper is aimed at critically investigating the function of those associations that have been allowed to operate in the agri-food sector since the last reform of the Common Agricultural Policy and the role that they can virtuously play in order to rebalance the bargaining power of agri-food entrepreneurs, specifically focusing on the consortia for the protection of local products. The nature and tasks of producer organisations (POs) and interprofessional organisations (IOs) will be described. Protection consortia in the wine sector will then be analysed, together with the delicate issue of *erga omnes* extension of rules laid down for IOs in the wine sector to non-members of the consortium. In addition, some tentative evaluation will be undertaken and the hypothesis of extending the interprofessional model to all consortia for the Protection of typical products will be assessed.

Keywords: agri-food market, *erga omnes*, interprofessional organisations, producer organisations, consortia

[1] This chapter is the result of a research undertaken jointly. However, Section 10.1, 10.4 and 10.6 are attributed to Lorenza Paoloni; Section 10.2, 10.3 and 10.5 to Marianita Gioia.

DOI 10.3920/978-90-8686-844-5_10, © Wageningen Academic Publishers 2017

10.1 Introduction

The 1st of January 2015 marked the adoption of five European regulations aimed to determine the Common Agricultural Policy (CAP) for the period 2014-2020. This last reform, as well as introducing important innovations regarding the tools to be used to implement it, contains general provisions and definitions that further emphasize the will of the legislator (which has existed, in truth, since 2003) to promote centralized regulation, formerly entrusted to individual Member States.

This decision excludes a 'renationalisation' of the CAP. The abandonment of a centralized price policy and the management of resources by Member States, which had characterized the CAP since its inception, seemed, in fact, to give back wider powers to each Member State.

In other words, the European Union, which has always operated through economic and regulatory mechanisms (Albisinni, 2015)[2], has in recent years given greater centrality and importance to the latter rather than the former, giving States the ability to implement economic measures upon the establishment of unified legal rules and necessary guidelines.

In this context, the role played by the bodies representing the interests of farmers, such as protection consortia (PC), producer organisations (POs)[3] and interprofessional organisations (IOs), becomes particularly important. These bodies constitute (together with the vertical integration agreements and the use of essential written procedural requirements for contracts between food business operators (for vertical integration contracts, see Costantino, 2013; Russo, 2015)) some of the models chosen by the legislator to improve the functioning of the supply chain, develop the competitiveness of the European agricultural sector and safeguard producers' income.

Unlike instruments used in the past (such as clusters, intervention prices and production quotas) that operated directly and automatically, these measures act indirectly on the exchange.

With this paper, we intend to critically investigate the function of the associations currently allowed to operate in the agri-food sector since the last CAP reform and the role that they can virtuously play for rebalancing the bargaining power of agri-food entrepreneurs, particularly focusing on the consortia for the protection of local products.

[2] The measures that have always been used to achieve the CAP's objectives can be summarized in two main macro-areas: economic and regulatory measures. The former has as their objective prices, production, and distribution requiring direct market intervention. Conversely, the latter entail actions that define and classify products and the contexts of intervention. In other words, the regulatory mechanisms form the legal boundaries within which Member States are allowed to operate. For further details, see Albisinni (2015).

[3] For more details on the regulation of producers organisations, see Jannarelli, 2004, 2008.

For this purpose, the second section will describe the nature and tasks of POs and IOs; the third will analyse PC in the wine sector. The fourth will tackle the delicate issue of *erga omnes* extension of the rules laid down for Interprofessional Organizations in the wine sector to non-members of the consortium; in the fifth, some primary evaluation of these issues will be undertaken. The sixth and final section will examine the hypothesis of extending the inter-professionalism model to all types of consortia for the protection of typical products.

10.2 Powers and functions of the bodies representing the interests of agri-food entrepreneurs in the new CAP

Collective bodies composed of operators acting, for various reasons, in the agri-food market, are essentially aimed at rebalancing the bargaining power of farmers. These associations, through multiple-level regulatory measures and the European Commission's proposals for the reform of the CAP for the period 2014-2020 (Costato, 2011), play today an important role not only in Europe but also internationally.

The opportunity that allows POs and IOs to extend, in certain circumstances, their rules to non-members belonging to the same sector, makes said entities potential 'regulators' of the market.

Firstly, it is necessary to clarify that POs, together with associations of POs, are horizontal, as they consist of producers in a particular sector. Their general supervision is currently regulated at European level by Regulation 1308/2013 and by specific provisions contained in the same legislation on individual sectors. While POs are top-level bodies, associations of organisations (APOs) are called second-level bodies as they consist of more producer groups.

POs may carry out operational and regulatory functions. In the first case, the organisation sells products produced by its members; this is called 'direct marketing' as the organisation operates in the market itself.

Conversely, when exercising the regulatory functions, the PO dictates the technical rules, namely those relating to the methods of production and marketing. These tasks are closely linked to the operational tasks, since the organisation must ensure that the production is as homogeneous as possible. The circumstance whereby the PO carries out only normative functions is known as 'indirect marketing', because acting in the market is the single producer, not the organisation.

The concentration of supply and marketing are non-essential tasks for the PO, which can also have only a normative role. This option, contained in Regulation 1308/2013, does not have an equivalent in the Italian legislation, which gives importance to the aspect of marketing so much as to consider it a major goal.

The content of the contract between the producer and the buyer originates from the 'framework agreement' concluded by the organisations to which they belong to. Farmers therefore enjoy the benefits of a collective bargain carried out by their organization using the aggregate volume of production as a lever. In other words, the more the members, the greater the volume of production available to the PO and the greater the bargaining power within the framework contract.

Regarding IOs, unlike the POs, it is necessary to underline that they are vertical bodies, designed to group all operators at different levels within the same supply chain (producers, processors, distributors). Moreover, Regulation 1308/2013 supervises the functioning of these bodies to facilitate communication within the supply chain, thus promoting good practices and market transparency.

The tasks that IOs can currently perform are those of research, market knowledge, and information activities but do not include marketing tasks, which, conversely, fall within the operational functions carried out by the POs.

This decision stems from the necessity for the European legislator to protect competition: damages to competition are in fact a risk to which associations of operators within the same supply chain are more easily exposed.

Regulation 1308/2013 lays out for the first time rules for IOs that can be defined as horizontal, as they address all sectors covered by said regulation. Conversely, the previous legislation on the single common market organization, Regulation No 1234/2007, was limited to the IOs producing tobacco, wine, olive oil, fruit and vegetables, milk and dairy.

Regulation 1308/2013 admits the possibility, for POs and IOs, to extend their rules to non-members. As a result, resolutions adopted by these organisations can be said to bear *erga omnes* effectiveness.

As explained earlier, this feature would allow such associations, which have sufficient representative power, to convert private rules into legally binding provisions for all sectors within the same economic area. Alongside this, States can demand that non-members operating in the same territory and enjoying the benefits brought by the extension of the rules contribute financially to the functioning of the organisation.

What described thus far suggests that the European legislator has only taken into consideration the POs and IOs, without addressing PC. There seems to be little doubt that such bodies play a fundamental role in the agri-food sector, despite not being mentioned explicitly by the CAP's five Regulations.

Indeed, because of their nature and their widespread presence, PC could be considered the type of association most likely to function as a PO or IO. However, unlike other

European countries, Italy has only recently granted PC belonging to certain sectors, the possibility to carry out the functions envisaged for POs and IOs.

The Italian legal system defines PC as voluntary, non-profit organisations. In the agri-food sector, the primary role accorded to such institutions is to safeguard products bearing protected designation of origin (PDO) and protected geographical indication (PGI) certifications.

With a measure of 1999, the Italian legislator accorded multiple competences to such bodies: protection, promotion, enhancement of geographical indications and consumer information. To date, due to a confused and fragmentary legislation, PC are denied the opportunity to control, plan and schedule the supply, which, conversely, is the case in other Member States. The wine sector represents an exception to this gap, as it benefits from a specific legislation that grants consortia of said sector further roles and powers.[4] As of today, wine consortia in Italy can perform functions that at European level are granted to POs and to IOs, including the possibility to extend *erga omnes* the effectiveness of their rules (for further details, see Paoloni, 2012).

Another area that has recently undergone changes in the Italian legal system is milk and dairy. A legislation dated July 2015[5] has dictated the rules concerning POs of this sector. The content of its provisions, however, suggests that they are also applicable to other agri-food products, thus further complicating the regulatory framework.

In 2012, the European legislator, with the enactment of Regulation 1151 containing provisions on quality schemes for agri-food products, while not specifically mentioning the PC, strengthened their role. Article 45, without prejudice to the specific provisions relating to POs and IOs, recognizes, indeed, ample prerogatives and responsibilities to producer groups, in order to contribute to the development and promotion of indications of origin. In addition, Member States are given the opportunity to contribute to the dissemination of producer groups in their territory.

Moreover, in the light of other regulations contained in the CAP reform, which runs until 2020, the central role of PC is inferred. Therefore, the Italian decision to deprive them of the key functions recognized to POs and IOs, to which, as shown later, they are often equated, is increasingly at odds with European standards.

[4] Legislative decree 8th April 2010, n. 61 pertaining to PDOs and PGIs of wine products, issued for implementation of art. 15 regulation 7th July 2009, n. 88, published in Official Gazette of the Italian Republic n. 96, 26th April 2010.

[5] Regulation 2nd July 2015, n.91, containing measures to boost and support sectors facing extreme economic conditions. Published in the Official Gazette of the Italian Republic n. 152, 3rd July 2015.

10.3 The protection consortia for PDO and PGI wine products

In 2010, the Italian legislator reshaped the broad subject of the PGI and PDO of wine products, at the same time providing a very articulate and detailed set of tools for the wine sector.

Particularly interesting is the new regulation on PC, which presents significant implications for its relevance (not only limited to this sector) on which some brief remarks can be made.

The two profiles that deserve a deeper analysis, since they are more specific and relevant to the peculiar dynamics typical of the agri-food market, touch the fundamental postulates of representativeness, within the PC, of the so-called *erga omnes* effectiveness in the implementation of government supply policies, alongside the interprofessional nature of such bodies.

These principles are directed to the preservation and protection of PDO and PGI product quality, but they also appear to pursue a better-coordinated market placement of the product and the definition of quality improvement plans.

The attention paid to such bodies is justified by the fact that PC have occupied, for a long time, a significant segment of the supply of quality products within national and European food markets and, consequently, their operations have often been monitored either by the legislators (national and EU) or by the competition and market authority.

The long and eventful life of PC has thus been marked by several legislative interventions over the past few years. In detail, these measures have established the separation between the control and the supervisory functions. The control function has been conferred to a third independent party, while for a long time it was carried out by the association itself, giving rise to the controversial figure of the consortium being the 'controller' and the 'controlled', at the same time. The supervisory function remains within the consortium structures, and is to be exercised once the certification body authorized by the Ministry (for further details, see Paoloni, 2000a,b) has completed its control activities.

Transferring the control for the recognition of the denomination to a third party agency, has deprived those consortia established before the reg. 2081/92 of the strongest power that tied necessarily all producers of the same product to the consortium (Giacomini, 2006).

Alongside this new regulation, the Authority for competition and the market intervened, its investigations resulting in the prohibition, for the major national consortia, such as those for *Parmigiano Reggiano*, *Grana Padano*, *Parma* and *San Daniele*, to control and plan the supply, therefore depriving them of the institutional and regulatory tasks that older consortia could perform in the past.

In this legal context, the qualification of PC as interprofessional bodies has also remained unaddressed; in fact, a Ministerial decree established the 'General provisions concerning the representation pre-requisites for the recognition of consortia' and 'the criteria for a balanced representation of categories of producers and processors interested in PDO, PGI and TSG within the consortia themselves' (Mancini, 2003).[6]

This open front in the redefinition of PC in the field of quality local products has raised heavy criticism (Giacomini, 2005) regarding the lack of effective operational tools in the national agricultural policy.

PC, partly because of the rigid stance taken by Antitrust authorities, 'more as a result of ignorance than a correct market analysis of protected products' (Giacomini, 2005) have not been able to regulate the supply of their associates, differently from what happened in other EU countries.

More generally, the above-mentioned issues constitute some of the reasons for the failure of our agricultural supply, alongside critical aspects of production structures, in need of innovation and investment that have recently suffered delays due to the country's general crisis.

This particular issue has not, however, gone unnoticed in the EU: the Commission's communication on agricultural product quality policy of 2009 appropriately pointed out that 'representatives of producers of geographical indication products have called for greater rights and control over their use – for example, to control production (using quotas), and to control the use of geographical indication names on the packaging of processed products'.

This issue, as highlighted, is particularly delicate: today, the task of adjusting quantities and prices of the products is still precluded to the PC (save for exceptional cases, and following authorisation by the public authority), with the risk of exposing the production of typical products to serious market crises. As a result, 'the claim of the Commission, which is not reflected in policy options, offers an opportunity to reopen the debate on the role that PC (and generally IOs representing the production of geographical designations) have in coordinating the product sales management, considering the possibility to use self-managed tools for market placement and supply planning in relation to expected market trends' (Arfini *et al.*, 2010).

[6] Ministerial Decree n.61413, 12th April 2000 (identifying requisites for representation), followed by Ministerial Decree 61414, 12th April 2000 (defining criteria for representation within the Protection Consortia), another decree of the same date, repealed by Ministerial Decree of 12th October 2000 (defining the limits for the cooperation between Protection Consortia and the Anti-fraud Inspectorate) and, finally, Ministerial Decree 12th September 2000 n. 410 (imposing the duty to contribute to some expenses to non-members). More details on said Decrees, in Mancini (2003).

10.4 *Erga omnes* and interprofessional organisations

The Italian legislator considers it necessary, firstly, 'to fully regulate the procedures necessary to recognize and accord PC for PDO and PGI of wine products the functions of protection, promotion, enhancement, consumer information and general safeguard of interests related to the designation, as well as market surveillance actions mainly at the sale stage'.

It is also essential to regulate the operations described above even when they are addressed to all 'winegrowers, winemakers and bottlers of the designation subject to the controls' contained in the above Legislative Decree and not just to the very members of the consortium.

The provision refers, as will be explained, to the well-known possibility for the PC's functions to be extended *erga omnes* to all producers and users of the same designation, that are regularly subjected to compliance checks in the use of the designation, despite not being members of the consortium (Paoloni, 2000a,b).[7]

The Italian legislation guarantees the use of consortium trademarks to all the producers complying with the approved product specification, accepting the controls by the selected inspection body and contributing to the management costs of the consortium. This procedure, in fact, gives PC a central role in the life of the designation (especially for protection and promotion) but at the same time does not compel producers to be members of the consortium avoiding, however, potential opportunistic behaviours of certain categories of producers (Gardini and Lazzarini, 2000).

Moreover Art. 1 of the Ministerial Decree, December 16th 2010, defines the type of subjects related to a particular designation allowed to form a protection consortium of geographical indications and designations of origin for wines by following (and therein confirming the regulation in force) the model contained in the Civil code. These are growers, winemakers and bottlers in the control system for the denomination composing the wine industry, this one being the essential prerequisite for an IO to exist.

This measure confirms, firstly, the functions that the PC, instructed by a Decree of the Ministry of agriculture, are expected to perform, alongside the regulation proposals and advisory duties. These are technical assistance, study, proposal, economic evaluation of the PDO or PGI, product enhancement, protection and preservation of the PDO or PGI against abuses, unfair competition, counterfeiting, misuse of designations, etc. Functions limited to associates are protection, promotion, enhancement, consumer information and general safeguard of the interests of the designation together with supervisory actions at the sale stage.

[7] For further theoretical considerations regarding the principles of *erga omnes* and representativeness, see Paoloni (2000).

A first important element contained in the Decree is art.5, which reiterates PC's possibility to perform *erga omnes* functions, provided that such associations represent at least 40% of winegrowers and 66% of the production of registered vineyards within their PDO or PGI, calculated on the declared amount produced in the preceding two years. At least 20% of the last figure must originate from production stages other than viticulture.

Such *erga omnes* functions are generally regulated by art.17, par. 4 of legislative decree 61/2010, and provide for PC to perform tasks in the interests of non-members, namely to define, given prior consultation with the representatives of the designation concerned, the implementation of supply government policies in order to safeguard, protect and define improvement plans of PDO and PGI product quality; to organize and coordinate the activities of the groups involved in the production and marketing of PDO or PGI products; to act at judiciary and administrative levels to protect and safeguard PDOs or PGIs and rights and interests of producers; finally, to supervise, protect and preserve the designation at the sale stage, namely by verifying that certified products meet the disciplinary requirements in order to prevent similar products from generating confusion in consumers, thereby harming the protected productions.

Such activities are different from the control tasks performed as per the guidelines given by the MIPAF, the national authority in charge of coordinating control activities.

The second prominent feature, outlined by the Ministerial Decree of December 16, 2010 hereby analysed, establishes that PC are IOs pursuant to regulation 1234/2007.[8]

Regarding the possibility for collective bodies in the agri-food market to be recognized as IOs in the wine industry, the European legislation requires them to be composed of representatives from economic activities related to the production, sale and processing of the products, and that they be established by either all or some of the above-mentioned representatives. When these associations meet such requirements, and apply for official recognition, they must demonstrate the ability to operate in one or more regions of the territory. In addition, they must represent a significant share of the production or sale of the products and must not be involved in the production, processing, and marketing of products of the sector concerned.

Notably, the purpose of IOs (Paoloni, 2003)[9] refers to the following aspects: to concentrate and coordinate the supply and marketing of their members' production; to adapt the production and processing to market requirements and improve the product; to promote the rationalisation and mechanisation of production and

[8] As of today, art. 125 pertaining to IOs has been transposed into art. 158, regulation 1308/2013.

[9] The definition hereby given is by Lorvellec (1988), which still today represents a frame of reference. IOs are 'institutions that gather the economic actors within the agri-food supply chain, such as producers and companies, with the aim of managing the specific market of a product or a group of products, in compliance with state economic policies'.

processing; to carry out research on methods of sustainable production and market developments. They represent, in essence, a sort of control room, which has also the power to affect the definition of standard contracts.

As the combination of said provisions suggests, the consortium/ IO emerges as different from POs, its functions being solely normative and not operational.

It seems appropriate to emphasize that the aforementioned Ministerial Decree takes shape in the wake of a national legal journey, which has marked the evolution of PC towards the Interprofessional model, a model celebrated by some and disregarded by others and subject of several and meaningful, though not fully effective, provisions of the recent past.

The status of inter-branch organisation had previously been assigned to bodies representing economic activities linked to the production, processing and marketing of agricultural products mentioned by the EU regulation, though limiting the freedom to define their legal entity to associations with legal personality.

Resorting to the parameter of approved associations led, therefore, to the exclusion of both consortia and associations thereof; this exclusion has today been neutralized thanks to the regulatory measures mentioned above and PC can thus aspire to become IOs in all respects.

Similarly, the establishment of the *erga omnes* effectiveness of collective bodies' decisions has distant origins and relates to the wider issue of extending interprofessional features to some actors essential to the functioning of the agri-food market and supply policies, which producers organizations, inter-branch organisations and PC indeed are.

The economic doctrine has long acknowledged that certain sectors of quality products are 'hybrid organizational forms' (Martino, 2007), where the governance power is devolved to a third party with interprofessional features, this normally being the protection consortium.

The decree under consideration has thus permanently accepted the idea that PC for typical products albeit limited to the wine sector, can take the form of inter-branch organisations and it also incorporates the guidance of some legal and economic theories, which have long advocated the establishment of such development in the agri-food market.

In this way, the supply of products with protected designation can take place through the adoption of common specifications between agricultural and industrial producers, and even within IOs representing different components of the production chain.

In addition, the internal structure of PC for quality products with a PDO changes, as it guarantees a level playing field for the individual components and an adequate number of farmers (possibly represented by their producer groups), preventing the achievement of dominant positions by groups with higher bargaining power.

10.5 Discussion

From this brief analysis, it can be inferred that no association model is better than the others are, but each form (POs, IOs, PC) needs to be transposed from theory into practice in order to assess their actual mode of operation. In other words, only by placing them in real legal, economic and social contexts will it be possible to identify the most suitable for each situation.

Specifically, with regard to the European experience, the particular relationship between competition rules and the CAP plays a fundamental role.

Article 3 of the TFEU states that the EU has exclusive competence in the definition of the competition rules necessary to the functioning of the internal market. Competition rules are contained in title VII, articles 101 and 102 of the Treaty.

Article 101 specifically prohibits agreements between companies, decisions by associations of companies and all those practices which may affect the trade between Member States and which have as their objective or outcome the prevention, restriction or distortion of competition, identifying a compulsory, non-exhaustive list of practices that restrict competition.[10] The formula used by the legislator plays two different roles: to prevent companies from pursuing practices aimed at distorting competition, hurting the market and the final consumer, and to prohibit, not just agreements between competitors, but also between companies operating at different levels within the same production or distribution chain.

In other words, there is a general approach, which dictates that each economic actor must independently determine the conduct to follow in the market, because only in doing so will the competitive effect be achieved. Competition rules therefore dictate that actors must operate in isolation on the market, regardless of others; the exchange of information between competitors can therefore be contrary to the rules of competition if it reduces or eliminates the degree of uncertainty in the market. The very concept of uncertainty, the heart of competition, requires actors to operate without revealing information to their competitors.

The basic idea underlying the economic and legal model for the decisions made by European legislators since the Treaty of Rome in 1957 has in fact always been the

[10] The limiting effect of competition is objective and not subjective, as the intent of contracting parties and the effects of each agreement have little importance, as the Court of Justice has confirmed.

autonomous principle of free competition by which the market must be regulated. For this reason, agreements between companies, decisions by associations of companies and practices that aim at or result in impeding competition in the common market, have always been prohibited.

Inelasticity of demand and supply of food products, the peculiar nature of the products themselves, and the involvement of several actors along the production chain, can however be considered some of the fundamental reasons underlying the preferential treatment that the European antitrust law has reserved for the agri-food sector.

The peculiarity of the agricultural market (mainly in the primary sector, characterised by a different speed from the secondary and tertiary) led the European legislator to adopt, since the inception of the CAP, a different regulation that waived the general competition rules, in order to adapt the tools used to build the common market to the peculiarity of the sector.

The purpose of the derogation from the General rules on competition envisaged for agri-food products (contained today in article 42 TFEU) reflects the intent pursued by antitrust regulations, namely economic efficiency and consumer interests.

In this context, competition is not an end but a means to achieve the objectives of the CAP. This interpretation was endorsed by the Court of Justice, according to which this provision 'recognized both the agricultural policy priorities in relation to the objectives of the Treaty in the field of competition and the power of the Council to decide the extent to which competition rules are to be applied in the agricultural sector' (EC, 1994).[11]

This explains the importance for operators in the agri-food market to negotiate agreements and benefit from convenient organizational forms, as in doing so they can stabilise the contractual asymmetry faced by producers.

If from a legal point of view, the establishment of organisations restricts producers' freedom to exercise their activities and structure their production, from an economic perspective it makes them more integrated into the production project of the food industry.

In other words, the loss of autonomy in production and organisational choices, both when operators become part of an organisation (be it of producers or interprofessional) or in case they are bound by business decisions of organisations they are not part of, is offset by the increased competitiveness in the market.

[11] Judgement of the EU Court of Justice, 5th October 1994, case 280/93 Federal Republic of Germany v. Council of the European Union.

It could be argued that, for some areas of the agricultural sector, the European legislator has chosen to apply the motto 'unity is strength', because associations of producers, or of different actors operating along the same chain, are considered a suitable means to balance the intrinsic structural weakness of the agri-food sector.

10.6 Extension of inter-professionalism to all protection consortia of typical products

At the end of this paper, it seems appropriate to recall an issue that has recently emerged in the legal and economic landscape regarding the agri-food sector.

A trend among scholars has hypothesized the possibility to extend the economic and legal features outlined by the legislative framework for IOs, to agri-food sectors other than wine.

This hypothesis has been drawn in an essay (Giacomini *et al.*, 2011) according to which, at the present state of agri-food economics, there seems to the basis to convert the legal structure of the consortium for *Parmigiano Reggiano* Cheese, one of the major representatives of the Italian food system, into an IO by amending its present statute, thereby conferring representative power to producers, processors and distributors, and providing clear rules to ensure a democratic decision-making process.

In this way, the consortium would reduce its active participation in the supply chain and could, as a body comprising all the components of the production line, secure the balance between the members of the *Parmigiano Reggiano* production chain, protect the interests of the sector and strengthen their performance in the long term.

On the regulatory front, the so-called 'milk package' (reg. 261/2012 today merged in reg. 1308/2013), was launched to respond to price volatility at the end of the guaranteed and serviced quota regime, by means of indirect instruments that enabled new procedures for contractual relationships such as written contracts between producers and processors, POs and IOs and production planning. This measure has introduced, in the specific market sector, inter-branch organisations with the task of: improving knowledge of production and the market; improving the marketing of milk and dairy products; promoting consumption and exploring potential export markets; drawing up standard contracts; coordinating research, innovation and product quality.

Member States may also establish, for a limited period, binding rules for the regulation of the supply of PDO/PGI cheese, at the request of POs, IOs or consortia. This means that reg. 26/2012 authorizes POs, IOs and consortia to plan the supply of PDO and PGI cheese. Such supply planning is subject to conditions necessary to avoid practices that may restrict competition (Prete, 2015).

These rules limiting the decision-making process of the aforementioned organizations stipulate that supply planning must be carried out upon agreement either between at least 2/3 of milk producers or representatives thereof, or between 2/3 of cheese producers representing at least 2/3 of the total cheese production in the given geographical area. Furthermore, supply planning must have the sole purpose of adapting cheese supply to meet the demand, must not include pricing and must have a duration of three years, renewable upon request.

Member States are given the task to render such agreements between producers and processors binding without incurring the restrictions imposed by EU competition rules, thus contributing to the activation of supply adjustment tools.

Of course, from a legal perspective, there is nothing to prevent the export of such organizational structures to other areas of the agri-food sector.

However, despite some openness to new (but old) interprofessional models essential to supply adjustment, the combination 'consortium / IO' is struggling to become a shared parameter throughout the agri-food market. In fact, despite a positive trend, even supported at the national level as part of the 'quality package' (containing measures for the protection and enhancement of typical products, recently approved by the EU Parliament) the amendment extending the production planning system for PDO cheese to certified ham, has not been accepted. As this paper has attempted to show, this is an age-old question, more heartfelt in times of crisis, on which our Country's competition authority has repeatedly debated.

It is clear that the 'scaffold' of contractual economics hereby erected and standing on such old foundations, will be able to stay in place only if farmers achieve effective organizational capacity and if Member States successfully encourage the involvement of supply chain operators (producers, processors distributors) and guarantee the mandatory use of contract agreements for supply planning schemes.

References

Albisinni, F., 2015. Strumentario di diritto alimentare Europeo, seconda edizione, UTET Giuridica, Turin, Italy, 148 pp.

Arfini, F., Belletti, G. and Marescotti, A., 2010. Prodotti tipici e denominazioni geografiche. Strumenti di tutela e valorizzazione. Gruppo 2013. Tellus, Rome, Italy, 59 pp.

Costantino, L., 2013. L'integrazione verticale per contratto nel settore agro-alimentare: fattispecie giuridica e disciplina applicabile. Contratto e Impresa, 1448 pp.

Costato, L., 2011. Regime disaccoppiato, trattato di Lisbona e obiettivi della Pac verso il 2020. Agricoltura Istituzioni Mercati 2: 13.

European Commission (EC), 1994. Judgement of the EU Court of Justice, 5th October 1994, case 280/93 Federal Republic of Germany v. Council of the European Union. EC, Brussels, Belgium.

Gardini, C. and Lazzarini, C., 2000. Il ruolo dei consorzi di tutela in relazione alle nuove norme. Terra e Vita 50.

Giacomini, C., 2005. L'agricoltura che vorremmo dopo la riforma della PAC. Available at: http://tinyurl.com/ya8zpb7w.

Giacomini, C., 2006. Consorzi di tutela registi delle DOP. Informatore Agrario.

Giacomini, C., Arfini, F., De Roest, K., 2011. Interprofession and typical products: the case of Parmigiano-Reggiano cheese. New Economic Papers – Agricultural Economics: 125-150.

Jannarelli, A., 2004. Le organizzazioni dei produttori: dalle fughe in avanti alla retromarcia? Diritto e Giurisprudenza Agraria, Alimentare e Dell'Ambiente 5: 312.

Jannarelli, A., 2008. Organizzazioni di produttori ortofrutticoli e loro associazioni dopo la recente riforma comunitaria del 2007: prime riflessioni. Rivista di Diritto Agrario 1: 25.

Lorvellec, L., 1988. Droit rural. Masson, Paris, France, 455 pp.

Mancini, M.C., 2003. Le produzioni alimentari tipiche. L'impatto economico e organizzativo della normativa europea. MUP, Parma, Italy, 39 pp.

Martino, G., 2007. Sicurezza degli alimenti, ambiente istituzionale ed economia delle strutture ibride. Available at: http://tinyurl.com/ycr7mnos.

Paoloni, L., 2000a. Gli accordi interprofessionali in agricoltura. CEDAM, Padova, Italy.

Paoloni, L., 2000b. I consorzi fra produttori agricoli tra passato e presente. Agricoltura e Diritto, Scritti in onore di Emilio Romagnoli 2: 908.

Paoloni, L., 2003. Organizzazioni interprofessionali. Digesto civ., Agg., Turin, Italy, 983 pp.

Paoloni, L., 2012. I consorzi di tutela ed i contratti per le politiche dell'offerta dopo il d.lgs. 61/2010. Rivista di Diritto Alimentare 3: 1.

Prete, F., 2015. Piani di regolazione dell'offerta di formaggi DOP e IGP e limiti alla concorrenza in agricoltura. Rivista di Diritto Agrario 1: 322.

Russo, L., 2015. I contratti della filiera agroalimentare nel quadro della nuova politica agricola comune. Rivista di Diritto Agrario 1: 470.

Part III – Case studies on drivers of organizational diversity

11. Value in territorial branding, agro-food production and governance patterns: a case study

R. Furesi, F.A. Madau and P. Pulina*

Department of Science for Nature and Environmental Resources, University of Sassari, Via Muroni 25, 07100 Sassari, Italy; famadau@uniss.it

Abstract

A territorial brand highlights the specific characteristics of goods and services derived from being produced in a particular place. This information can be considered important by consumers and can allow producers in a specific location to increase their profitability. Examples of territorial brands are those related to an environmentally sensitive area or to a natural park. The objective of this study was to determine the governance patterns more suitable for the introduction and promotion of the Porto Conte Regional Natural Park brand by wine growers producing near the protected area in the northwest of Sardinia (Italy). We found that the governance solutions adopted allowed larger companies to manage the territorial brand efficiently. Furthermore, the brand was found to be able to create (new) shareholder value. Finally, brand adoption would be more suitable for enterprises with hierarchical governance rather than those characterized by different market transactions. This implies that a hybrid form of governance might be the more suitable solution for small firms promoting brand adoption and the related environmental, territorial and quality attributes.

Keywords: wine producing, natural park, organizational patterns

DOI 10.3920/978-90-8686-844-5_11,

11.1 Introduction

A 'territorial brand' is a form of location-related brand that highlights the specific characteristics of goods and services derived from being produced in a particular place (Lorenzini *et al.*, 2011; Neto, 2007). More specifically, a territorial brand identifies a certain product as inherently related to its territorial origin in that producers operating in a given territory can make a product or products which can only be created there and cannot be replicated elsewhere (Charters *et al.*, 2011). This is because the products can be made in a distinctive style in that place due to highly specific territorial factors, such as the climate, topography, soil, history, cultural heritage, etc. (Kavaratzis and Hatch, 2013).

These features contribute on the whole to defining a territorial brand and the lack of certain characteristics recalls other sorts of brands. For example, a 'regional brand' is commonly related to products situated in a specific geographical region, but the environment does not necessarily affect its nature (Perrouty *et al.*, 2006; Van Ittersum *et al.*, 2003). In other words, not all regional brands can be identified as territorial brands whereas a territorial brand is always a type of regional brand.

Examples of territorial brands in the agro-food sector are those related to an environmentally sensitive area or to a natural park. Little attention has been paid by the economic literature to these sorts of brands, which have been introduced throughout the world albeit in different modalities. Generally speaking, they can be defined as 'umbrella' brands aimed at creating and sharing additional value for food products made in or near a natural park. A highly specific environment such as a natural park – often distinguished by a high presence of endemic species, landscape amenities and location-specific characteristics – has particular features in terms of intrinsic and extrinsic quality.

In the case of brands related to a natural park, value creation might be associated with the promotion of both the geographical and environmental characteristics of that area. Furthermore, some agro-food products can inherently be characterized by the environmental features of the highly specific region in which the natural park is located.

This sort of territorial branding belongs by nature to the category of the collective brands. A crucial point is the definition of the 'creation of economic value'. According to some authors, economic value regards the possible creation of shareholder value (Charters *et al.*, 2013). This means that value is created by relationships between producers, intermediaries and consumers (Lindgreen and Wynstra, 2005).

The creation and sharing of value is a cooperative process that requires some degree of governance (and coordination) among actors in the chain and at any point in the chain (Helm and Jones, 2010; Humphrey and Schmitz, 2002). Understanding which governance model is appropriate depends on the participation of producers within the value chain and the capabilities required for managing transactions and relationships and the complexity related to a given chain. Gereffi *et al.* (2005) affirm that governance models essentially depend on three variables: (1) the complexity of transactions; (2) the ability to codify transactions; (3) the capabilities in the supply base. However, enterprises often do not have a concrete definition of value and are uncertain how to quantify and share it among the local operators involved. Different branding strategies can lead to different sorts of governance in vertical transactions and therefore the most appropriate form of governance needs to be adopted according to the brand introduced and the branding strategies as a whole (Raynaud *et al.*, 2009).

In light of these considerations, the objective of this study was to determine the appropriate governance model related to the brand introduction and promotion of the Porto Conte Regional Natural Park by wine growers. This is a territorial brand recently introduced by agro-food producers close to a protected area in the northwest of Sardinia (Italy), which has specific environmental, landscape and cultural characteristics. The study was carried out by adapting the conceptual frameworks for value chains to take into account the value created by the eventual introduction of this sort of territorial brand. In essence, we aimed to determine whether governance patterns might be expected to change after the introduction of the territorial brand and the main critical areas related to the brand introduction from a managerial perspective.

Figure 11.1. The Porto Conte Regional Park area (Sardinia).

11.2 Description of the case study

The Porto Conte Regional Park is a protected area located in north-western Sardinia (Italy) (Figure 11.1). It was set aside by the Sardinian Regional Authority in 1999 and is managed by the Alghero City Council. The park area amounts to 5,350 ha of land and 60 km of coastline and it is a rich collection of landscapes, with a wealth of vegetation, fauna and karstic features that are among the most important in the Mediterranean basin (Dettori, 2013). The botanical elements consist of lush maquis with an abundance of Mediterranean vegetable species and fascinating endemic species. The landscape also benefits from a sinuous coastline that penetrates deep into the landmass, giving rise to an utterly unique land/sea spectacle.

The main objective of the Porto Conte Regional Park is to conserve and enhance the area's natural, environmental, historical and cultural resources, as well as to promote scientific research and teaching, working towards the ecological regeneration of the settlements and developing compatible economic activities with a particular focus on traditional activities of farming, animal husbandry, crafts and tourism.

Agriculture, extensive breeding and agro-food processing (mainly wine, olive oil and agro-tourism) are typical activities of the region in which the park is sited, but many farm and food processors operate in the protected area. However several agro-food activities are located close to the borders, creating a strong connection between environmental and agro-food amenities, with useful implications for tourism. Agro-food products also show intrinsic characteristics provided by the environmental, soil and coastal characteristics of the area observed; specifically some types of wine are uniquely produced in this particular region.[1]

On the basis of the strong environmental and agro-food vocation of the area, a few years ago a territorial brand related to the park was introduced to enhance agro-foods and other goods produced close to the protected area. It was conceived by the Porto Conte Regional Park management, which is now also the brand authority. The aims of this territorial brand are very ambitious and are directed at creating value through the promotion of the environmental and territorial characteristics of the area. Specifically, they relate to: improving the environmental performance of territory, appreciating those producers and service providers whose activities have positive effects on the environment; making visible the environmental contribution of producers and the services provided; promoting the sustainable development of certain economic sectors; promoting the use of low environmental impact techniques; helping to improve the local population's quality of life; diffusing and increasing the reputation of the park territory in the market and among consumers.

[1] For example, a commercial wine derived from a fermentation process developed in the sea. This is an original technique practised only by a wine maker located close to the Porto Conte Regional Park.

Since its institution, a few enterprises have introduced the territorial brand. These operate in the field of agro-food, handcrafts, cosmetics and pharmaceutics and tourism services. Concerning the agro-food sector, seven farming and processing enterprises have adopted this territorial brand: three wine processors, two olive oil processors and two farms that produce foods and other connected services, such as agro-tourism.

Generally, food and agro-food service producers should subscribe to a quality control protocol to avail themselves of the Porto Conte Regional Park brand; this contains a set of rules focused on strengthening the environmental value of production, as well as establishing minimal quality production requirements. As it stands today, a quality control protocol has been drawn up for olive oil, honey and wine production and processing.

An import point is the possibility of adopting the brand for a unique product or for a whole range. In any case, the typology of branding can be applied jointly to other brands or labels, such as geographical characteristics and traditional specialty labels (e.g. PDO, PGI and TSG for food) or organic labels. This implies that the authority intends the brand to be perceived as additive and not as an exclusive brand, thus allowing producers to expand their branding strategies.

In our case study, we focused on the three wine companies close to the park that had recently introduced the Porto Conte Regional Park brand. The wine sector was selected for several reasons. First, it represents the most important agro-food sector in the area from an economic perspective. Second, the wine firms played a crucial role in supporting the authority in the phase of elaborating the brand. Finally, wine is the agro-food product that clearly exhibits the inherent characteristics of the environmental features of the area.

The firms observed present notable differences in terms of size, marketing, organization and the motivations that led them to adopt the brand. Therefore, this allows us to provide evidence concerning different behavioural models on the part of wine companies in terms of value creation and governance. The main characteristics of the three wine companies (we use the generic terms Firm A, Firm B and Firm C) are summarized in Table 11.1 and elaborated thereafter.

Firm A is a large private company, one of the market leaders in the Italian wine sector. It is an old firm, founded in 1899. Since 2012, Firm A has been a subsidiary of the Campari Group, which holds 100% of its shares.[2] The company appears vertically integrated as more than 550 ha of grape cultivation close to the park area is owned by the firm. The strong integration between grape growing and wine processing allows Firm A to internalize many production functions, specifically those related

[2] The Campari Group is today the sixth largest player worldwide in the premium spirits industry and spans three main segments: spirits, wines and soft drinks.

Table 11.1. Main general characteristics of the wine companies analysed.

Characteristic	Firm A	Firm B	Firm C
Legal status	private	cooperative	private
Ownership	joint-stock company	cooperative of grape growers	wine maker
Size	large (market leader)	medium	very small
Vertical integration	yes	partially	yes (not fully)
Size of market	global	regional/global	regional
Number of commercial products	41 (wines and grappas)	28 (wines and grappas)	4 wines

to relationships between suppliers of raw materials and processors. The grapes are partially supplied by growers throughout Sardinia. Firm A is the company that mostly processes grapes in the region, producing more than 40 types of wine and grappa and selling them in local, regional and global markets. The level of capital investment amounts to around €100 m and the firm's revenues in 2014 were more than €30 m.

Firm B is a cooperative that comprises more than 300 grape growers, mainly operating in the area of the case study. The cooperative produces around 30 commercial types of wine and grappa and provides services to members, such as technical services, marketing and farm accounting. Firm B not only supplies some commercial wines that date back to its foundation in 1959, but also periodically introduces some innovative and selected wines appreciated by consumers worldwide. Its reference markets are traditionally regional, but in recent years the management of the cooperative has explored new market opportunities, especially with respect to its more innovative wines. Capital investment and revenues amount to around €20 m and €10 m, respectively.

Firm C is a very small private firm, founded recently. It is a family-run company that fully devoted itself to grape growing early on. A few years ago, the owners decided to integrate grape production using 9 ha of land with wine making. Today, Firm C produces four commercial types of wine. Therefore, it is a vertically integrated firm, although some functions are acquired from the market because of the small-scale level of its activity. Its reference market is at the regional level. Its capital investment is around €150,000 and its revenues amounted to less than €100,000 in 2014.

The three companies have adopted the territorial brand since 2013, when the quality control protocol for wine was provided by the brand authority. Essentially, the Porto Conte Regional Park brand represents a very young opportunity for these firms, implying that many effects (e.g. commercial, income, export) cannot fully be valued due to the short time since its introduction. However, the effects on firm management patterns can be verified more concretely, even if the information collected might provide indications only from a short-term perspective.

Table 11.2. Level of adoption of the Porto Conte Regional Natural Park brand.

Products	Firm A	Firm B	Firm C
Number of commercial wines	3	1	all
Innovative product	no	yes	no

Thus, we evaluated the effects of brand implementation on firm management patterns in terms of expectations. We formulated a semi-structured questionnaire administered to each firm in July 2014. The interviews aimed to establish the motivations that led the firms to adopt the brand and their expectations concerning value creation, markets, governance and the quality of products and processes. The interviews allowed us to collect sufficient and appropriate information to evaluate the issues identified above.

11.3 Results and discussion

Different patterns were found among the firms regarding the level of application of the brand, i.e. the range of goods produced which could benefit from the brand (Table 11.2). Firm A applies the brand for only three commercial wines. All these wines represent traditional products in the companies' supply. This is because Firm A aims to understand the opportunities and risks of brand implementation before expanding its application to a wider range of goods, especially to wines recently introduced to the market.

In contrast, Firm B has chosen to apply the brand only to a unique wine that was first commercialized in 2015.[3] This indicates that Firm B has strongly linked its promotion of environmental and territorial issues to an innovative and highly specific product, rather than extending the brand to its traditional wines. Unlike Firm A, Firm B considers the territorial brand more suitable for new wines introduced to the global market than for products already supplied.

Vice versa, Firm C has adopted an integrated perspective, applying the brand to all four wines it produces. In this case, the small size of the company and the youth of its enterprise in wine making have enabled it easily to adopt the brand for its entire production without notable effects on its management and governance, at least according to the firm's perspective and expectations.

[3] This is the wine described in footnote 1. When undertaking our investigation, this wine was yet to be realized and bottles were lying under the surface of the sea for fermentation.

11.3.1 Motivation

Concerning the motivations that led the companies to adopt the territorial brand, we found similar patterns in the two largest, Firms A and B (Table 11.3). For both these firms, the Porto Conte Regional Natural Park brand is a commercial tool able to promote sustainability because it should transmit to consumers the idea of environmentally friendly wines and recall territorial amenities related to the park. At the same time, the adoption of the brand should create shareholder value in the supply chain, especially in the case of the cooperative, Firm B, which is strongly based on the relationship and trust between management and wine growers and relationships with territorial food and tourism operators. Concerning the latter issue, we found that these large wine makers believed that adopting the territorial brand could facilitate the creation of networks between local enterprises based on horizontal cooperation and functionalities to reduce some vertical transaction costs (e.g. the provision of raw materials).

However, we found differences between Firm A and Firm B in terms of their perceptions concerning the role of the brand in generating higher commercial value. Firm A, which produces a wider range of wines and generates greater economic value, considered that adopting the brand could contribute little to determining a 'premium price', whereas Firm B, which has applied the brand to an innovative product, was of the view that more favourable market conditions could be created by adopting a territorial brand for selected top wines.

In contrast, Firm C was driven by a different motivation in choosing to introduce the territorial brand for all its products. This smaller wine company considered that the brand would result in significant premium prices for its wines and it would be able to improve its relationships with local operators, with anticipated benefits in terms of minimizing its transaction costs. The promotion of sustainability and environmentally friendly goods is not a priority for Firm C. This is important because it suggests that the adoption of the brand is exclusively associated with creating marketable value and increasing collaboration with other economic operators, especially to obtain advantages in reducing transaction costs.

Table 11.3. Main motivations leading firms to adopt the territorial brand.

Motivation	Firm A	Firm B	Firm C
Increase in commercial value	yes (limited)	yes (strong)	yes (strong)
Creation of shareholder value	yes	yes	yes (limited)
Promotion of sustainability	yes	yes	no
Relationship among local enterprises	yes	yes	yes (strong)

11.3.2 Market expectations

The adoption of the territorial brand was considered to have the potential to contribute to expanding the market size for each wine producer, with the level of expectations tending to be inversely proportional to the scale of the firm (Table 11.4).

It seems that smaller firms might primarily trust the Porto Conte Regional Natural Park brand as useful tool for expanding market possibilities and thus profitability. It is important to note that the brand is also viewed as an opportunity for selling wines not only in the regional market, but also national and world markets. Finally, all firms considered that their reputation would be enhanced after brand introduction.

11.3.3 Expectations concerning governance

We found that the governance solutions adopted allowed larger companies to manage the territorial brand efficiently (Table 11.5). Indeed, both Firms A and B considered that significant effects would be generated in management because of the introduction of the brand. On the other hand, these firms adopted the brand for a limited number of products and no other remarkable effects were expected.

Table 11.4. Expected effects on market dimensions.

Effect	Firm A	Firm B	Firm C
Market size	enlarged (+)	enlarged (++)	enlarged (+++)
Extra-regional market	yes	yes	yes
Firm reputation	enhanced	enhanced	enhanced

Table 11.5. Expected effects on governance.

Effect	Firm A	Firm B	Firm C
Conflicts among stakeholders	large vs small firms	large vs small firms	large vs small firms
Management	no significant effects	no significant effects	logistics quality requirements input provision
Line of product(s) organization	no significant effects	limited effects	to be verified
Complexity of transactions	no significant effects	no significant effects	with providers with purchasers

In contrast, Firm C took the view that some adjustments were needed in terms of logistics, input provision and quality dimensions. More specifically, Firm C adopted the brand with the aim of increasing the firm size and/or operational scale, thus making it possible to manage the expected increase in demand. Furthermore, the intention to adopt the brand was based on gains in bargaining power with input providers, especially concerning the purchase of technical inputs and extension services. The producer expressed a preference for outsourcing the provision of some technical inputs to a 'supervisory' entity able to coordinate the activities of the firm. The park authority was considered the most effective institution in playing this role. Finally, Firm C considered that handling the brand would confer improvements in quality requirements, determining a general cost increase. More generally, this means that governance changes would be necessary in the small firm to exploit the brand efficiently and minimize transaction costs both downstream and upstream.

However, all else being considered, wine producers agreed that conflicts among stakeholders could occur in the future due to adoption of the Porto Conte Regional Natural Park brand. In the case of inefficacy in terms of oversight by part of the authority disciplining the brand, there is a risk of exacerbating the differences between large and small wine makers with regard to market opportunities and costs. Therefore, each firm wishes to strengthen the role of the authority in enhancing brand use.

Hence the empirical results suggest that brand adoption may be more suitable for wine enterprises with hierarchical governance, rather than those characterized by different market transactions. According to Gereffi *et al.* (2005), this implies that a hybrid form of governance involving multiple actors might be a more appropriate solution for small firms in promoting brand adoption and exploiting the related environmental, territorial and quality attributes. This is because the complexity of transactions was high in the small wine company observed and therefore the optimal option might be to implement a hybrid form of management model. In this case, the Natural Park Authority is an institutional entity that can directly manage some functions, thus relieving the small firm of certain costs (e.g. marketing, communication, provision of some technical inputs required by the brand).

11.3.4 Expectations concerning value creation

The Porto Conte Regional Natural Park brand was found to be able to create firm and (new) shareholder value (Table 11.6). However, only the larger firms had a clear idea of the type of value that could be created. In terms of the motivations that led them to adopt this territorial brand, they considered it might serve to intensify their relationships with local enterprises and promote sustainability. The brand has a strong common identity, based on its highly specific territory, with many environmental, cultural and heritage amenities and being in the location of a natural park.

Therefore, the inherent nature of the brand allows local stakeholders to promote their territorial identity with a specific focus on environmental issues. This sort

Table 11.6. Expected effects concerning creation of value.

Effect	Firm A	Firm B	Firm C
Creation of value	higher	higher	to be verified
Creation of shareholder value	higher	higher	to be verified
Description	promoting relationships between firms	promoting relationships between firms	uncertain possibility of creating value
	promoting sustainability	promoting sustainability	territorial brand as strategic for firm development
	only part of the intangible value of a commercial brand		

of promotion represents in and of itself a new economic value created due to the institution and introduction of the brand. From this perspective, the adoption of this territorial brand leads to an improvement in the quality of relationships between producers, intermediaries and consumers, by enhancing a common and well-defined identity. This suggests that the motivations inspiring the Natural Park Authority to institute a territorial brand have found validation in firms' expectations as the brand is viewed as a tool for creating value by several large wine makers.

However, it must be highlighted that Firm A considered this specific brand to be only part of the intangible value of a commercial brand and more generally that local agro-food firms should promote the geographical and territorial characteristics of the area considered irrespective of the introduction of this specific brand.

Regarding Firm C, the possibility of creating shareholder value is far from being a concrete opportunity. From its perspective, the current benefits related to brand adoption are generally lower than the burdens it imposes. Specifically, the adoption of the brand has resulted in an increase in bureaucracy without producing benefits in terms of commercial and transaction costs.

In light of this result, the risk is that compliance with the standards required could be a constraint for smaller firms. This is a paradox when one considers the goals of the initiative promoted by the Porto Conte Regional Natural Park and by the stakeholders. However, Firm C is convinced that brand adoption can provide an opportunity for firm development in the medium- to long-term perspective.

In summary, the firms' motivations and expectations related to adoption of the territorial brand and constraints in terms of governance varied significantly among the wine companies examined in this case study. However, such contrasting perceptions based on firm characteristics have recently been found in the literature on the introduction of voluntary and umbrella territorial and/or environmentally friendly brands in the agro-food sector (Higgins *et al.*, 2008; Spielmann and Williams, 2016).

Some studies on this theme have specifically concerned the wine sector (Aranda *et al.*, 2015; Charters *et al.*, 2013).

In essence, the findings of various studies have highlighted that the perceived value linked to the adoption of a territorial brand varies among stakeholders. Some operators report that new value derives from improving the inherent quality of their own products, while others are of the opinion that value is related to the extrinsic qualities of their products (e.g. image) and yet others identify the creation of value with the expansion effects for the territory as a whole. This presumes the presence of potential or real conflicts – also related to different interests – that need to be managed opportunely.

Although it is not possible to generalize our findings, our analysis shows results similar to those found in other research. In particular, the creation of value derived from the introduction of the Porto Conte Regional Natural Park brand is not uniformly perceived by the wine producers observed, implying that opposing expectations can be generated and adequate management should be promoted to minimize conflicts.

11.4 Conclusions

The purpose of the paper was to verify how the introduction of a territorial brand, the Porto Conte Regional Natural Park brand, might generate changes in local agro-food firms and particularly their governance patterns. By analysing the perspectives of three wine producers located close to the park that have adopted the brand, we found that expectations concerning governance changes and value creation vary according to firm size, structure and organizational structure. This suggests that a common governance model cannot be determined; the most appropriate form depends on firm characteristics and small firms are expected to greatly adjust their governance and management due to brand introduction.

As these findings are based on perspectives – the brand was created very recently and only introduced in 2013 – they represent preliminary results that need to be tested in future research when the brand has been applied at operating speed. However, this research highlights that governance might be a critical issue in the case of small firms when a territorial brand is implemented. In large firms, governance and patterns might vary depending on the range of goods benefiting from the territorial brand, but this is a distant prospect as things stand today, at least concerning the wine makers considered in this study.

At the same time, firms' capabilities in terms of creating long-term firm and shareholder value derives not only from ownership of the territorial brand, but also from having the ability to deliver consistently on environmental, territorial and sustainability issues that are inherently related to this sort of brand. In other words, the territorial brand associated with a natural park is not sufficient in and of itself to create new value; rather, firms must be conscious that addressing environmental and sustainability issues through effective governance can be efficiently promoted by applying the brand.

Our findings also suggest that value creation related to a natural park brand can be enabled by managing a virtuous cycle of stakeholder expectations to overcome the organizational and structural burdens present in smaller firms. From this perspective, the authority called on to safeguard and enhance the brand can play a relevant role in ensuring firms use the brand efficiently to allow them to create shareholder value and at the same time minimize possible shocks in governance based on adoption by only some local enterprises.

References

Aranda, E., Gómez, M. and Molina, A., 2015. Consumers' brand images of wines: differences between two leading Spanish denominations of origin. British Food Journal 17: 2057-2071.

Charters, S., Menival, D., Senaux, B. and Serdukov, S., 2013. Value in the territorial brand: the case of champagne. British Food Journal 115: 1505-1517.

Charters, S., Mitchell, R. and Menival, D., 2011. The territorial brand in wine, paper presented at the 6th AWBR InterRegional Conference. June 9-10, 2011. Bordeaux, France.

Dettori, S., 2013. Sardinia. In: Agnoletti, M. (ed.) Italian historical rural landscapes, environmental history. Vol.1. Springer, Dordrecht, the Netherlands.

Gereffi, G., Humphrey, G. and Sturgeon, T., 2005. The governance of global value chains. Review of International Political Economy 12(1): 78-104.

Helm, C. and Jones, R., 2010. Extending the value chain – a conceptual framework for managing the governance of co-created brand equity. Journal of Brand Management 17: 579-589.

Higgins, V., Dibden, J. and Cocklin, C., 2008. Building alternative agri-food networks: certification, embeddedness and agri-environmental governance. Journal of Rural Studies 24: 15-27.

Humphrey, J. and Schmitz, H., 2001. Governance in global value chains. Institute of Development Studies Bulletin 32(3).

Kavaratzis, M. and Hatch, M.J., 2013. The dynamics of place brands; an identity-based approach to place branding theory. Marketing Theory 13: 69-86.

Lindgreen, A. and Wynstra, F., 2005. Value in business markets: what do we know? Where are we going? Industrial Marketing Management 34: 732-748.

Lorenzini, E., Calzati, V. and Giudici, P., 2011. Territorial brands for tourism development: a statistical analysis on the Marche region. Annals of Tourism Research 38: 540-460.

Neto, P., 2007. Strategic planning of territorial image and attractability. In: Matias, A., Nijkamp, P. and Neto, P. (eds.) Advances in modern tourism research. Physica Verlag, New York, NY, USA.

Perrouty, J.P., d'Hauteville, F. and Lockshin, L., 2006. The influence of wine attributes on region of origin equity: an analysis of the mode rating effect of consumer's perceived expertise. Agribusiness 22: 323-341.

Raynaud, E., Sauvée, L. and Valceschini, E., 2009. Aligning branding strategies and governance of vertical transactions in agri-food chains. Industrial and Corporate Change 18(5): 835-868.

Spielmann, N. and Williams, C., 2016. It goes with the territory: communal leverage as a marketing resource. Journal of Business Research 69: 5636-5643.
Van Ittersum, K., Candel, M.J. and Meulenberg, T.J., 2003. The influence of the image of a product's region of origin on product evaluation. Journal of Business Research 56: 215-226.

12. Functioning of the dairy value chain in Slovenia

Š. Bojnec

University of Primorska, Titov trg 4, 6000 Koper – Capodistria, Slovenia; stefan.bojnec@fm-kp.si

Abstract

This paper provides an overview of the functioning and patterns in development of the dairy value chain in Slovenia. The focus is on horizontal and vertical market integration, the regulatory framework, government policy, dairy value chain business practices and dairy value chain market structures. The overview shows a considerable restructuring of the dairy value chain with the exit of smaller farms, an increase in average farm size and yields, the establishment of contractual relations of farms with producer associations or cooperatives and domestic and foreign dairies, a transformation in milk processing, and the entry of foreign retail chains. Institutional innovations include the setting up of e-cooperatives that have low operational costs and reduce information costs with aim to improve the bargaining position in horizontal and vertical dairy chain relations.

Keywords: dairy value chain, market structure, institutional and organizational innovation, Slovenia

12.1 Introduction

The dairy chain in Slovenia plays a crucial economic role in Slovenia considering its value of production and employment. Its importance is recognised at different value chain levels: at farm-gate, milk processing and at retail level.

The Milk Board was important in setting the milk price and milk purchasing conditions after the Slovenian independence at the beginning of the 1990s. It includes representatives of government, farmers, dairies, and trade. In the early days of the evolution of the dairy value chain, the main focus of the Milk Board was on setting a basic price of milk, while the actual market price of milk was driven by other criteria, particularly milk quality. The Milk Board lost its importance with the macroeconomic and agricultural policy reforms because relations between farmers-farmer associations/cooperatives-dairies have increasingly become an issue of contracts and business relations. The milk price has largely become driven by milk quality and the stable supply of milk. Therefore, the actual farm-gate price depended on quality and purchasing conditions already before the Slovenian entry in the European Union (EU).

The entry of Slovenia in the EU has brought adjustments to the dairy value chain due to the Common Agricultural Policy (CAP) and the single EU market with its regulations on market functioning and competition, including regulations for quality and other standards. Market arrangements in the dairy value chain between farmers, producer associations/cooperatives, dairies and retailers have become more market driven with greater importance of economies of scale and milk quality. The entry of new retailers has caused a greater role for intra-industry trade with simultaneous export and import of similar dairy products in order to use economies of scale in production and to match demands by consumers for different milk and dairy products.

There have been two important changes at the farm level: first, exit of small scale dairy farms, and second, growth of surviving dairy farms with a greater level of commercialization. Changes have also occurred at the processing level, where the number of dairies has declined. Main changes at the retail level have been the entry of foreign supermarkets and hypermarkets, which has increased competition in local markets and has diversified products for consumers.

A new challenge for dairy farmers and the dairy value chain is the abolishment of the milk quota on 1 April 2015, which will lead to new arrangements and adjustments in the functioning of the dairy value chain. The rest of the paper is organized in five parts: a brief review of previous literature on Slovenian agri-food value chains, the regulatory framework, government policy and dairy value chain business practices, dairy value chain market structures and conclusions.

12.2 Previous literature

Earlier studies on Slovenian agri-food marketing and value chains have focused on the econometric analysis of vertical market integration and competition in the pork and meat sectors (Bojnec and Peter, 2005), on monetary impacts and overshooting of agricultural prices (Bakucs *et al.*, 2012), on nonlinearities in apple price transmission (Hassouneh *et al.*, 2015a), on spatial wheat product market integration between the Slovenian and Hungarian economies (Bakucs *et al.*, 2015) and on modelling price transmission and volatility spill-over in the wheat value chain between the farm-gate and the consumer level (Hassouneh *et al.*, 2015b).

Another strand in the literature is based on case studies and interviews. Van Herck *et al.* (2013) investigated regulations and business practices in the dairy value chain in Hungary, Slovakia and Slovenia. The focus of investigation was on how price transmission occurs in the presence of imperfect markets and institutional constrains. Other attempts to assess food industry competitiveness and government policy based on interviews for a larger number of countries, including the Slovenian dairy chain include a study by Stojanovic *et al.* (2014).

Despite the great importance of the dairy value chain in Slovenian agriculture and rural development, there are few studies that analyse this sector in detail. Volk *et al.* (2014) present some evidence on the dairy sector in Slovenia in terms of market balance and their evolution. Bojnec and Fertő (2008, 2014) analysed the competitiveness of dairy products in Slovenia in comparison with other neighbouring countries and EU-27 member states. The results suggest a potential for improved competitiveness, which should not be focused on raw milk and its exports, but on higher value-added processed products. However, this requires competitive milk processing and international dairy products marketing.

12.3 Regulatory framework

The regulatory framework and the quality of institutions can contribute to the transparency of transactions, which can in turn reduce imperfect and incomplete information and the uncertainty in exchange in business transactions (e.g. North, 1990). Moreover, the quality of institutions can have an influence on the enforceability of contracts, reduce transaction costs in exchange and encourage growth (e.g. Van Herck *et al.*, 2012).

In Slovenia, increased quality standards for purchased milk were introduced at the beginning of the 1990s. This has increased the demanded quality of purchased milk and linked the price of milk to the quality of milk and a stable milk delivery. With the entry of Slovenia in the EU further adjustments have occurred as a result of regulations on market functioning and competition in the single EU market and on higher quality standards.

Changes in market arrangements in the dairy value chain from farmers, milk processors to retailers have been caused by the entry of retailers, particularly foreign retail chains, which have created a greater role for intra-industry trade with simultaneous export and import in similar milk and dairy products.

Van Herck *et al.* (2013) investigated entry regulations, price controls and regulation of promotional activities, operational restrictions and the competition regulatory framework. Entry procedures for the establishment of a new company depend on the size of the farm/company and registration in the commercial register is required. It is mandatory to comply and fulfil different environmental requirements and hazard analysis and critical control point (HACCP) regulations at the local level. In general, HACCP is as a system that helps dairy business operators with the management of how to handle milk and dairy products and introduces procedures to make sure that dairy products are safe for consumers. There are no price controls in dairy value chain, while regulation of promotional activities exists from time to time. The latter can relate to domestic dairy market actions and to export promotion activities. The operational restrictions are largely related to the opening hours in the retail sector, and are less applicable to other stages and/or activities in dairy value chain.

Similarly as in other EU countries, the Slovenia Competition Office is responsible for the application of competition rules and competition policy. They include compliance with antitrust regulation, merger regulation and cartel agreements. Relevant to competition policy in the dairy value chain is the calculation of different concentration ratio measures in milk processing and in the retail sector. At the farm level, such concentration measures are of less relevance as there are still more than 5,000 farms, which daily supply milk to milk processors in Slovenia and for exports abroad (particularly to Italy).

12.4 Government policy and dairy value chain business practices

The CAP reforms in the EU have shifted government transfers from market-price support to direct payments and rural development payments. The reform of the CAP in 2007 has transformed subsidy payments into payment entitlements to milk producers. The majority (80%) of premiums for milk claimed by farmers before the CAP reform in 2007 was transformed into an individual (historical) addition to payment entitlements, while the remainder was included in the regional single payment per hectare that is different for arable land and permanent grassland. In 2010, the mid-term review of the CAP for Slovenia has introduced two new measures. The first is aimed at mountain and steep slope farms producing milk, and the support is bound to the annual individual reference quantity of milk of the individual farmer. The amount of payment per farm depends on the sum of all quantities of farmers who claim these payments. In addition, the holdings with milk cows, like other farms with cattle, sheep, goats, and horses, could also claim support to keep livestock on grassland. This support takes the form of an addition to the rights of the individual holding. In 2013 direct payments represented an important measure to supplement the income of cow farmers. These different support measures should assure more sustainable agricultural and dairy sector development in terms of economic, social and environmental sustainability. Similar to other livestock activities cow farmers are also entitled to payment for protecting livestock against disease as a percentage of the insurance premium, and they may exercise the entitlement to a refund of part of the excise duty on fuel used in agriculture. Farmers are also eligible to apply for European regional, cohesion and rural development funds for the restructuring of the dairy sector.

During the last two decades the dairy value chain in Slovenia has changed rapidly due to the exiting from business of small scale farms and the concentration of surviving dairy farms. A greater farm concentration with fewer dairy farmers, yield increases per cow and a larger milk production are the outcome of dairy farm restructuring (Volk *et al.*, 2014). Growth in dairy farm size and cow productivity and a greater commercialization of milk production with sales of milk from farms have contributed to different dairy value chain business practices (Blatnik and Bojnec, 2015). Stricter requirements for high milk quality and stable milk delivery have been introduced at the farm level. This has led to competition between dairy farmers and unwillingness for cooperation. Input costs increases have further increased pressures to reduce costs, while there has been some decline in farm-gate milk price, which has been consistent with global and regional developments in the milk market. Consequently, this has led to a decline in dairy farm profitability, lack of finance and liquidity problems with implications for dairy farm restructuring.

Vertical coordination and restructuring of dairy value chains in Central and Eastern Europe have been investigated in the literature (e.g. Dries *et al.*, 2009). In addition to econometric studies, which have investigated vertical and horizontal as well as spatial agri-food market and price integration, Van Herck *et al.* (2013) investigated the dairy value chain based on the degree of competition. This study distinguished between horizontal interactions and agreements between actors at the same level of the dairy value chain and vertical interactions and contract types between actors at different levels of the dairy value chain, including farm/service cooperatives, milk processors and retailers.

Regarding the presence of horizontal organizations in the dairy value chain, different farm associations exist in Slovenia and around half of the dairy farmers are a member of a farm association for which membership is not compulsory. Farm associations are used for lobbying the government, the provision of extension services, training and seminars, and the preparation and conducting of public tenders. At the milk processing level, the Dairy Association of Slovenia is a milk processor association and most of the larger dairies are members of this association. The Dairy Association of Slovenia is lobbying the government and assisting the government in the preparation of legislation. There is also a retail association, which is not dairy specific.

Collective action occurs only at the farm level. Activities are related to the selling of milk, where producer organizations engage in joint selling activities and joint buying of inputs, mainly for fertilizers, but the latter is becoming less important. The imposition of supply restrictions for members of a producer organization in case of oversupply or financial difficulties is very rare.

Vertical value chain relations more often rely on marketing contracts than on production contracts. Written marketing contracts are common between milk producers (farmers), farmer associations/cooperatives and milk processors for the duration of at least one year. The contract specifies milk sale conditions particularly on price, milk quality and delivery terms (e.g. timing, temperature and packing). The importance of long-term written contracts has increased. The milk price between the farmer and the processing company or producer organization/service cooperative is renegotiated on a monthly basis and largely depends on the milk quality delivered by the farmer. While contracting may contribute to efficiency gains, long-term contracts may also lead to a collusion of interests between different partners in the dairy value chain. For example, in some regions of Slovenia, farmers or producer organizations/service cooperatives have implemented selective milk sale agreements between domestic or local milk processors and foreign dairies.

There is a need for a greater horizontal cooperation between farmers and their organizations, as well as for their vertical integration in the dairy value chain. In order to reduce asymmetric information and transaction costs, new initiatives have occurred in business practices. For example, an innovative institutional and organizational development regarding the use of advanced information and communication technologies is the setting up of e-cooperatives for better cooperation and exchanges of information between farmers. Namely, cooperation of farmers within such cooperatives can assure greater farmer participation in increasing the quantity of milk supply, which can increase economies of scale and improve negotiation power on input and output (milk) markets resulting in better contractual arrangements and farm-gate milk price stabilization.

12.5 Dairy value chain market structures

Product and market differentiation and market segmentation have increased in the dairy value chain. Dairy production plays an important role in Slovenian agriculture and rural areas. The sector accounts for about 15% of the total value of agricultural production and about one-third of the value of livestock production (Volk *et al.*, 2014). Structural changes in milk production have been substantial since the entry of Slovenia in the EU in 2004. The number of farms rearing dairy cows has declined by almost two-thirds, while the number of milking cows fell by around a quarter. The average number of milk cows per holding has more than doubled to around 10 cows per farm. In 2010, only 15% of all agricultural holdings were engaged in cow milk production, while in 2000 this proportion was 33%. Interestingly, despite of the decline in the number of dairy farms and in the number of milk cows, yield per cow has increased and the level of milk production has not declined. The smaller number of farms is now producing a similar level of milk due to the increase in dairy farm size in terms of the number of cows per farm and the increase in the yield per cow. This pattern in restructuring can have positive economic effects due to the possible increase in international competitiveness as well as potential environmental effects due to the smaller number of cows, which produce similar levels of milk as in the past.

Larger milk processors in Slovenia have established a dairy association. The following Slovenian dairies are members of the Slovenian Dairy Association: Ljubljanske mlekarne, Mlekarna Celeia, Pomurske mlekarne, Mlekarna Planika, Mlekarna Škofja Loka and Mlekarna Krepko. They together employ almost a thousand people and buy more than 320 million litres of milk from almost 5,000 Slovenian farmers. They sell most milk on the domestic market and are regular suppliers of milk and dairy products to Slovenian schools and kindergartens. Some problems, underlined by the Slovenian Dairy Association, lie in the disconnectedness of Slovenian dairies and their associations. It is suggested to increase activities for purchasing home produced milk and for better sales of dairy products in Slovenia.

Milk production in Slovenia has exceeded domestic consumption. The rate of self-sufficiency is around 116%. The surplus of milk and dairy products is exported. The structure of milk and dairy product exports has changed due to milk processing restructuring and the entry of foreign retailers. On the one hand, raw milk from famers is exported because part of the raw milk is bought directly by foreign dairies (mostly Italian and a small amount also by Croatian). Since the entry of Slovenia in the EU, exports of raw milk have grown rapidly to more than 188 thousand tons in 2013, which is more than 36% of purchased milk from farmers in Slovenia. In spite of the fact that milk production has been rather stable and that dairy farm commercialization has increased, the purchase of milk by Slovenian dairies has declined (Table 12.1) due to increasing exports of raw milk directly from farms and/or their associations / service cooperatives. On the other hand, changing patterns in development can be seen in the export of dairy products as a result of milk processor specialization towards certain dairy products to utilize economies of scale and increased import competition of dairy products. Imports of dairy products have increased, which generate substitution effects in the demand for domestic dairy products. The increased intra-industry trade with simultaneous exports and imports of similar milk and dairy products is consistent with production and export specialization towards products with comparative advantages and demand for varieties by consumers.

Changes in farm-gate milk prices in Slovenia mainly follow the changes in the EU market. After strong fluctuations in farm-gate milk prices between 2007 and 2011, there has been a slight increase in farm-gate milk prices, which were slightly higher in the EU than in Slovenia. On the Italian market raw milk prices have increased less, but were still on average higher than in Slovenia.

Table 12.1. Purchase of milk in Slovenian dairies (Valjavec, 2014).

Year	Purchase of milk (in million litres)	Year	Purchase of milk (in million litres)
1992	365	2003	484
1993	346	2004	486
1994	375	2005	449
1995	388	2006	378
1996	383	2007	372
1997	387	2008	351
1998	420	2009	346
1999	435	2010	353
2000	446	2011	354
2001	459	2012	345
2002	474	2013	319

Slovenia as a small country and thus also relatively a smaller global milk and dairy product producer is a price taker in international trade. Main focus should be on milk quality as the driver of higher farm-gate milk prices and higher quality of dairy products. While abolishment of the milk quotas on 1 April 2015 has not had a substantial shock on the Slovenia milk and dairy market, in the long-term it is possible that a slight reduction in Slovenian milk production will occur, particularly in hilly and mountainous areas. Yet, also a decline in raw milk exports and in exports of dairy products is possible, if no further improvements in international competitiveness are achieved. Due to the nature of perishable raw milk and dairy products, most trade is expected to remain with countries in the region, particularly those, which are bordering Slovenia (Austria, Croatia, Hungary and Italy). Therefore, the role of proximity factors can have an important role in further milk and dairy sector developments.

12.6 Conclusions

Dairy production is the most important sector in Slovenian agriculture. The number of dairy farms has declined by almost two-thirds and the number of milk cows by a quarter compared to 2000, while the average number of milk cows per farm has doubled. The decline has been seen also in the number of milk processors. Milk production in Slovenia is greater than domestic consumption and more than one third of total milk purchase is bought directly by foreign dairies (mostly from Italy). Changes in farm-gate prices for milk mainly follow the changes in the EU market. Dairy farms receive also different subsidies from CAP direct payments, payments from the Rural Development Programme and other subsidies.

Considerable changes in the Slovenian dairy value chain have occurred since the entry in the EU. Main changes are related to the exit of small-scale farms, dairy farm concentration with growth of the surviving dairy farms and a greater commercialization. Milk production has stabilized and after the milk quota abolishment in April 2015 it is more likely that similar patterns in development will continue in the short to medium term. Slovenia is a price taker in regional and global trade. Therefore, focus is on milk quality as a driver of higher farm-gate milk prices to mitigate competition problems, which are caused by higher costs due to structural limitations and economy of scale problems, technology and innovation gaps. Selling of raw milk to Italy is more likely to continue due to proximity factors and a lack of competitiveness in higher value-added dairy products.

Challenges are to improve cooperation and contractual relations in the dairy value chain. Competition between dairy plants has targeted the domestic market and has been less focused on international competition problems caused by high production costs due to limited economies of scale, innovation problems and the need to create new added value products.

Acknowledgements

This publication was generated as part of the COMPETE Project, Grant Agreement No. 312029 (http://www.compete-project.eu), with the financial support from the European Community under the 7th Framework Programme.

References

Bakucs, L.Z., Bojnec, Š. and Fertő, I., 2012. Monetary impacts and overshooting of agricultural prices: evidence from Slovenia. Transformations in Business and Economics 11: 72-83.

Bakucs, Z., Bojnec, Š. and Fertő, I., 2015. Spatial product market integration between two small, open neighbouring economies. Agribusiness 31(2): 171-187.

Blatnik, P. and Bojnec, Š., 2015. Dairy market in Slovenia. COMPETE working paper. Available at: http://tinyurl.com/yaz8d3l6.

Bojnec, Š. and Fertő, I., 2008. Dairy trade competitiveness: crucial role of processed products. Milchwissenschaft 63(4): 434-438.

Bojnec, Š. and Fertő, I., 2014. Export competitiveness of dairy products on global markets: the case of the European Union countries. Journal of Dairy Science 97(10): 6151-6163.

Bojnec, Š. and Peter, G., 2005. Vertical market integration and competition: the meat sector in Slovenia. Agricultural and Food Science 14(3): 236-249.

Dries, L., Germenji, E., Noev, N. and Swinnen, J., 2009. Farmers, vertical coordination, and restricting of dairy supply chains in Central and Eastern Europe. World Development 37: 1742-1758.

Hassouneh, I., Serra, T. and Bojnec, Š., 2015a. Nonlinearities in the Slovenian apple price transmission. British Food Journal 117: 461-478.

Hassouneh, I., Serra, T., Bojnec, Š. and Gil, J., 2015b. Modeling price transmission and volatility spillover in the Slovenian wheat market. IAAE Congress, Milan, Italy.

North, D., 1990. Institutions, institutional change, and economic performance. Cambridge University Press, Cambridge, UK.

Stojanovic, Z., Ognjanov, G. and Filipovic, J., 2014. The EU food industry competitiveness and government policy: exploration of findings obtained by in-depth interviews. COMPETE meeting, Prague, Czech.

Van Herck, K., Noev, N. and Swinnen, J.F.M., 2012. Institutions, exchange and growth: evidence from Bulgarian agriculture on the impact of hold-ups and contract innovations. European Review of Agricultural Economics 39: 29-50.

Van Herck, K., Bojnec, S., Ferto, I. and Pokrivczak, J., 2013. Regulations and business practices in the dairy supply chain in Hungary, Slovakia and Slovenia. Appendix 1. Report prepared for the FP7 TRANSFOP project. Centre for European Policy Studies, Brussels, Belgium.

Valjavec, I., 2014. Purchase of milk, processing and exports in Slovenian dairies up to 2013. Dairy Association of Slovenia, Ljubljana, Slovenia.

Volk, T., 2014. Poročilo o stanju kmetijstva, živilstva, gozdarstva in ribištva v letu 2013 (Report on situation in agriculture, food, forestry and fisheries in 2013). Ministry of Agriculture and Food, Agricultural Institute of Slovenia, Ljubljana, Slovenia.

13. Agricultural cooperation and corporate social responsibility: alternative approaches in a literature review

D. Toccaceli

Department of Economics and Management, University of Florence, Via delle Pandette 9, 50127 Florence, Italy; daniela.toccaceli@unifi.it

Abstract

The paper attempts to shade a light on the relationship between corporate social responsibility and agricultural cooperation in the French institutional environment, through the case study of the French cooperative Fermes de Figeac. Why corporate social responsibility (CSR) could be a successful choice for a medium-size agricultural cooperative is the question viewed through alternative approaches whose I propose a literature review. The first one is the French mainstream literature to interpret what kind of cooperative Fermes de Figeac is. Some threats from CSR literature to highlight the possible advantages the coop could gain follow. Regional development and rural network creating in the proximity approach try to highlight the interaction with economic and social dynamics. The fourth approach try to set the problem in the New Institutional Economic standing to attempt organizational reasoning. I discuss the four approaches having regard to the peculiar and different aspects they allow to focalise.

Keywords: cooperative pattern, proximity, stakeholder, hybrids, transaction costs

DOI 10.3920/978-90-8686-844-5_13,

13.1 Introduction

The relationship between agricultural cooperation and corporate social responsibility (CSR) is not so obvious in the present changing scenario, even taking into account that the stated international values of the cooperation already include social responsibility.

About CSR there is a patchy corpus of theories relating to multiple approaches and purposes (Garriga *et al.*, 2011; Lockett *et al.*, 2006; Margolis and Walsh, 2003), whose analysis is beyond the ambition of this paper. Without tackling the hard task of providing a definition – Carroll (1999) supplies an interesting historical review of its evolution- it is relevant to remember that in such a broad class of actions classifiable as CSR, also those rising from sustainable development issue – especially the triple bottom line of economic, social and environmental sustainability – take place (Garriga *et al.*, 2011). They are those I deal with by focalising on the case of a French agricultural cooperative through alternative approaches, in a literature review.

The agricultural cooperation in France seems to be standing at a crossroads between to change the classical cooperative pattern for surviving in a more competitive global market, or maintain the same pattern surrendering to lower performance pathways.

The paper analyses the case of the cooperative Fermes de Figeac (FdF), in the south-west France, which is a medium-size cooperative adherent to the international cooperative values. The case is worthy of attention because the cooperative was achieving economic goals while accomplishing environmental and social sustainability standards. In addition, it was contributing to social initiatives according to its program of CSR actions. Furthermore, it was engaging in the development of the region together with local institutions and other enterprises.

Why CSR should be a profitable choice for an agricultural cooperative is the unavoidable question I tackle.

I propose a literature review of several alternative approaches able to deal with the question, as each ones highlight different mechanisms at work.

Section 13.2 sketches the case of the FdF cooperative in its evolutive path. Section 13.3 grounds on French mainstream literature to understand what kind of cooperative FdF is. Section 13. 4 is about CSR literature approaches that fit with the case. Section 13.5 takes into account the French proximity view. Section 13.6 attempts a New Institutional (NIE) approach by highlighting both the institutional environment and how it influences the special organizational arrangement that the cooperative is. In Section 13.7, I discuss how the alternative approaches focalise different aspects able to answer the main question and Section 13.8 proposes some final remarks.

13.2 The case study

13.2.1 The FdF Group

What nowadays is known as FdF '*Un coopérative au service des hommes et du territoire*' (a cooperative to serve the local community in the region) is the type of cooperative Society of Collective Agricultural Interest (SCAI, that in French is *SICA*, *Societé d'Intérêt Collectif Agricole*) operating in the areas of *Ségala* and *Limargue*, thus the name *SICASELI*. It is a cooperative group that has 650 adherent members, 96 direct employees – 160 if the 5 subsidiaries are taken in the account – and a turnover of seventeen millions of euro, only 55% of which comes from agricultural provision to the members, the rest from the other activities undertaken over the time. The turnover is increasing by 5% per year since 2011. In the cooperative group the subsidiaries are to provide services to farmers (*Agri Quercy* Service and *CUMA Lot Environment*), raw materials (*Delbard* and *Big MAT Lafargue*), and produce energy (*Ségala Agriculture et énergie solaire*, 120 farmers managing 60,000 m^2 of photovoltaic panels). The shared firms are *SARL Chaux de Montmurat* (providing materials to reduce soil acidity), *GIE de 2 Vallées* (squeegee for mais and cereals) and *La Quercynoise* (palmiped slaughterhouse).

Members are mostly stockbreeders to which FdF offers the equipment procurement and technical services. To this early core activity, over the thirty-year time others were added to enhance the value of food production in the local short chain, to supply of stores for the local community (food, gardening, bricolage, building raw materials and clothes) and to produce green energy.

13.2.2 The region where it operates

FdF was set up in 1985 having the form of a simple agricultural cooperative of the *Ségala-Limargue* area, in northeast of the Department of *Lot*, *Midi-Pyrénées* Region. It has identified its own territorial circumscription (the *ancrage territorial* (AT) imposed by the *Code Rural*) in the cantons[1] of *Figeac*, *Lacapelle-Marival*, *Latronquière*, *Souscyrac*, a rural region at the foot of the *Massif Central*.

Agriculture in the region is typical of medium mountain areas, with medium size farms (60-70 hectares) that are employing the 9% of active population, even though employees are reducing by 30% every ten-year time. Bovine breeding is dominant with 25,000 cows and 66 million litres of milk per year.

The region is coming from an history of marginality (Cariou, 2013) where agriculture has been the main activity for many time, and the most relevant agricultural activity is still the ovine and bovine breeding. While the region is still keeping the rural features, there is also a well-known industrial cluster of aero industry, stemmed from

[1] It refers to the administrative system in force before the reform applied in 2014.

the helicopter's inventor at the early twentieth century, thanks to the wide woods too. Moreover, in the last decades, tourist activity has been also emerging, because several cities and historical places are in the 'Road of Saint James of Compostela', one of the major Christian pilgrimage route. Such a particular development pattern has entailed a deep change in the workforce market so that employment has been decreasing in agriculture (Pham and Aubert, 2013a, 2014) causing a big problem the cooperative have been facing, early in the interest of own adherents, then looking at the territory's interest too.

13.2.3 The case history

The emerging of agriculture workers dropping was the beginning of something of new: a reflexion shared with the local stakeholders that pushed the cooperative to look forward. On this base, the general cooperative 'social responsibility' principle took a new and more specific meaning: self-commitment for sustainability of production system whose co-operators are actors; self-commitment for the local (social, environmental and economic) sustainable development.

The choice of social responsibility has played a key-role because has allowed the role of 'territory' changed the cooperative's vision and action regarding the local sustainable development. Its own involvement together with the local stakeholders followed.

The coop took a peculiar path leading to focalise caring for the 'territory' (let me use the jargon word in the 'social building construction' sense whose I refer to in Section 13.5) such as a basic component of the value creating process.

To simplify, I present the path in three phases.

- In the first phase, the leading idea was the commitment to improve technical services offered to the adherent members, as well as for the community rural development. From 1985 to 2002, CASELI aimed to enhance the typologies of services offered to the members, mostly by improving technical services and the access to new technologies at the less cost. In 1989, for example, it took an action for improving the acid soil in *Ségala* region. Anyway, from the beginning CASELI was deeming itself a 'cooperative of the territory' and feeling the need to enhance the services both for the cooperative members and for all others in the region. As far as, in 1991, it adopted the new statute of SCAI, taking the name SICASELI. Consequently, the opening to the territory became possible, relationships with stakeholders were intertwined and major actions were undertaken (Cariou *et al.*, 2006). In 1994, it started up a point of sale of regional products in the '*Gamm vert de Figeac*'. In 1995, a new plant for transforming the agricultural products of the territory became operative. The awareness of the sustainability of development was rising, so that in 2002 SICASELI promoted the setup of the cooperative of services

CUMA[2] Lot Environment. This new body was also able to provide technological innovative tools and services for everyone in the region.

- In the second phase, from 2003 to 2010, FdF has been engaging in social responsibility actions for the community and for facing the challenge of sustainability development. In 2003, SICASELI adopted the *Bilan Societal* (BS) as the specific instrument to manage its own engagement for the community sustainable development. The BS is a peculiar tool provided by the Centre of Young Directors and Players of the Social Economy to the firms of the social economy in France. It is a questionnaire conceived for both evaluating corporate responsibility in terms of social and environmental development, and managing the process to achieve the goals (Bodet and Picard, 2006; Cariou, 2003; Cariou *et al.*, 2006; Heslouin, 2011; Mangin, 2001; Persais, 2004). It answered also to the need to distinguish the firms of the social economy from the ones of other sectors equally engaging in CSR. Coop de France has adapted the BS to the agricultural cooperatives (Mangin, 2001). Several aims moved that choice: valorising the externalities produced by the agricultural cooperatives on terms of economic, social and environmental development; reinforcing the AT; promoting the participative approach to open the agricultural cooperatives to the territory. It serves as a tool to manage the process of stakeholder involvement. So reshaped, BS consisted in a guideline to which a public project -the program Equal, financed by the European Social Fund – has been adding a consultant service to favour the early implementation of that device by some innovative agricultural cooperatives. SICASELI has been one of the few in that group (Cariou, 2003; Cariou *et al.*, 2006). The BS applied by SICASELI has relied upon a survey based on 450 indicators and conducted with a participative approach involving the main stakeholders and local players upon five domain: internal democracy, human promotion, solidarity, economic action, and relations with the territory. The board of directors chose to engage its members to conduct the interviews, with the assistance of a consultant (Cariou, 2003). The findings of the BS survey allowed SICASELI to identify a complex action plan with new goals in each relevant area: internal governance, economic performance, environment and social sustainability, rural network creating. That was the start of a strategic plan committed to the 'stakeholder management', declined in agricultural and cooperative stance (Cariou *et al.*, 2006).The pathway has taken multiple integrated directions. The one for pursuing a higher agricultural value-added. So that, in 2003, SICACASELI gained also the certification ISO 9001 for the quality of services management and opened the first butcher shops of the local ovine breeding. The purpose was to add new value to the local short-chain, grounding in an enhanced value of the territory (Mollard and Pecqueur, 2007; Perrier-Cornet *et al.*, 2000) and of the social value of the producers (Marotta and Nazzaro, 2012). Another one for environmental sustainability, so that it promoted the project of a photovoltaic park, realised in 2010 and able to provide renewable energy for the need of the entire region. In 2009, it gained the certification ISO

[2] CUMA, acronym of *Coopérative d'Utilisation des Materiels Agricoles* that is a kind of cooperative for commonly using agricultural material.

14000. Another one for supporting rural population, so it arranged an agreement with the municipality of *Felzin* for keeping in that area the store and the mail service. The cooperative representatives have affirmed that they have realised in that period 'the utopia of a territorial republic' re-elaborating in a territorial key the Draperi's concept of 'cooperative-republic' (Draperi, 2012; Draperi and Corroller, 2015) where, instead, the sole cooperative is central (Cariou, 2013). In the period, it focused also on seeking the answer to the previously mentioned problem of reduction of agricultural activities and workforce availability. That was raising a new way – the GTEC process[3] (Pham and Aubert, 2014; Sol et Civilisation, 2010; Toccaceli, 2016) – since that effort was possible just by working together with other players of the territory.

- Relying on such an enhanced system of territorial relationships, in the third period, the leading idea evolved in that of becoming an 'innovative territory' able to project and collectively act for renewing futures perspectives. In 2010, the new cooperative project was grounding on the idea of enlarging agricultural activity up to include each aspect of the rural life that resulted in the purpose to make agriculture *gestionnaire du vivant*. It happened in parallel with a collective initiative to develop the first community reflexion for the expected future promoted by the *Pays de Figeac* (Pham and Aubert, 2013a, 2014). That was the push for the jump into the more recent phase, when SICASELI changed in FdF with the purpose of keeping agricultural activity in the territory and face the loss of skilled workers (Thomas, 2009). The last evolution points out the cooperative's convincement to be on the right way, as far as it decided to present its own candidature for ISO 26000 certification for the social responsibility, which was gained in 2015.

13.3 What kind of cooperative Fermes de Figeac is

In France, cooperation has given a connective structure to agricultural organizational system, so that nowadays three farmers on four are represented by a cooperative (Frey, 2013) and the World Cooperative Forum lists six French cooperatives among the first twenty-five in the world, that have a highest level of turnover (INEA, 2014).

French mainstream cooperative literature provides many contributions to understand what kind of cooperative FdF is, and how the multi-annual commitment for local community sustainable development takes sense. Scholars' contributions take pregnant sense in reference to the evolution of the French law.

A peculiar French rule that has conditioned the cooperative evolution is the AT that is regional or 'territorial' – let me use again the jargon term herein and in the follow-

[3] *Gestion territorial des emplois et des compétences (GTEC)*, is a tool established by the French Ministry of Labour to maintain skilled labour forces in rural areas instead of leaving for the urban labour market. Implemented by the *Pays de Figeac* for the territorial forward planning of skills and employment, it consists in a social process, of which FdF has been a key-actor. The case of Fermes de Figeac has resulted one of the few having implemented both the *Bilan Societal* and the *GTEC* process.

anchorage. Stated by the Code Rural, it is the juridical obligation to freely choice a bounded area where the cooperative can operate. AT contributes to limit competition among cooperatives in the same field and area, for reinforcing them. It allows also supporting competitiveness of cooperatives in other various way, and preserving the root in the local dimension (Chomel, 2013; Filippi, 2013b). Filippi (2013a) argues that AT refers to that complex and relational sense, which the dynamic of proximity approach (Colletis *et al.*, 2005; Filippi *et al.*, 2008; Pecqueur and Zimmerman, 2004; Rallet and Torre, 2004; Torre and Beuret, 2012; Torre and Rallet, 2005) gives to the concept of territory, to be intended as a collective construction.

From a juridical point of view, the link between territory and cooperatives emerges also in the more recent cooperative statute, the SCAI. It is a cooperative oriented to manage plant of transformation and offer services for farms and rural population, which have to associate farmers and at least twenty per cent of non-farmers, as far as one may deem it a 'cooperative of the territory'. It is considered suitable for the multifunctional agriculture.

In the same direction, in 2001 was established another new pattern, the Cooperative Society for Collective Interest (SCIC). Even if it is not specific for the agriculture, it has resulted to be innovative and suitable for developing innovative solutions in rural milieu, dealing with multiple challenges, through projects having social, cultural or environmental aims (Cariou *et al.*, 2006; Draperi and Corroller, 2015; Thomas, 2009).

Albeit well-fitting for a classical cooperative, in the more recent time AT is thought of as a limit to the growth of the cooperative (Filippi, 2013a; Filippi *et al.*, 2013). French agricultural cooperatives can develop as business groups of cooperative (BCGs), a new pattern to adapt both to the global change and to the progressive revision of the European Common Agricultural Policy (Chomel, 2013; Filippi *et al.*, 2013). BCGs have a cooperative statute and can concentrate together cooperatives and other kind of firms to achieve larger dimension and conditions that are more competitive. BCGs are also present in the territory by their agencies that are not cooperatives. Developing several different patterns, they often lead to very complex structures of interlinked societies – cooperative and not, often including a financial holding – that make possible the process of concentration, vertical integration, internationalisation, and raising financial capital and partners (Koulytchizky and Mauget, 2003). As the new pattern was succeeding, it became clear that the classical cooperative were changing in a more hybrid structure, method and values as well (Filippi *et al.*, 2008, 2013).

Over the last decades, both tendencies – the one towards global and the other towards local dimension – have been working together so that many different types of cooperatives developed. However, BCGs have shown a so growing dynamic, especially after the French Laws of 1991 and 1992 (Filippi, 2013b), that seeking a new cooperative paradigm turned out to be necessary (Côté, 2000, 2005, 2007), but not easy.

Côté (2000) have represented the evolution of the cooperative pattern according to the intensity of cooperative principle, and the intensity of market values expressed by the cooperatives. Complementary, Koulytchizky and Mauget (2003) have deepened the question developing an ideal space in which all types of cooperative are placeable, taking into account the cooperative adherence to values of solidarity and position in the market. In a so defined space, (Figure 13.1) they consider as limits the path of utopia (exclusively adherence to solidarity values) and the path of banalisation (exclusively adherence to market values). Between the two, they place the virtuous area where the observable cooperatives are representable, as the small biological local, as the mega-cooperative nearer to a multinational company than to a classical cooperative in a rural region.

The banalisation problem represents the risk of cooperative-identity loss, due to the hybrid nature of the BCGs that put together cooperatives, industrial firms, financial partners, banks (Filippi, 2013b). Nonetheless, facing that risk and trying to answer the question whether the running evolution is so deep as to undermine the basic cooperative principle is worth. In fact, BCGs meet the need of adaptation to global and more competitive markets, while protective measures have no longer put in place by European agricultural policy. Therefore, several strategic reasons seems to be for continuing to create BCGs and consequent holding structures (Filippi *et al.*, 2013). It seems very much true after considering that the financial, juridical and of management improvements are typical points of weakness of the agricultural cooperative pattern.

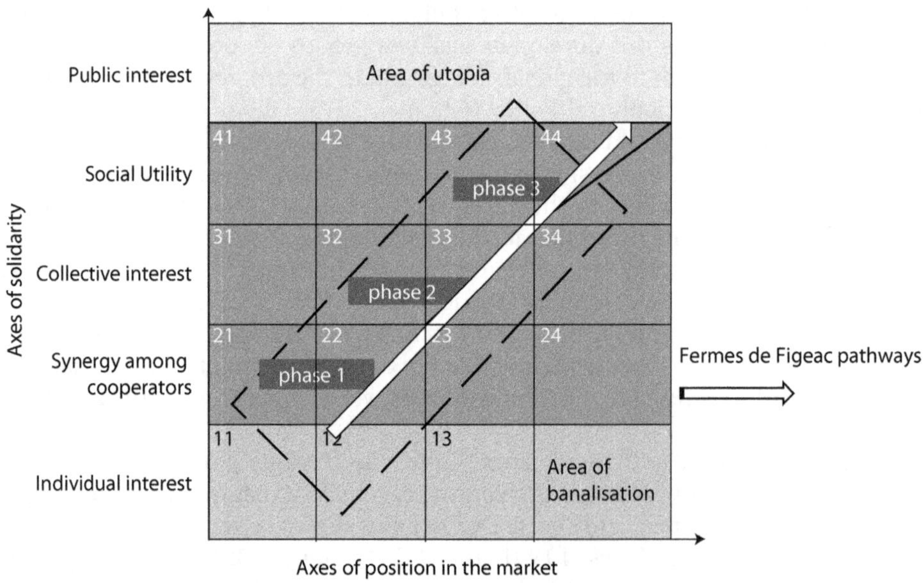

Figure 13.1. The pathway of Fermes de Figeac placed in Koulytchizcky-Mauget schema.

Contrariwise, in such a changing scenario, the case in point shows an alternative way to face those points of weakness and develops a different kind of Group, which is able to underpin cooperative growth and meet the new needs of rural areas with new arrangements (Toccaceli, 2016).

The above quoted Koulytchizky and Mauget schema (2003, p. 36) has been adapted to discuss the case in point. According to the scholars, the schema (Figure 13.1) allows to put a generic cooperative pathway in relation to both economic and solidarity performance, and to identify an area where virtuous evolutive path is possible. The area of virtuous performance is contained between the two extreme tensions: that towards purposes of general interest without economic aims, and that towards purpose of economic performance without solidarity.

In the virtuous area of this schema, we can place the pathway of FdF (the white arrow in Figure 13.1) because the aims of stimulating local sustainable development and growing economic performance have run together. The white arrow points out the three aforementioned phases. In the first – the square 21 –, technical support and synergy among co-operator have prevailed, enhancing the position in the market. Therefore, the line starts from the square 21 and runs diagonally through the 22. In the second, the interest for the local community arises and gives rise to new economic activities well performed. To represent it, the arrow moves through the square 33. It indicates the beginning of the interest for the environmental sustainability and for producing public goods through the new investments in renewable energies. For the third phase, the arrow moves yet diagonally through the square 44 to represent the adjunctive actions performed in favour of local community and social interest. Finally, one can observed that, remaining in its own perimeter, FdF has widely differentiated the economic activities by adding as affiliated new cooperatives or participating to others of interest. The difference respect to usual BCGs is that the main aim has been remaining to add value to the cooperative-mother – hence to the adherent members- instead of the group. In this new perspective, the economic and financial growth of the cooperative make a deeper sense.

13.4 Corporate social responsibility approaches

Although the focus is on the relationship between CSR and business performance, even in this narrower field there are numerous theoretical streams, wherefrom I withdraw the ones closer to explain the problem at stake. In lacking of a general theory, it need arrange together the pieces of the mosaic to reach the purpose of understanding the mechanisms at work. For the case in point, it need arrange together three pieces to fit the relevant characteristics. The first is that maximizing shareholders value fits the need to reach financial benefits to coop-adherents, so that they become available to support the innovation process and the related investments. The second is that maximizing competitive advantages fits the need to interplay in a specific (local) competitive context. Finally, maximizing the social value produced within the firm's value-chains fits the case of multifunctional agriculture.

13.4.1 The first piece of the mosaic

It aims to clarify why firms can reach financial benefits through several kinds of CSR strategy. The common basic idea is that firms engage themselves in producing social benefits because it results also in firms' benefit. How it is possible is object of a number of theories. The most attuned for the case in point are the following. The first – perhaps one of the most famous – is the stakeholder theory based upon the Freeman's (1984) idea that taking care for the core interests of the stakeholders can lead to financial benefit. As this approach raised some criticisms, to reinforce it, Jones (1995) developed an ethical stakeholder theory based on the idea that if managers contracts with their stakeholders on the base of mutual trust and cooperation, they will have a competitive advantage, because an ethical behaviour can largely reduce transaction costs (TCs) borne from opportunism. Its approach aim at justifying the existence and the success of sincerely altruistic behaviour. That is also relevant because underpins the stakeholder theory with the transaction cost economics, as Barnett (2007) has observed. Yet referring to the stakeholder theory, Barnett (2007) has developed a new paradigm to explain the great variety of financial performance stemming from CSR, among firms and over time. He established the criteria to distinguish CSR – the orientation to social welfare and to stakeholder – from other corporate investments or direct influence tactics that aim at stakeholder influence without bring social welfare increasing, or vice versa. He underlines that the link between CSR actions and the financial return is indirect. Speculating on this link, he formulated the stakeholder influence capacity (SIC) paradigm as the ability of a firm to identify and profit from opportunities to improve stakeholder relationships through CSR. Hence, he argues, 'a firm should not engage in an act of CSR that is unlikely to offer a compensating increase in stakeholder favour or stakeholder influence capacity.' (p. 812). As this paradigm points out, that relation has a path-dependent nature, on whose base Barnett builds a contingent model of the business case for CSR. The path-dependent nature of stakeholder relationship means that a given investment in CSR may prompt different stakeholder's reactions and yield different financial results for different firms at different points in time. However, by this way he places the CSR beside other management issues – especially material investment – that are in the direct interest of the firm, not beyond.

The role of manager in this dynamic is crucial and is suspected to become easily opportunistic. Jensen (2002) criticises managerial approach and method to apply stakeholder theory and evaluate their own performance, deeming it somewhat opportunistic because based on manifold measure. By contrast, he attempts to clarify the relation between shareholders value maximization and stakeholder theory through the concept of 'enlightened value maximization'. The long-term value maximization is the one the manager ought to seek as firm's object, therefore is proposed as the true criterion to evaluate managers performance.

McWilliams and Siegel (2011) ground their cause-related marketing approach on an extension of the resource-based theory to CSR. They deem CSR a co-specialized asset such as to make other assets more valuable. The best example they refer to is firm's reputation for quality or for a particular brand of its product, in that able to enhance the value of the firm. Consequently they introduce the role of the media in enhancing (or damaging) the reputation of firms. They consider several ways to capture the social value created, all of which relies upon the reputation and advertising as key concept. Therefore, they argue that a good reputation can result in premium pricing or consumer loyalty, which increase revenue, and may decrease capital and personnel costs. The former may be lower if CSR actions dampen the subjective risk profile of the firm; the latter may be lower if employees draw up motivations from the firm's CSR actions. On this base, a firm can develop a strategic plan involving CSR as co-asset.

13.4.2 The second piece of the mosaic

It focuses on the importance of choosing strategically how to better the external environment or, as Porter and Kramer (2002, 2006) put it, the competitive context of a firm, or a group, in order to reinforce its own business. They adopt an integrated and strategic approach grounding on the same idea lying under the cluster concept, that of a deep and functional link between the firms and their competitive context. Therefore, they deem that corporations can focus on contributions able to enhance their competitive context, by this way increasing their competitive advantages (Porter, 1990). This interplay of actions – creating social value and improving its own performance –, in their opinion, can be felt as 'a sincere commitment to bettering society', so CSR actions result credible and can actually reinforce the firm's reputation. Porter and Van der Linde (1995) argue also that firms can enhance competitive advantages through implementing regulation to protect the environment, when they can pick up the opportunity to innovate technologies or make the use of inputs more effective.

13.4.3 The third piece of the mosaic

It starts from a different approach, strictly focused on the European model of agriculture multifunctional and social responsible. The European Commission (2001, 2011) has considered CSR[4] as a field of private actions, though allowing the policies can play a role in creating 'an environment more conducive'. At the same time the concept of multifunctionality were permeating the modern agriculture (OECD, 2001, 2005) and related policy (OECD, 2003). In the new perspective, agriculture provides not only food and commodities but, jointly, public goods and positive externalities able to pursue the social goal of reaching sustainable development. Thus, the concept

[4] EU Commission has provided in 2001 this definition: 'A concept whereby companies integrate social and environmental concerns in their business operations and in their interaction with their stakeholders on a voluntary basis' and in 2011 this integration: 'Corporate social responsibility concerns actions by companies over and above their legal obligations towards society and the environment. Certain regulatory measures create an environment more conducive to enterprises voluntarily meeting their social responsibility.'

of conventional agriculture was coming through and the new model has risen that encompasses manifold ways a farm can create value (Van der Ploeg *et al.*, 2002).

In this concern, Marotta and Nazzaro (2011, 2012) develop their approach which thinks of: (1) multifunctional farm as 'an ensemble of governance structures optimizing the creation of value'; (2) the products of the farm as embedding the social value crated. On this basis, they advance the concept of 'portfolio of values'. It encompasses the material and immaterial values produced by the farms, which can ground on it their positioning in the market and social role. Therefore, CSR and multifunctionality together are sources of competitive advantages. The scholars argue that multifunctional farm can reach major economic and financial values (through a premium price), at once the consumers their benefit (through a minus price), meanwhile the society achieves the goal of sustainable development, especially in the case of short-chain and strategy to valorise regional products. This approach is suitable for agricultural cooperatives too. CSR actions coincide with the process of social value creation. Farm's products, services and processes, embed the social value created; so it becomes a part of the transactions in which products and services are involved.

13.5 Proximity approach

Understanding how the spatial and social dynamics work together is the task to perform for placing the economic and financial dynamics in a concrete context.

I take into account those theories stating that the space is not neutral. The dynamic of proximity approach refers to several French scholars (Belletis *et al.*, 1993; Pecqueur *et al.*, 2004; Rallet and Torre, 2004; Torre and Beuret, 2012; Torre and Rallet, 2005). It has been broadly applied to the rural governance issue (Berriet-Solliec and Trouvé, 2013; Bertrand *et al.*, 2004; Lardon *et al.*, 2008; Torre, 2007) and to cooperative studies (Colletis *et al.*, 2005; Filippi, 2004; Pecqueur and Itçaina, 2012). The idea that 'territory' – in the French pregnant meaning – means a bounded region and a complex dynamic of social relation relies upon the decomposition of the concept of territory in two[5] complementary components. According to Torre and Rallet (2005), the spatial component is the geographic proximity that relates to the compresence of subjects in a same delimited area. The second component is the organised proximity that rise from the union of two distinct having relational nature. One is cultural and relates to the 'feel of belonging' to a group, anywhere defined. The other is functional, relates to institutional issues and imply a collaborative behaviour to reach common projects. Together, they represent the territorial proximity as a complex notion in

[5] Or three, in the institutionalist way: geographic, organised, and institutional that is the third component mentioned herein as 'functional' (Belletis *et al.*, 1993; Colletis *et al.*, 2005). Anyway, the three together identify the 'territorial proximity'. Herein we do not present the institutionalist position for not giving rise to misunderstandings about the different means and use of the term 'institutional'.

which the spatial, relational, institutional and cultural interactions are able to identify the territory as a localised social construction.

The specific organised proximity sorts four effects. Firstly, the concrete spatial dimension affect the ways – formal or informal – the involved subjects exchange information. Secondly, referring to stakeholders, it allows finding 'who and what really counts' (Mitchell *et al.*, 1997) in the specific local system. Thirdly, it indicates that several firm's initiative for local sustainable development may stay into shared local project in which local players and stakeholders mix the twofold nature. Fourthly, it points out that formal or informal local institutions are involved, and that local public or mixt authorities play a role. Thus, it serves to characterise hardly the concept of institutional environment for the case in point.

13.6 New institutional economics approach

In the NIE approach, transaction is the unit of analysis. It tries to explain how economic players organize for minimizing TCs which are rising for both behavioural (opportunism and bounded rationality) and environmental (uncertainty and small numbers) causes (Williamson, 1975). TCs are also positively related to the increasing of asset specificity (Riordan and Williamson, 1985). According to North (1990), institutions are to reduce uncertainty by shaping human interactions. Thus, the institutional environment has to be analysed to understand which are the rules that establish how the game to play, since institutions allocate the property and decision rights to the players that organise transactions. Consequently, players choose alternative organizational arrangements – the 'make or buy' decision – in order to minimize TCs.

To deal with the issue of agricultural cooperatives and CSR, I need to focus on: the institutional environment; what kind of organizational arrangement the coop is; at what institutional level CSR rules of the game are established; how they affect the choice about whether or not stay with the same institutional arrangement (the play of the game).

13.6.1 The institutional environment

International principles sketch what a cooperative is and inspire the national laws: A co-operative is an autonomous association of persons united voluntarily to meet their common economic, social, and cultural needs and aspirations through a jointly owned and democratically controlled enterprise. According to Iliopoulos (2014), in the most of Countries typical agricultural cooperative property rights are: '(1) ownership rights are restricted to member-patrons; (2) residual income rights are non-transferable, non-appreciable, and redeemable; (3)residual income is distributed to members in proportion to patronage; (4) residual decision rights are distributed to member-patrons on the basis of either the one-member-one-vote rule or in proportion to patronage.'

Without contradict this basic framework, in France a number of laws are concerning with cooperation. Among others, the 10th September 1947 that establishes distinction between investor-owned firms and cooperatives. The 27th June 1972 law concerning cooperatives and their union. The 3rd January 1991 law n. 91-5 that allows the Business cooperative Groups creation (or subsidiarization, Filippi, 2012).

Referring to the case of FdF, the institutional environment is strongly characterised by the previously analysed, AT, SCAI and BCGs. AT establishes a spatial limit to the coop's right of operating. The SCAI statute enlarges the previous possibilities to operate, to the transactions concerning with the multifunctionality of agriculture. That make allowance to act in the environmental and social field. The BCGs rules delineate the field in which the relationships among the mother-cooperative and the subsidiaries takes place according to different arrangements. In this field, each mother-cooperative can freely choose to maximize BCG benefits or its own.

13.6.2 At what institutional level corporate social responsibility rules are placed

The European Commission (2001, 2011) addressed the firms to voluntary behave according to CSR. In fact, there are no law enforcing firms to be virtuous, but citizen-consumer opinion matters in this concern as far as to 'constraints' some strategic and organizational firms' choice. The public sensitiveness for environment protection that entail firms, farms and cooperatives too are choosing to behave beyond law enforcement, to meet citizen-consumer preferences is an example. The instrument to make citizens and consumers aware about the firm adoption of such a virtuous behaviour is the voluntary certification. It belongs to the private order. Concerning the case in point, I refer to ISO 14000, ISO 26000 certifications. The International Standard Organization (ISO) is a non-governmental body that develops standards for social responsibility and environment management (among many others), having effect on 163 countries.

According to Ménard (2014), the ISO bodies may be considered as meso-institutions. They are the devices that provide the guidelines by which the general rules established in the institutional environment become applicable in each specific 'sector and/or geographic areas, thus framing and delineating the domain of activities of actors (individual as well as organizational arrangements) operating within these rules'. Actors can affect their efficiency in two ways: by their degree of acceptance of the rules of the game 'which is rooted in the mental maps that frame their behaviour'; by their 'capacity to monitor the diverse organizational arrangements', as they need to adapt to changing conditions and require 'rules flexible enough' (p. 578).

Namely, the ISO Bodies frames the transactions and provides guidelines to implement general standards in each Country. ISO, through the Committee on Conformity

Assessment (CASCO), interact also with the Certification bodies (again private bodies) whose role is to assist firms willing to gain the ISO certification and arrange audits according to the CASCO procedure. They are M-I too, as well as the ISO Bodies. These M-I have established limits, possibilities, and monitoring system to FdF's action in order to make both citizen-consumers and member-owners sure about the coop contribution to enhance environment quality and social responsibility.

Contrariwise, the BS is not to be considered an M-I. It is not an audit, but a device to identify and plan CSR action in a way studied to fit to agricultural cooperatives in France.

13.6.3 What kind of organizational arrangement the coop is

Agricultural cooperatives are governance arrangements that meet the characteristics of the hybrids (Karantininis, 2007; Ménard 2004, 2007), the intermediate forms that Williamson (1991, 1996) places between markets and hierarchies. Recent contributions have enriched the early definition (Ménard, 2008, 2012, 2014). According to Royer *et al.* (2016, p. 107) we can recognize that agricultural cooperatives involve subjects that 'remain legally and economically autonomous, holding distinct property as well as decision rights, but also pooling subsets of rights on which they abandon part of their control, in some cases in a substantial way'. As Ménard (2008) synthesises, Williamson provides the reasons to explain why players can chose to stay with the hybrid arrangements. They are depending on asset specificity and uncertainty. If the first is specific 'enough to generate contractual hazard without justifying integration' and the second is 'enough to require tighter coordination than market can provide' there is the incentive to choose the hybrid form.

Pooling resources implies to use the ability to coordinate decisions and safeguard against free riding and other opportunistic behaviours. Cook (1995) listed five problems as source of TCs: free rider, horizon, portfolio, control and influence cost problems. (We refer to a reach literature to deepen the five problems: Cook and Iliopoulos, 2000; Iliopoulos and Cook, 1999; Sykuta and Cook, 2001). They are depending on what Cook identified as the distinctive characteristic of the cooperative, i.e. the 'vaguely defined property rights', which are descending from two reasons. The first is the twofold nature of user/producer of the 'member-owner'. The mechanism to determine the price for their bestowing (or using) is not the market but the result of management activity of the cooperative. Therefore, there is a trade-off between the short-run expectation of the producer – who aims at claiming money refund as higher as possible – and the long-run perspective of the owner that aims at enhance the capital value by right investment. Manager instead of the producer-owner usually represents the latter position, having the interest to stay with the cooperative arrangement.

The second reason is that neither the capital shares are marketable, nor there is a market to assign them a value. The combined effect results in horizon and portfolio problems and in free rider as well. The members' preference to invest is typically lower and the hazard felt is higher.

Another reason comes from the structural feature of delegating the decision rights to the manager. Grounding on informational asymmetry, this characteristic generate both the classical costs of agent control, and the influence costs problem that are associated with the multiple activity of a cooperative. This point highlights that any element able to reduce/increase asymmetric information and affect TCs, can reduce/increase the preference to invest. Consequently, the way to manage the five problem affects both the propensity of the owners to support cooperative long-term investments and their loyalty.

13.7 Discussion: why corporate social responsibility could be a good choice for FdF?

The FdF CSR choice has resulted in implementing the consequent CSR action plan. A few of facts emerges from the territorial and cooperative point of view.

On the territory side, perceived positive effects on environmental, social and economic sustainability are the more obvious. Directly, the region has benefitted from the investments realised in the interest of the environment and the social community, e.g. the photovoltaic park that has involved a number of farmers. Another example is the coop's engagement for keeping a store and mail service in a more remote area of the region. Indirectly, the major economic value created has contributed to underpin the economic development of the region and, not last, some reasons for people to live there and make agricultural activity. All the process has been possible through the large involvement of the local stakeholders as well as the *collectivités territoriales*, which in France can play a crucial role in the effort of promoting local development and delivering public policy for rural development.

On the cooperative side, FdF members have benefitted from the major profits coming from the new undertakes. A more complex organisational arrangement resulted, without contradict the initial one.

13.7.1 The French mainstreaming approach

This approach allows underlying the difference between FdF and the usual BCGs. The main aim has been remaining to add value to the cooperative-mother – instead that to the group – hence it was not to bypassing the market limit imposed by AT. In the growth and development of FdF there is something of more complex. The many investments have yielded to multiple value chains, quite each one having its specific organizational arrangement, most again in the cooperative form. One can observe that from the SICASELI stemmed a peculiar form of BCGs, whose rationale mostly grounds on CSR action plan, whose new partners are mostly from the local system, whose business scope mostly coincide with the mother's AT. FdF has become a player of the local green economy, able to benefit its own adherent (remember that in the SCAI statute at least 20% are people, not farms) from the profit so gained. In this perspective of the economic and financial growth of the cooperative, the CSR plan implemented makes a deeper sense.

13.7.2 The corporate social responsibility multiple approaches

From the three 'piece of the mosaic' proposed, as many conceptual frameworks apply.

- The first is the stakeholder management. SICASELI has achieved the rural networking creation goal by means of stakeholder management, which also sorted a number of effects, direct and indirect. The involvement of the stakeholders led to develop collective projects and shared investment. In turn, that led to the provision of products or services embedding both social value and social dynamics. So that the engagement for enhancing local sustainable development results embedded.
- The creation of a rural network is the second. The intentional and direct involvement with the other player-stakeholders has sorted several effects. The cooperative action has reinforced the territory-building process and has also reinforced the replication of social capital (Parodi, 2005). It has bettered the rural governance process and the delivering of public policies (Bertrand and Moquay, 2004) by the public or mixt bodies charged of the local administration (such as the Pays of Figeac). It has meant more cohesion on shared values able to sketch the line for the future, which has enhanced the general trust on the collective capabilities to reach better future perspectives for the local system (Toccaceli, 2016). Besides, each type of action has developed within the competitive context of the cooperative, which is operating in the local system, via AT. Thus, working with, and in the interest of local community, in Porter's words, has offered positive competitive advantages to the cooperative. Enhanced economic returns from several new investments have come. Furthermore, the introduction of new technologies for improving environmental performance (Porter and Van der Linde, 1995) has also reached competitive advantages in respect of the competitive context.
- The third is the arrangement of a Portfolio of values (Marotta and Nazzaro, 2011, 2012) whose value- chains embed social value in multiple products or services, so that objective of sustainability are embedded too. From the beginning, SICASELI has continued to progressively 'shift the boundaries' (Van der Ploeg *et al.*, 2002) of its initial activity and has created multiple value-chains each one has sought its optimal organizational arrangement, and has managed all of them. The Portfolio of values has progressively incremented its monetary and social value. Those values have developed some effects through the market in several ways. One is valorising local food products, thanks to qualification labels, like Protected Designations of Origin, Protected Geographical Indications or collective labels (Mollard, 2001; Tregear *et al.*, 2007). Another one is composing baskets of integrated products and services of the territory on which to set up local strategies (Mollard and Pecqueur, 2007; Pecqueur, 2001; Perrier-Cornet and Sylvander, 2000) or collective marks. For example, the creation of the point for selling of the *agneau du Quercy* for adding

value to the short-chain, reinforcing the ties with a number of co-operators and enhancing economic performance.

13.7.3 The proximity approach

The proximity approach (Torre and Rallet, 2005), especially the concept of territorial proximity (Belletis *et al.*, 1993; Filippi, 2004, 2013a), helps to explain the social dynamics that 'frame' the 'territory'. By this point of view, the cooperative CSR action (plan and implementation) is part of a broader process, the social building of the rural space. In this respect the constraint coming from the AT obligation can give rise to cooperative strategies able to transform it in a regional advantage. According to Filippi (2014) such strategies can aim at enhancing the value of local resources, reinforcing members loyalty or building interdependencies with consumers in the supply chain. All can apply to the case in point and try to explain why the cooperative governance can result enhanced.

13.7.4 The new institutional economics approach

The success of the cooperative could be viewed in terms of loyalty of its members, capability to invest and decision to stay with the same arrangement. Albeit it could seem to be closer to the manager's standpoint than to the owners' one, I think it could be the right hypothesis to verify, by applying the NIE conceptual framework.

For the case in point, making a CSR choice concerning the triple bottom line of environmental, social and economic sustainability, have some implications. Firstly, member-owners pool resources to carry out CSR action plan. Secondly, managers use decision rights and related property rights to implement CSR action plan that needs for additional investments and new plants. Thirdly, the set of CSR related transactions involves asset specific and uncertainty, which are 'enough' to justify TCs are minimized by staying with the same organizational arrangement.

The new CSR initiatives have been taking the form of a group of cooperative, suggesting the idea that a satisfying level of coordination has been reached.

13.8 Final remarks

What is the relation between CSR and agricultural cooperation in the current challenging scenario and why CSR should be a profitable choice for an agricultural cooperative are the early questions. CSR has been understood in terms of the triple bottom line of environmental, social and economic sustainability.

The proposed alternative approaches have highlighted different profile of analysis and implications. The changing pattern of the cooperative, challenged by the more competitive market, has shown a 'third view' is possible thanks to the CSR engagement.

As better explained by the CSR and proximity approaches, a CSR set of actions can start multiple new ways – and related chains – to create value. In this regard, the 'territory' became endogenous in the value creation process and the 'territory', in the meaning of social building, seems able to explain why positive effects on stakeholder opinion would be internalised in a stronger loyalty of cooperative members and governance.

The NIE approach has put the basis to highlight the internal side of the cooperative, the institutional layers to be focalised in order to comprehend how uncertainty, asset specificity and moral hazard together push toward the hybrid arrangement as the better organizational solution for the farms.

Each one of the proposed approaches allows highlighting different mechanisms at work, profiles of analysis and implications. Each one would need of as many direct surveys to verify the emerging hypothesis.

References

Barnett, M.L., 2007. Stakeholder influence capacity and the variability of financial returns to corporate social responsibility. Academy of Management Review 32(3): 794-816.

Belletis, C.G. and Lung, Y., 1993. Economie de proximités. Revue d'Économie Regional et Urbaine 3: 357-361.

Berriet-Solliec, M. and Trouvé, A., 2013. Développement des territoires de projet. Quel enjeux pour les politiques rurales? Économie Rurale 335: 7-19.

Bertrand, N. and Moquay, P., 2004. La gouvernance locale, un retour à la proximité. Économie Rurale 280: 77-95.

Bodet, C. and Picard, D., 2006. Le Bilan Sociétal: de la prise en compte des intérêts contradictoires des parties prenantes à la responsabilité sociétale. Développement durable et territoires, Dossier 5: économie plurielle, responsabilité sociétale et développement durable. Available at: http://tinyurl.com/yc96mky9.

Cariou, Y., 2003. Le bilan sociétal dans la coopération agricole : une démarche participative pour s'ouvrir au territoire. Recma- Revue Internationale d'Economie Sociale 290: 15.

Cariou, Y., 2013. Nous, cooperateurs. l'Aventure des éleveurs du pays de Figeac. Bécherel, France.

Cariou, Y., Fournie, S. and Wallet, F., 2006. Le bilan sociétal : un outil de management pour renforcer l'ancrage territorial et la responsabilité socialedes coopératives agricoles. Développement durable et territoires, 5. Available at: http://tinyurl.com/ycgkgabq.

Carroll, B.A., 1999. Corporate social responsibility. Business and Society 38(3): 268-295.

Chomel, C., 2013. Le cadre juridique et la gouvernance des coopératives agricoles. In: Chomel, C., Declerck, F., Filippi, M., Frey, O. and Mauget, R. (eds.) Les coopératives agricoles. Identité, gouverance et stratégies. Larcier, Brussels, Belgium, pp. 65-140.

Colletis, G., Gianfaldoni, P. and Richez-Battesti, N., 2005. Économie sociale et solidaire, territoires et proximité. Recma-Revue Internationale d'Economie Sociale 296: 8-25.

Cook, M.L., 1995. The future of U.S. Agricultural cooperatives: a neo-institutional approach. American Journal of Agricultural Economics 77: 1153-1159.

Cook, M.L. and Iliopoulos, C., 2000. Ill-defined property rights in collective action: the case of US agricultural cooperatives. In Ménard, C. (ed.) Institutions, contracts, and organizations. Perspectives from new institutional economics. Edward Elgar, Cheltenam, UK, pp. 335-348.

Côté, D., 2000. Cooperative in the next millenium: the emergence of a new paradigm. Recma-Revue Internationale d'Economie Sociale 275-276: 149-166.

Côté, D., 2007. Fondements d'un nouveau paradigme coopératif. Recma – Revue Internationale d'Economie Sociale 305: 72-91.

Côté, D., 2005. Loyauté et identité coopérative. Recma-Revue Internationale d'Economie Sociale 295: 50-69.

Draperi, J., 2012. La republique cooperative. Larcier, Brussels, Belgium.

Draperi, J. and Corroller, C., 2015. Cooperatives and regions in France : specific and complex relationships. Recma – Revue Internationale d'Economie Sociale: 34-48.

European Commission (EC), 2001. Green paper: promoting a European framework for corporate social responsibility. COM 366: 32.

European Commission (EC), 2011. Communication from the Commission to the European Parliament, the Council, the European Economic and Social Committee and the Committee of the Regions: a renewed EU strategy 2011-14 for corporate social responsibility. COM 681: 1-15.

Filippi, M., 2013a. L'ancrage territorial des enterprises coopératives agricoles et le développement durable. In: Chomel, C., Declerck, F., Filippi, M., Frey, O. and Mauget, R. (eds.) Les coopératives agricoles. Identité, gouverance et stratégies. Larcier, Brussels, Belgium, pp. 205-262.

Filippi, M., 2013b. Les relations des coopératives avec leurs associés coopérateurs et la gouvernance en coopérative. In: Chomel, C., Declerck, F., Filippi, M., Frey, O. and Mauget, R. (eds.) Les coopératives agricoles. Identité, gouverance et stratégies. Larcier, Brussels, Belgium, pp. 141-203.

Filippi, M., 2004. Réorganisations dans la coopération agricole: proximités et solidarité territoriale. Economie Rurale 280(1): 42-58.

Filippi, M., 2012. Support for farmers' cooperative. Countryreport France. In: Bijman, J., Iliopoulos, C., Poppe, K.J., Gijselinckx, C., Hagedorn, K., Hanisch, M., Hendrikse, G.W.J., Kulh, R., Ollila, P., Pyykkönen, P. and Van der Sangen, G. (eds.) Support for farmers' cooperatives. Final report for European Commission. Available at: http://tinyurl.com/ycczjl8l.

Filippi, M., 2014. Using the regional advantage: French agricultural cooperatives' economic and governance tool. Annals of Public and Cooperative Economics 85(4): 597-615.

Filippi, M., Frey, O. and Mauget, R., 2013. Les coopératives agricoles face à l'internationalisation et à la mondialisation des marchés. Recma – Revue Internationale d'Economie Sociale 310: 31-51.

Filippi, M., Frey, O. and Torre, A., 2008. Les stratégies d'ancrage territorial des groupes coopératifs agricoles français, mesures de la diversité. Journées de Recherches en Sciences Sociales, pp. 1-17.

Freeman, R.E., 1984. Strategic management: a stakeholder approach. Pitman Publishing Inc., Boston, MA, USA.

Frey, O., 2013. La genèse du mouvement coopératif et les facteurs de développement des coopératives agricoles. In: Chomel, C., Declerck, F., Filippi, M., Frey, O. and Mauget, R. (eds.) Les coopératives agricoles. Identité, gouverance et stratégies. Larcier, Bruxelles, Belgium, pp. 19-63.

Garriga, E. and Melé, D., 2011. Corporate social responsibility theories : mapping the territory. Journal of Business 53(1): 51-71.

Heslouin, M., 2011. Le Bilan sociétal: un outil pour mettre en oeuvre l'ISO 26000? Recma – Revue Internationale d'Economie Sociale 321: 13-16.

Iliopoulos, C., 2014. Ownership, governance and related trade-offs in agricultural cooperatives. Dovenschmidt Quarterly 2(4): 159-167.

Iliopoulos, C. and Cook, M.L., 1999. The efficiency of internal capital markets in customer-owned firms : the influence costs problem. In: 3rd Annual Conference of the International Society for New Institutional Economics. Washington, DC, USA.

Innovation and Networks Executive Agency (INEA), 2014. La cooperazione: una nuova centralità nello sviluppo del sistema agroalimentare Italiano. Edagricole, Bologna, Italy.

Jensen, M.C., 2002. Value maximization, stakeholder theory, and the corporate objective function. Business Ethics Quarterly 12(2): 235-256.

Jones, T.M., 1995. Instrumental stakeholder theory: a synthesis of ethics and economics. Academy of Management Review 20: 404-437.

Karantininis, K., 2007. The network form of the cooperative organization – an illustration with the Danish pork industry. In: Karantininis, K. and Nilsson, J. (eds.) Vertical markets and cooperative hierarchies: the role of cooperatives in the agri-food industry. Springer, Dordrecht, the Netherlands, pp. 19-34.

Koulytchizky, S. and Mauget, R., 2003. Le développement des groupes coopératifs agricoles depuis un demi-siècle. Recma – Revue Internationale d'Economie Sociale 287: 14-40.

Lardon, S., Chia, E. and Rey-Valette, H., 2008. Introduction: dispositifs et outils de gouvernance territoriale. Norois 209(4): 7-13.

Lockett, A., Moon, J. and Visser, W., 2006. Corporate social responsibility in management research: focus, nature, salience and sources of influence. Journal of Management Studies 43(1): 115-136.

Mangin, P., 2001. Le Bilan sociatal, un outil novateur d'évaluation de l'impact global des coopératives agricoles. Recma – Revue Internationale d'Economie Sociale 281: 101-107.

Margolis, J. and Walsh, J., 2003. Misery loves rethinking companies : social initiatives by business. Administrative Science Quarterly 48(2): 268-305.

Marotta, G. and Nazzaro, C., 2011. Verso un nuovo paradigma per la creazione di valore nell'impresa agricola multifunzionale. Rivista di Economia Agro-Alimentare 3: 1-38.

Marotta, G. and Nazzaro, C., 2012. Responsabilità sociale e creazione di valore nell'impresa agroalimentare: nuove frontiere di ricerca. Economia Agro-Alimentare 14(1): 13-54.

McWilliams, A. and Siegel, D.S., 2011. Creating and capturing value: strategic corporate social responsibility, resource-based theory, and sustainable competitive advantage. Journal of Management 37(5): 1480-1495.

Ménard, C., 2007. Cooperatives: hierarchies or hybrids? In: Karantininis, K. and Nilsson, J. (eds.) Vertical markets and cooperative hierarchies: the role of cooperatives in the agri-food industry. Springer, Dordrecht, the Netherlands, pp. 1-17.

Ménard, C., 2008. A new institutional approach to organization. In: Mènard, C. and Shirley, M.M. (eds.) Handbook of new institutional economics. Springer, Dordrecht, the Netherlands, pp. 281-318.

Ménard, C., 2012. Hybrid modes of organization. Alliances, joint ventures, networks and other 'strange' animals. In: Gibbson, R. and Roberts, J. (eds.) The handbook of organizational economics. Princeton University Press, Princeton, NJ, USA, pp. 1066-1108.

Ménard, C., 2014. Embedding organizational arrangements : towards a general model. Journal of Institutional Economics 10(4): 567-589.

Mitchell, R.K., Agle, B.R. and Wood, D.J., 1997. Toward a theory of stakeholder identification and salience: defining the principle of who and what really counts. Academy of Management Review 22(4): 853-886.

Mollard, A., 2001. Qualité et développement territorial: une grille d'analyse théorique à partir de la rente. Économie Rurale 263(1): 16-34.

Mollard, A. and Pecqueur, B., 2007. De l'hypothèse au modèle du panier de biens et de services histoire succincte d'une recherche. Economie Rurale 300: 110-114.

North, D.C., 1990. Institutions, institutional change and economic performance. Cambridge University Press, Cambridge, UK.

Organisation for Economic Co-operation and Development (OECD), 2001. Multifunctionality. Towards an analytical framework. OECD Publishing, Paris, France.

Organisation for Economic Co-operation and Development (OECD), 2005. Multifunctionality in agriculture. What role for private initiatives. OECD Publishing, Paris, France.

Organisation for Economic Co-operation and Development (OECD), 2003. Multifunctionality, the policy implications. OECD Publishing, Paris, France.

Parodi, M., 2005. Économie sociale et solidaire et développement local. Recma - Revue Internationale d'Economie Sociale 296: 26-41.

Pecqueur, B., 2001. Qualité et développement territorial: l'hypothèse du panier de biens et de services territorialisés. Économie Rurale 261(1): 37-49.

Pecqueur, B. and Itçaina, X., 2012. Economie sociale et solidaire et territoire: un couple allant de soi? Recma - Revue Internationale d'Economie Sociale 325: 48.

Pecqueur, B. and Zimmerman, J.-B. (eds.), 2004. Ѐconomie de proximités. Hermés, Paris, France.

Perrier-Cornet, M.P. and Sylvander, B., 2000. Firmes, coordinations et territorialité une lecture économique de la diversité des filières d'appellation d'origine. Economie Rurale 258: 79-89.

Persais, E., 2004. Bilan sociétal : la mise en œuvre du processus rse au sein du secteur de l'économie sociale. Recma - Revue Internationale d'Economie Sociale 294: 14-39.

Pham, T. and Aubert, F., 2013a. Diagnostic des enjeux territoriaux de ressources humaines en milieu rural : retour d'expériences des territoires de Fougères et de Figeac. 7èmes Journées de Recherche en Sciences Sociales: 1-14.

Pham, T. and Aubert, F., 2014. Gerer les ressources humaines au niveau territorial en Pays de Figeac. Available at: http://tinyurl.com/ycg3ta9x.

Porter, M.E., 1990. The competitive advantage of nations. Harvard Business Review 68: 73-93.

Porter, M.E. and Kramer, M.R., 2002. The competitive advantage of corporate philantropy. Harvard Business Review 80(12): 56-68.

Porter, M.E. and Kramer, M.R., 2006. Strategy and society: the link between competitive advantage and corporate social responsability. Harvard Business Review 84(12): 78-92.

Porter, M.E. and Van der Linde, C., 1995. Green and competitive: ending the stalemate. Harvard Business Review 73(5): 122-133.

Rallet, A. and Torre, A., 2004. Proximité et localisation. Économie Rurale 280(1): 25-41.

Riordan, M.H. and Williamson, O.E., 1985. Asset specificity and economic organization. International Journal of Industrial Organization 3: 365-378.

Royer, A., Ménard, C. and Gouin, D.-M., 2016. Reassessing marketing boards as hybrid arrangements: evidence from Canadian experiences. Agricultural Economics 47(1): 105-116.

Sol et Civilisation, 2010. Gestion territorial des emplois et des compétences. Reseau Rural National, Paris, France.

Sykuta, M.E. and Cook, M.L., 2001. A new institutional economics approach to contracts and cooperatives. American Journal of Agricultural Economics 83(5): 1273-1279.

Thomas, F., 2009. Scic et agriculture : le temps des défricheurs. Recma – Revue Internationale d'Economie Sociale 310: 17-30.

Toccaceli, D., 2016. From farm cooperation to territory building: a French case study. Rivista di Economia Agraria 71(3).

Torre, A., 2007. The intriguing question of regional and territorial development in rural areas. Analytical Variations and Public Policies 1: 1-26.

Torre, A. and Beuret, J.-E., 2012. Proximités territoriales. Economica, Paris, France.

Torre, A. and Rallet, A., 2005. Proximity and localization. Regional Studies 39(1): 47-59.

Tregear, A., Arfini, F., Belletti, G. and Marescotti, A., 2007. Regional foods and rural development: the role of product qualification. Journal of Rural Studies 23(1): 12-22.

Van der Ploeg, J.D., Long, A. and Banks, J., 2002. Living countryside: rural development process in Europe: the state of the art. Elsevier, Doetinchem, the Nederlands.

Williamson, O.E., 1975. Market and hierarchies: analysis and antitrust implication. Free Press, New York, NY, USA.

14. Concepts in European policy intervention and food exchange: a Delphi approach

G. Martino[1*], J. Tóth[2] and A. Frascarelli[1]

[1]Department of Agricultural, Food and Environmental Sciences, University of Perugia, Borgo XX Giugno 74, 06100 Perugia, Italy; [2]Department of Agricultural Economics and Rural Development, Corvinus University of Budapest, Fovam ter 8, 1093 Budapest, Hungary; gaetano.martino@unipg.it

Abstract

The paper focuses on stakeholder beliefs in the recent changes of the Common Agricultural Policy. The new policy requires farmers to set up organisational forms dedicated to the exchange of agricultural products. These forms are required to replace the pattern of intervention established by the European Union and based on a series of specific common market organisation. Also the private solutions that farmers will be able to implement should provide sustainable income for the farming sector, food availability for European society and environmental protection. Huge changes are then expected in terms of agricultural exchange. From an institutional perspective, the paper investigates stakeholder beliefs in such a change, assuming that these beliefs are an important driver of effective changes. The empirical analysis is based on the Delphi approach and indicates that respondents recognise the potential impact of the policy change on the organisational change. The role of individual characteristics and the focus on some organisational forms would suggest the importance of a broader public debate.

Keywords: Common Agricultural Policy, governance structures, beliefs, institutional analysis, Delphi approach

DOI 10.3920/978-90-8686-844-5_14, © Wageningen Academic Publishers 2017

14.1 Introduction

The new Common Agricultural Policy (CAP) regulation emphasizes the importance of the organisational arrangements and critically posits the necessity to create and implement effective governance structures that are able to support the exchange of agricultural products replacing the role that public authorities have held for several years (Martino *et al.*, 2016). The new regulation deeply innovates the context of farmers' decisions and calls for the effective role of organisational agreement along the chain in order to achieve the goal that public European authorities aim to pursue. This study assumes that an articulated set of actors create a framework around the decisions made by farmers through a complex set of institutional relationships and restrictions (Saccomandi, 1998).

The study focuses on stakeholder beliefs in the new European Union (EU) regulation approach of agricultural product exchange in the Agro-Industry Supply System (AISS) and underlines the possibility that these beliefs contribute to shaping the new decision context of farmers. The objective of the study is to identify the beliefs of a specific type of stakeholder type with respect to the main focuses of the new CAP approach and of the potential organisational solution the farmers may implement in the new scenario.

European policy makers are becoming aware of the importance of governance modes in improving the functioning of the AISS (EC, 2009). Moreover, integration across regions and countries relies on the emerging of structures and norms (Varzakas and Jukes, 1997, p. 507) that, in turn, influence the adoption of the organisational choices (Williamson, 2005). European integrations also based on the convergence of the institutional arrangements toward a common European framework in which the governance modes – the ways the parties of a transaction organise it (Williamson, 2005) – are able to provide equal or comparable opportunities to resource mobilisation and valorisation in all European productive contexts. On the other hand, the European integration process influences the patterns of labour division in AISS (Chevassus-Lozza and Galliano, 2009, p. 139; Levchenko and Zang, 2012, p. 586) and the cohesion between economic processes. In this context, the CAP reform still connects the AISS activities with a general view of integration and cohesion, but emphasizes the private capabilities of organising the exchange of agricultural products.

Compared with the previous market regime characterised by systematic EU intervention, the new approach depicts scenarios characterised by market uncertainty. Moreover, there is a lack of experience in many EU countries on the possibilities of organising the market process privately. The challenge is thus to identify and select effective organisational solutions. In the uncertain scenarios, the beliefs of external stakeholders on farm opportunities may provide useful information on creating and managing an effective organisational arrangement. This study thus aims at eliciting stakeholder beliefs, to contribute to the knowledge of the resulting uncertain scenario. The paper is organised as follows. Section 14.2 illustrates the conceptual framework of the study. The strategy of the empirical analysis is illustrated in Section 14.3. Section

14.4 illustrates and discusses the results. Section 14.5 provides the final remarks of the study.

14.2 Motivation and conceptual framework

14.2.1 Study motivation: the new CAP reform and private orderings

The CAP has been profoundly transformed over the 2014-2020 programming period. The new CAP aims at enhancing agricultural production, stabilising income and providing support to family-owned businesses. The CAP reform has introduced new goals, such as competitiveness, external sustainability, public goods supply and rural development, to be achieved through new tools. In these reforms, while market measures were scaled down and simplified, the role of spontaneous economic initiatives of farmers is expected to be prominent. Accordingly, twenty Common Markets Organisations (CMOs) – the label assigned to an institutional set of instruments of the EU that regulates the exchange of specific products – have been therefore joined under a *single CMO* which is thought to be general enough to include all classes of organisational solutions that farmers – and their unions or sustaining organisations – may conceive in order to cope with the new uncertain context. The European reform – namely, Reg. 261/2012 and Reg. 1308/2013 – strongly relies on indirect tools by fostering market transparency and the setting up of relationships negotiated among the AISS value chain partners (Martino *et al.*, 2016). In principle, this new approach for agricultural product exchange in Europe calls for all the possibilities allowed by private organisational solutions.

Political governance plays a critical role in ensuring the essential prerequisite of a market economy (Coase, 2013; Dixit, 2009). The institutional environment creates a framework around the governance of private decisions influencing, on a very general level, the choices of individual actors aimed at organising transactions (Williamson, 2005). As an innovation of the institutional environment, the new EU regulation focuses on the capability of private actors to set up viable market relationships and it is expected to cause a huge change in market relationships. The CAP reform is inducing the AISS to react to market pressures by adopting effective organisational forms (Pascucci *et al.*, 2015). In this context, we contend that the CAP reforms call for posing the problem of new regulation of agricultural markets emphasizing the role of private arrangements (Martino *et al.*, 2016). In other words, from a conceptual view, the role of private ordering becomes prevalent in predicting and assessing the choice made by AISS agents. As stated by theory, the policy intervenes in the institutional environment to contribute to the establishing of the system public ordering. Private ordering, on the other hand, concerns 'how the game is played':

> [...] Thus even if the polity has fashioned good rules of the game, transaction of idiosyncratic kind where immediate parties have deep and non-verifiable knowledge and for which continuity of the exchange

> is important are ones for which simple market exchange is poorly suited. Private ordering effort by the parties to realign incentives and embed transactions is more protective governance structures, have the purpose and effect of mitigating the contractual problems that would otherwise arise (Williamson, 2002, p. 38)

Private orderings are efforts to infuse order, thereby to mitigate conflicts and better realise the mutual advantage of voluntary exchange (Williamson, 1998). Private ordering is established through the alignment between the transaction and the governance structure (Williamson, 2002, p. 441). Conceptually, this transaction ordering provides the supporting basis for the finalised and intentional exchanges along the agricultural value chains and for the consequent, expected achievement of the EU policy goals.

14.2.2 Conceptual framework

14.2.2.1 The increasing uncertainty of agricultural transactions

The new CAP reform raises the uncertainty of the environment in which farming and processing activities of AISS take place. Actually, the radical change of the policy approach posits, for the farmers, the problem of facing the market forces directly. The ability of the farmers to cope with open competition processes is challenged because of the abolition of any form farms of direct regulation of the market, such as price interaction. Moreover, volume uncertainty (Walker and Weber, 1984) and price volatility (Bakucs and Fertö, 2005; Gilbert and Morgan, 2010; Rezitiz and Stavropoulos, 2010; Tadesse *et al.*, 2014) are strongly affecting the agricultural product supply. Therefore, the farmers' lack of experience and adequate tacit knowledge of the markets and the inherent uncertainty of the agricultural markets call for institutional and organisational plan.

Private ordering inspires farmers to seek for support in achieving two objectives: to guide the farmers in coping efficiently with uncertainty, and guiding private organisations toward the CAP's general, public objectives (Martino *et al.*, 2016). Both objectives can be achieved by the choice of the governance structures. There are many questions to be addressed in the field. Firstly, private orderings are thought of as being able to foster public interest: in this study we assume this case, even though future research should investigate the mode and extent to which this outcome could actually be achieved. Secondly it invokes a great role for both individual and collective organisational arrangements such as producer organisations, consortia, interbranch arrangements and so forth (Martino *et al.*, 2016) which can finely address the multiplicity of organisational issues arising in an open exchange of products entailing a large number of relevant qualitative characteristics in the context of increasing technological uncertainty.

14.2.2.2 Governing the agricultural transaction in the new CAP context

Theory shows that the agricultural exchange is organised through a complex variety of forms, spanning from the simple spot exchange to vertical integration and including many hybrid forms (Ménard, in press). In this context, also the existing forms such as cooperatives, producer associations and consortia, are likely to evolve. The central question to be addressed in this field concerns those types of organisational arrangements that can make private solutions able to regulate the agricultural markets, steering agent behaviours toward the achievement of public objectives (Cacchiarelli *et al.*, 2016; Martino *et al.*, 2016). The theoretical inquiry into the complex variety of the Agri-food industry (Ménard, in press) lays the basis to answer this question to be addressed. Moreover, it is necessary to investigate further dimensions of the AISS and take into account the fact that the recent CAP reform entails a significant transformation, from a public role-centred intervention to a regulation based on private capabilities and effective decisions. In turn, this transformation entails a change in farmer and stakeholder attitudes. The farmers' new perspectives have to be supported by the awareness of the main role that their private choices have on how the agricultural markets are organised. Following this rationale, it is necessary to point out that farming activity is, however, art of a framework of densely woven relationships entailing public (e.g. local regulatory agencies, research centres, certification bodies, etc.) and private (e.g. firm associations, farmers unions, etc.) actors (Saccomandi, 1998). This net of relationships tends to channel and restrict individual choice, particularly in the case of the adoption of governance structures requiring the participation of many farms, as in the case of producer organisations, cooperative firms, consortia or similar organisations. In this field, the role of farmers unions is usually relevant. This analytical perspective tackles recent developments in agricultural policy analysis. Scholars paid attention to preferences for agricultural policy intervention. On the one hand, these studies concentrated on specific aspects of intervention (Hall *et al.*, 2004; McEachern and Willock, 2004; Sydorovych and Wossink, 2008); on the other hand, attempts were also made to explain these preferences (Bennet *et al.*, 2002; Kallas *et al.*, 2007). Daniel and Perraud (2009) investigated the way multifunctionality is addressed in France and the Netherlands, in the CAP perspective, and found differences explained by the agents' conception of farming activities. Dedeurwaerdere *et al.* (2015) investigated the relation between CAP payments for the agro-environmental services and the diffusion of the related practices. They found that the network of connections among farmers and bridging organisation favours both cooperation and organisational learning. Giordano *et al.* (2015) analyse the farmers' behaviour with respect to water management and investigate the impact of CAP of water management practices. The study proves the importance that farmer attribute to policies and, at the same time, the interaction between different policies. Focusing on social preferences, Rocamora-Montiel *et al.*(2014) explored the way the CAP is perceived by citizens. The empirical analysis indicates that CAP reform may present characteristics relevant to policy legitimisation. Burton and Schwarz (2013) highlighted the role of the institutional and cultural dimensions in setting environmental intervention. Mittenzwei *et al.* (2016) concentrated on policy preferences and compared knowledge-based and cultural background factors influencing those differences. Notably the authors found that not only the cultural factors seen to have a prominent influence but that the knowledge factors tend to focus on specific policy elements.

This study aims to contribute to this literature, investigate stakeholder beliefs as a central element of the attitudes (Fazio, 1986) and develop the analysis recognising that the human beliefs elaborated by the human agents lead to their choices and these, in turn, shape the change of the human environment (North, 2005). CAP reform requires a change in perspective of the stakeholders: their beliefs thus become relevant in depicting the potential paths of change. Actually, as stakeholder views could represent a framework for farmer views (Saccomandi, 1998), then investigating stakeholder beliefs can provide information on the potential path they are likely to undertake in creating a framework of CAP reform implementation at the private ordering level.

14.2.4 CAP Reform and change in farmers' beliefs

According to Fazio (1986) and Ajzen and Fishbein (1980), attitudes and norms interact at the basis of behaviour. The CAP reform intervenes on this level as it is imposing a transformation of the institutional framework of the agricultural sector and the transformation required to undertake an adaptive change, in turn centred on the beliefs of the actors (North, 2005).

The institutional analysis provides the basic categories for investigating the change in the beliefs in an economic system (North, 2005). The general process of economic change acts, as its main basis in the interplay between the individual beliefs and the role of institutions, as a coping device for uncertainty. In the case of the CAP reform we are addressing in this study, we identify the need for 'institutionalisation' of the new perspectives. For the purposes of this study we define this institutionalisation as the 'process of allowing the emerging of a framework of the farmers' organisational choice that is able to promote the effective private orderings relying on efficient governance structures'. This process requires aligning the stakeholder perspective with the new EU vision of the exchanges in AISS. The public understanding of the CAP reform is the results of the complex processes conducted in political arenas. However, the context of the farmers' choice is often set up by a multiplicity of actors defining institutional relations around the farmers (Saccomandi, 1998). In order to increase the efficiency of private governance choices, this institutional level should evolve (North, 2005) in the directions identified by the EU. Therefore, we expect the institutional context acting as the framework for the farmers (Saccomandi, 1998) and the policy views to be aligned to favour the development of governance structures chosen by the farmers. This alignment can be addressed in the terms of the institutional analysis. Aoki (2011) showed that the emerging of a new institutional dimension is the outcome of a recursive interaction among behaviours, public representation of the relevant phenomenon – in our case climate change – and the consequent behaviours of the actors. Substantially, the author contends that a public proposition mediating the stable states of play (strategic interactions) and individual beliefs in recursive ways may be referred to as the 'substantive form of an institution'. As indicated in Figure 14.1, the recursive cycle connects the 'public proposition' (in our case, the new view on the exchange of the agricultural products) with the 'beliefs' of agents/stakeholders and then with the 'strategic choice' (the choice of the specific governance structures adopted to organise the relevant transactions in the chains) and gives rise to the 'State of play' (i.e. the set of governance choices reflecting the necessity of the new regulation), in turn this outcome consolidates the 'public proposition'. As an institutional dimension, the connection between beliefs and public proposition may guide the farmers' choice (Saccomandi, 1998) and favour or limit the search for effective reaction to the new EU regulation.

14.3 Research questions and method of empirical analysis

The introduced conceptual framework encapsulates the agents' beliefs in a recursive institutional circle that connects policy intervention, public understating of this intervention, agent beliefs and practices. The resulting analytic perspective integrates policy interventions, public perception, beliefs and practical behaviours into a unique institutional causal series. As discussed above, these aspects were addressed separately in the literature. Focusing on expert beliefs on the basis of our conceptual framework therefore has two implications on the study. Firstly, it provides information on the experts' view of the uncertain context determined by the new CAP reform. Secondly, it can be allocated in the conceptual framework that highlights the relationship between beliefs and policy: this, in turn, provides opportunities for the practical creation of new organisational arrangements.

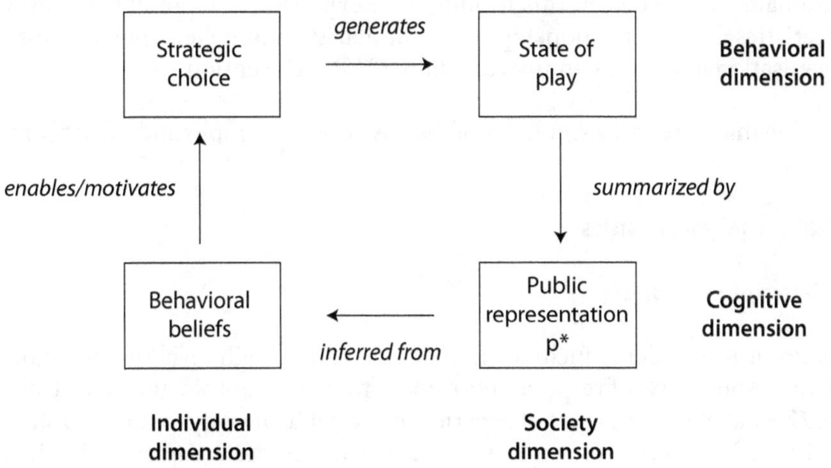

Figure 14.1. The recursive model of institutionalization.

Based on the proposed conceptual framework, this study is aimed at identifying a common general framework understanding of EU policy intervention in the exchange activities in AISS. The activity was carried out by contacting a selected sample of experts. The Delphi method (Adler and Ziglio, 1996) helps capture a preliminary, but comprehensive, understanding of the relationships between organisational choices and policy planning.

The first step was preparing the questionnaire. The main research questions addressed with the Delphi approach were the following:
- How does policy intervention take the organisational issues into account?
- To what extent do private orderings in AISS in Europe sustain the achievements of the objectives of agricultural and food policy intervention?
- What are the probable effects of European integration on the coordination of vertical stages in AISS?

The questionnaire was administrated by Surveymonkey. We availed of the personal contacts of one member of the research team who had the opportunity of personally contacting the potential respondents. All respondents are active in large Italian farmers unions. This strategy allowed us to collect 96 completed questionnaires. The sample is of course biased (personal network) and relates only to Italy. Table 14.1 illustrates the variables specified by the questionnaire items.

The data were analysed by describing the collected answers. In order to provide a synthesis of the beliefs of the respondents about the organisational forms we carried out a multidimensional scaling analysis (MDS) (Borg and Groenen, 2005; Kruskal and Wish, 1978). MDS is a method that represents measurements of similarity (dissimilarity) among pairs of objects as a distance between points of a low-dimensional space. We used this technique to explore and represent the individual characteristics and the relationship of the similarity with beliefs they express about the organisational forms as an instrument of CAP intervention.

In the following we report a summary of the answers with graphs and a brief comment.

14.4 Data analysis results

14.4.1 Descriptive analysis

The group of respondents includes individuals with a high level of education- and experience. About 54% of respondents is more than 50 years old and has a University degree. 77% has at least 16 years of experience in agribusiness dynamics and structures (Table 14.2). However, this experience varies across the sample: 38.3% has only regional level experience, 53.2% national, but only 17.0% has international experience. We also investigated the topic of experience through the professional responsibilities of the respondents. A high percentage of them have been responsible in the past in

Table 14.1. Variables description.

Variable definition	Label	Values
Age	Age	
≤25 years		1
26-45 years		2
≥45 years		3
Education	Edu	
Primary school		1
Secondary school		2
High School		3
University degree or higher		4
Professional experience	Prof_exper	
Low (≤5 years)		1
Intermediate (6-15 years)		2
Large (16-25 years)		3
Very large (>25 years)		4
Field of activity		
Local	Local	0, 1
Regional	Region	0, 1
National	Nation	0, 1
International	Intern	0, 1
Knowledge of the EU regulation		
Low	Reg_low	0, 1
Good	Reg_good	0, 1
Very good	Reg_vgood	0, 1
Excellent	Reg_exc	0, 1
Farmer contractual power	Farm_pow	0, 1, 2, 3, 4
Contractual enforcement	Enforc	0, 1, 2, 3, 4
Decision rights on the distribution of the gains	Gains	0, 1, 2, 3, 4
Certainty of the demand for the final products	Certan	0, 1, 2, 3, 4
Farmers costs	Power	0, 1, 2, 3, 4
Market power of the retailer	Farm_Cost	0, 1, 2, 3, 4
Processors costs	Proc_Cost	0, 1, 2, 3, 4
Retailers costs	Ret_Cost	0, 1, 2, 3, 4
Logistic issues	Logis	0, 1, 2, 3, 4
Market uncertainty	Mark_Unc	0, 1, 2, 3, 4
Technological uncertainty	Tech_Unc	0, 1, 2, 3, 4
Exchange development due to right choice	Exch_grow	0, 1, 2, 3, 4
Exchange reduction due to wrong choice	Exch_red	0, 1, 2, 3, 4
Short term production contract (max 1 year)	ST_Contr	0, 1
Long term production contract (more than 1 year)	LT_Contr	0, 1
Producer organization	PO	0, 1
Processing and trading consortia	Cons	0, 1
Protection consortia	Prot_Cons	0, 1
Voluntary certification	Cert_Vol	0, 1
Geographical indication	GI	0, 1
Private brand agreements	Brand	0, 1
Cooperatives	Coop	0, 1

Table 14.2. Respondents general characteristics.

	n	Frequency (%)
Age		
30 years or less	4	4.2
31-50 years	40	42.1
More than 50 years	51	53.7
Education		
Primary school	0	0.0
Scuola Media inferiore	2	2.1
High School	28	29.5
University degree or higher degree	66	69.5
Experience about Agribusiness		
5 years or less	5	0.05
6-15 years	17	0.18
16-25 years	31	0.33
more than 25 years	41	0.44

setting up many organisational forms: producer organisation (62.5%), chain contract (58%), cooperation (55.7%), but also processing contracts. A small number of the responses pertained to the current experience, but producer organisations (45.3%), chain contracts (41.9%) and cooperation (50.0%) are again the organisational forms in which the respondents are currently involved.

Table 14.3 summarises the respondents' evaluation of the role of the current CAP approach on key organisational elements. The respondents believe that the CAP is not properly considering the need to enforce the contract (23.9%), the distribution of critical decision rights (34.8%), and the role of demand certainty (18.3%). Accordingly only 11.9% believes they can expect development in exchange. The most frequently selected positive answer is 'In some cases', indicating that the respondent recognises the attention the CAP is paying, but the picture is still unclear.

Uncertainty and trust are the general drivers of organisational choice (Williamson, 1985, 1993). Table 14.4 shows that 99.6% of respondents think that the behavioural uncertainty may strongly affect the organisational agreement plan. 53.7% also believes that formalisation may guarantee the effectiveness of the agreement and almost 95% feels that trust is a critical factor in agricultural exchange organisation. We then investigated the beliefs about the potential role of the private organisational agreement in relation to agricultural and food policies (Table 14.5).

We selected a number of organisational forms and asked the respondents to assign a score to each of them (Table 14.6). While voluntary certification is mostly considered

Table 14.3. Respondents experiences in agribusiness.

	n	Frequency (%)
What is the territorial scope of your activity?		
Local	11	11.7
Regional	36	38.3
National	50	53.2
International	16	17.0
Have you ever managed the setting up of one of following forms?		
Producers organization	55	62.5
Processing and trade consortia	30	34.1
Protection consortia	17	19.3
Cooperatives	49	55.7
Chain contracts	51	58.0
Processing contracts	34	38.6
Local distribution contracts	12	13.6
Large retailer contracts	21	23.9
Do you have any professional role in one of the following forms?		
Producers organization	28	45.2
Processing and trade consortia	10	16.1
Protection consortia	5	8.1
Cooperatives	31	50.0
Chain contracts	26	41.9

Table 14.4. Do you think that the new CAP approach take rightly into account the following aspects?

Answer options	n	Frequency (%)				
		No	In some cases	Often	Fully	I do not know
Farmer contractual power	94	11.70	70.21	17.02	1.06	0.00
Contractual enforcement	92	23.91	61.96	9.78	1.09	3.26
Decision rights on the distribution of the gains	92	34.78	64.13	1.09	0.00	0.00
Certainty of the demand for the final products	93	18.28	63.44	17.20	0.00	1.08
Market power of the retailer	91	9.89	41.76	40.66	2.20	5.49
Farmers costs	95	22.11	62.11	11.58	1.05	3.16
Processors costs	92	10.87	48.91	28.26	2.17	9.78
Retailers costs	92	11.96	48.91	26.09	1.09	11.96
Logistic issues	89	15.73	57.30	13.48	2.25	11.24
Market uncertainty	93	7.53	59.14	26.88	3.23	3.23
Technological uncertainty	93	10.75	65.59	12.90	1.08	9.68
Exchange development due to right choice	92	11.96	51.09	22.83	1.09	13.04
Exchange reduction due to wrong choice	88	22.73	44.32	13.64	0.00	19.32

Table 14.5. Respondents evaluation of uncertainty and trust.

	n	Frequency (%)
What is the importance of the climate uncertainty to the organizational agreement design?		
Low	33	0.35
High	41	0.44
Very high	20	0.21
What is the importance of the behavioural uncertainty to the organizational agreement design?		
Low	5	5.4
Medium	58	62.4
High	30	32.3
What is the influence of the formal contracting on the agreement effectiveness?		
Low	17	17.9
High	51	53.7
Very High	27	28.4
What is the importance of trust to the agricultural exchange enhancement?		
Low	5	5.3
High	47	49.5
Very high	43	45.3

Table 14.6. Do you think that private agreements may support the achievement of agricultural and food policies?

Organizational forms	n	Frequency (%)				
		No	In some cases	Often	Fully	I do not know
Short term production contract (max 1 year	94	6.38	37.23	47.87	6.38	2.13
Long term production contract (more than 1 year)	95	3.16	30.53	53.68	10.53	2.11
Producers organization	93	5.38	15.05	63.44	16.13	0.00
Processing and trading consortia	92	5.43	16.30	66.30	9.78	2.17
Protection consortia	91	9.89	37.36	40.66	10.99	1.10
Voluntary certification	92	3.26	53.26	35.87	7.61	0.00
Geographical indication	92	2.17	34.78	56.52	5.43	1.09
Private brand agreements	93	4.30	38.71	50.54	5.38	1.08
Cooperatives	92	5.43	30.43	51.09	11.96	1.09

as being effective in some cases, notably private brand (54%) and cooperation (21.1%) were scored as 'Often' at the same percentage (50.5 and 51.1%). Processing and trade consortia (66.3%), producer organisation (63.4%) and geographical indication (56.5%) are often considered as able to support the achievement of the expected goals.

The respondents think that the European integration has an effect on the European agri-food systems (Table 14.7): To enhance the coordination among the stages (Often: 58.1%), To increase competition among various systems (Often: 68.4%), but less in case of farmers income improvements (In some cases: 42.1 and Often: 42.1%). Table 14.8 shows how the respondents believe that each type of agreement may contribute differently to the organisational solutions in different sectors. The variability can be explained by both the difference in uncertainty and by the direct knowledge of the sector. However, the differences observed are primarily related to the respondent's beliefs and it is worth pointing out that some arrangements – individual contract for bovine and ovine milk, and producer organisation contract with protection consortia for grains are not considered suitable at all. The respondents consider the price as the most important contractual term (Table 14.9). Also relevant are quality indicators and

Table 14.7. Do you think that the European integration may influence the European agri-food systems?

Answer options	n	Frequencies (%)				
		No	In some cases	Often	Fully	I do not know
To enhance the coordination among the stages	93	4.3	32.3	58.1	0.0	5.4
To enhance the food safety	94	3.2	17.0	64.9	13.8	1.1
To enhance the quality of the products	95	7.4	27.4	55.8	8.4	1.1
To increase the farmers income	95	9.5	42.1	42.1	4.2	2.1
To increase competition among various systems	95	1.1	22.1	68.4	5.3	3.2
To reduce the efficiency of the supply systems	92	18.5	47.8	19.6	2.2	12.0

Table 14.8. What are the best three forms to organize chain relationships in the following sectors?

Answer options	n	Frequencies (%)					
		Individual contracts	Contractual frameworks	Producer organization contracts	Producer organization contracts, with protection consortia	Chain agreement	Cooperatives
Cereals	88	6.82	6.82	13.64	0.00	59.09	13.64
Oil production	84	4.76	16.67	16.67	3.57	45.24	13.10
Fruits and vegetables	88	3.41	1.14	37.50	12.50	25.00	20.45
Olive oil	84	10.71	2.38	16.67	30.95	21.43	17.86
Wine	86	15.12	3.49	3.49	33.72	16.28	27.91
Bovine milk	83	0.00	12.05	16.87	7.23	20.48	43.37
Ovine milk	79	0.00	10.13	20.25	17.72	20.25	31.65
Bovine meat	83	2.41	7.23	15.66	22.89	43.37	8.43
Pig meat	83	1.20	8.43	12.05	22.89	44.58	10.84
Ovine meat	81	4.94	8.64	13.58	17.28	39.51	16.05

Table 14.9. What are the most important contractual terms in the following sectors?

Answer options	n	Frequencies (%)							
		Adequate price	Time of payment	Quality indicators	Certainty of the product delivery	Possibility of renegotiating	Free choice of production technique	No intermediary	Minimum prices
Cereals	91	50	0	5	12	4	0	4	16
Oil production	84	48	1	6	9	4	0	3	13
Fruits and vegetables	89	33	5	10	25	3	0	8	5
Olive oil	81	40	1	17	11	2	0	6	4
Wine	83	37	2	18	12	2	0	11	1
Bovine milk	82	51	3	8	4	2	0	4	10
Ovine milk	79	48	3	9	7	2	0	4	6
Bovine meat	82	50	3	8	10	1	0	5	5
Pig meat	79	44	3	9	11	1	0	5	6

certainty of product delivery, but only in the fruits and vegetables, olive oil and wine sectors. This concentrated pattern of answers illustrates how there is low awareness of the multiple content of the organisational forms. Actually no importance is assigned to the possibility of allocating rights about the technology to the farmers' decision.

14.4.2 Multidimensional scaling

We carried out a classical metric MDS computing on the following variables: age, education, geographical scope of professional experience and knowledge of regulation (Table 14.1, for details). We identified the satisfying number of dimensions by the stress index (Kruskal and Wish, 1978). Figure 14.2 reports the stress value associated to the increasing number of dimensions.

The stress value falls from 0.1094 to 0.0759 when the model passes from 4 to 5 dimensions. We then adopt a 4 dimension representation. Figure 14.3 illustrates the distribution of the respondents units in the space of the first two dimensions. We can see that there are two main groups of respondents' associated with five smaller groups and a few single units. This result indicates that there is no convergence among the respondents in relation to the possibilities opened by the new CAP. In light of the conceptual framework, the implication of this result is that we should expect a more intense interaction between the field of the policy intervention, public discourse, beliefs and practices. However, convergence between the units within the group seems more intense and this may suggest that opportunities exist to the emerging closer perspectives. Moreover, we contend that the internal diversification observed

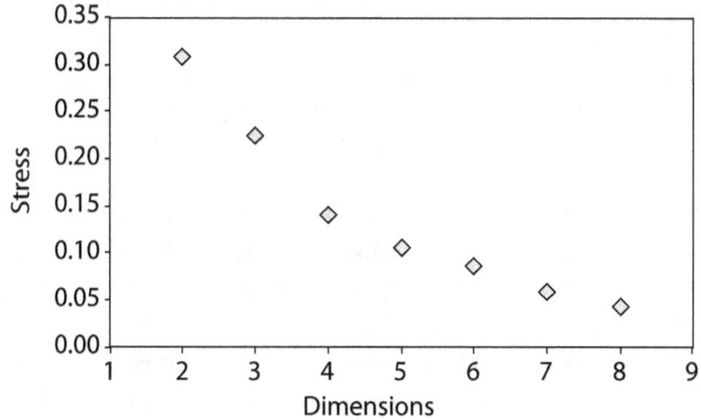

Figure 14.2. Multidimensional scaling analysis; stress and dimensions.

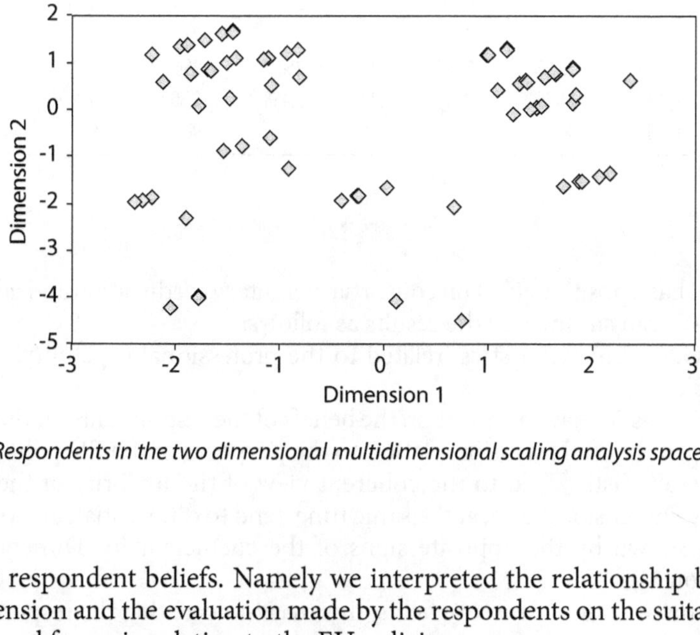

Figure 14.3. Respondents in the two dimensional multidimensional scaling analysis space.

may affect respondent beliefs. Namely we interpreted the relationship between the MDS dimension and the evaluation made by the respondents on the suitability of the organisational forms in relation to the EU policies.

For this purpose we regressed the score assigned to each form on the MDS dimensions (Kruskal and Wish, 1978). As we are dealing with dichotomic variables (1 if the form is considered, 0 if not), we used a logistic regression. Table 14.10 summarises the results. Short-term contract, private brand and cooperative scores are not explained by the respondents in the representation we made. Dimension 1 is, instead, negatively associated to all the remaining forms except consortia. Dimension 3 has no effect.

Table 14.10. Interpretation of the multidimensional scaling analysis dimensions by logistic regressions (standard errors in parentheses).[1]

	Short term contracts	Long term contracts	Producers organization	Consortia	Protection consortia	Voluntary certification	Geographical indications	Private brand	Cooperatives
Dim_1	-0.0486	-0.0979*	-0.0884*	-0.0725	-0.114*	-0.0940**	-0.0791*	-0.0716	-0.0285
	-(0.07)	-(0.06)	-(0.05)	-(0.05)	-(0.05)	-(0.04)	-(0.04)	-(0.06)	-(0.06)
Dim_2	0.00802	0.00184	0.00511	-0.183**	0.023	0.0454	-0.029	0.00558	-0.111
	-(0.08)	-(0.05)	-(0.06)	-(0.09)	(0.07)	-(0.06)	-(0.08)	-(0.05)	-(0.07)
Dim_3	0.0124	-0.036	-0.0409	-0.0471	-0.042	-0.0386	0.0291	0.0796	-0.0403
	-(0.09)	-(0.06)	-(0.07)	-(0.08)	(0.08)	-(0.05)	-(0.06)	-(0.07)	-(0.08)
Dim_4	0.0105	-0.00354	0.0933	0.178**	0.047	0.142*	0.15*	0.125	0.106
	-(0.10)	-(0.07)	-(0.09)	-(0.08)	(0.12)	-(0.08)	-(0.08)	-(0.09)	-(0.09)
Constant	2.552***	2.750***	2.812***	2.750***	2.43***	2.375***	2.573***	2.521***	2.615***
	-(0.09)	-(0.08)	-(0.09)	-(0.09)	(0.10)	-(0.08)	-(0.08)	-(0.08)	-(0.10)
R^2	0.015	0.069	0.06	0.129	0.06	0.093	0.074	0.066	0.041
AIC	253	233.6	249.7	254	277.7	237.8	241.2	239.8	266.5
BIC	265.9	246.4	262.5	266.8	210.6	250.7	254	252.6	279.3
F	0.17	1,279	1,185	2,874	1.37	2,458	2,309	0.969	0.994
N	96	96	96	96	96	96	96	96	96

[1] *$P<0.10$; ** $P<0.05$; *** $P<0.01$.

Dimension 4 has a positive effect on consortia, voluntary certification and geographical indication. We can summarise the results as follows:

- the individual characteristics, related to the professional capability, vary across the sample;
- the characteristics have an effect on the beliefs of the respondents on the suitability of the organisational forms in relation to the purposes of the EU policy;
- these characteristics lead to the coherent view of the majority of the forms (as shown by Dimension 1 and, at the same time, tend to differentiate the respondents view (as shown by the opposite signs of the coefficient for Dimension 1 and Dimension 2);
- some forms are out of the score fields, this may indicate that the respondents tend to focus only on some of them;
- some of these characteristics do not pertain to the organisational forms.

14.5 Final remarks

This study focused on some aspects concerning the relation between the CAP reform and the consequent organisational issues arising at AISS level. We considered these issues in the context of Transaction Costs Economics emphasizing the role of private ordering in achieving the outcomes expected by the EU and the farmers. The analysis emphasized the importance of the institutional framework surrounding the farm (Saccomandi, 1998) and focused on stakeholder beliefs. Our approach in particular showed the heterogeneous perceptions of a sample of stakeholders with respect to the organisational potential solution and the related outcomes. The results also show that a relation exists between the specific stakeholder beliefs on EU regulation and organisation and the assessment they assign to the usual organisational arrangements. This evidence points out the importance of stakeholder beliefs and of their institutionalisation.

References

Adler, M. and Ziglio, E., 1996. Gazing into the oracle: the Delphi method and its application to social policy and public health. Jessica Kingsley Publishers, London, UK.

Ajzen, I., and Fishbein, M., 1980. Understanding attitudes and predicting social behaviour. Prentice-Hall, Englewoo Cliffs, NJ, USA.

Aoki, M., 2011. Institutions as cognitive media between strategic interactions and individual beliefs. Journal of Economic Behavior & Organization 79: 20-34.

Bakucs, L.Z. and Fertö, I., 2005. Marketing margins and price transmission on the Hungarian pork meat market. Agribusiness 21(2): 273-286.

Bennett, R.M., Anderson, J. and Blaney, R.J., 2002. Moral intensity and willingness to pay concerning farm animal welfare issues and the implications for agricultural policy. Journal of Agricultural and Environmental Ethics 15(2): 187-202.

Borg, I. and Groenen, P. J., 2005. Modern multidimensional scaling: theory and applications. Springer Science & Business Media, New York, NY, USA.

Burton, R.J. and Schwarz, G., 2013. Result-oriented agri-environmental schemes in Europe and their potential for promoting behavioural change. Land Use Policy 30(1): 628-641.

Cacchiarelli, L., Lass, D. and Sorrentino, A., 2016. CAP reform and price transmission in the Italian pasta chain. Agribusiness 32(4): 482-497.

Chevassus-Lozza, E. and Galliano, D., 2009. Intra-firm trade in the context of European integration: evidence from the French multinational agribusiness. Agribusiness 25(1): 128-143.

Coase, R.H., 2013. The federal communications commission. Journal of Law and Economics 56(4): 879-915.

Daniel, F.J. and Perraud, D., 2009. The multifunctionality of agriculture and contractual policies. A comparative analysis of France and the Netherlands. Journal of Environmental Management 90: S132-S138.

Dedeurwaerdere, T., Polard, A. and Melindi-Ghidi, P., 2015. The role of network bridging organisations in compensation payments for agri-environmental services under the EU Common Agricultural Policy. Ecological Economics 119: 24-38.

Dixit, A., 2009. Governance institutions and economic activity. American Economic Review 99(1): 3-24.

European Commission (EC), 2009. A better functioning food supply chain in Europe. Communication COM(2009) 591. EC, Brussels, Belgium.

Fazio, R.H., 1986. How do attitudes guide behaviour? In: Sorrentino, R.M. and Higgins, E.T. (eds.) The handbook of motivation and cognition: foundations of social behavior. Guilford Press, New York, NY, USA, pp. 204-243.

Gilbert, C.L. and Morgan, C.W., 2010. Food price volatility. Philosophical Transactions of the Royal Society B: Biological Sciences 365(1554): 3023-3034.

Giordano, R., D'Agostino, D., Apollonio, C., Scardigno, A., Pagano, A., Portoghese, I., Lamaddalena, N., Piccinini, F. and Vurro, M., 2015. Evaluating acceptability of groundwater protection measures under different agricultural policies. Agricultural Water Management 147: 54-66.

Hall, C., McVittie, A. and Moran, D., 2004. What does the public want from agriculture and the countryside? A review of evidence and methods. Journal of Rural Studies 20(2): 211-225.

Kallas, Z., Gómez-Limón, J.A. and Arriaza, M., 2007. Are citizens willing to pay for agricultural multifunctionality? Agricultural Economics 36(3): 405-419.

Kruskal, J.B. and Wish, M., 1978. Multidimensional scaling. Vol. 11. Sage Publications, Thousand Oaks, CA, USA.

Levchenko, A.A. and Zhang, J., 2012. Comparative advantage and the welfare impact of European integration. Economic Policy: 567-602.

Martino, G., Frascarelli, A., Ventura, F. and Chiodini, G., 2016. Decision and property rights patterns in organizing Italian dairy sector: case studies evidence for the new EU market regulation. Paper presented at the 153th Seminar of the European Association of the Agricultural Economists. June 11-12, 2016. Gaeta, Italy.

McEachern, M.G. and Willock, J., 2004. Producers and consumers of organic meat: a focus on attitudes and motivations. British Food Journal 106(7): 534-552.

Ménard, C., in press. Organization and governance in the agrifood sector: How can we capture their variety? Agribusiness. DOI: https://doi.org/10.1002/agr.21539.

Mittenzwei, K., Mann, S., Refsgaard, K. and Kvakkestad, V., 2016. Hot cognition in agricultural policy preferences in Norway? Agriculture and Human Values 33(1): 61-71.

North, D.C., 2005. Understanding the process of economic change. Princeton University Press, Princeton, NJ, USA.

Pascucci, S., Dries, L., Karantininis, K. and Martino, G., 2015. Regulation and organizational change in the governance of agri-food value chains. British Food Journal 117(10), https://doi.org/10.1108/BFJ-07-2015-0268.

Rezitis, A.N. and Stavropoulos, K.S., 2010. Modeling beef supply response and price volatility under CAP reforms: the case of Greece. Food Policy 35(2): 163-174.

Rocamora-Montiel, B., Colombo, S. and Salazar-Ordóñez, M., 2014. Social attitudes in southern Spain to shape EU agricultural policy. Journal of Policy Modeling 36(1): 156-171.

Saccomandi, V., 1998. Agricultural market economics: a neo-institutional analysis of the exchange, circulation and distribution of agricultural products (European perspectives on rural development). Royal Van Gorcum, Assen, the Netherland.

Sydorovych, O. and Wossink, A., 2008. The meaning of agricultural sustainability: evidence from a conjoint choice survey. Agricultural Systems 98(1): 10-20.

Tadesse, G., Algiei, B., Kalkuhl, M. and Von Braun, J., 2014. Drivers and triggers of international food price spikes and volatility. Food Policy 47: 117-128.

Varzakas, T. and David Jukes, D., 1997. Globalisation of food quality standards: the impact in Greece. Food Policy 22(6): 501-514.

Walker, G. and Weber, D., 1984. A transaction cost approach to make-or-buy decisions. Administrative Science Quarterly 29: 373-391.

Williamson, O.E., 1985. The economic institutions of capitalism. Firms, markets, relational contracts. Free Press, New York, NY, USA.

Williamson, O.E., 1993. Calculativeness, trust, and economic organization. The Journal of Law and Economics 36: 453-486.

Williamson, O.E., 1998. Transaction cost economics: how it works; where it is headed. The Economist 146(1): 23-58.

Williamson, O.E., 2002. The lens of contract: private ordering. The American Economic Review 92: 438-443.

Williamson, O.E., 2005. Transaction cost economics. In: Menard, C. and Shirley, M.M. (eds.) Handbook of new institutional economics. Springer, Dordrecht, the Netherlands, pp. 41-65.

15. Collective arrangements in the agro-food supply chain: the case of the interbranch organization *Tabacco Italia*

S. Ciliberti and A. Frascarelli*

Department of Agricultural, Food and Environmental Sciences, University of Perugia, Borgo XX Giugno 74, 06122 Perugia, Italy; steciliberti@gmail.com

Abstract

The revision process of the tobacco Common Market Organization was characterized by the objective to reduce the production that caused an increasing uncertainty in supply. To confront such a dangerous situation, shareholders took advantage of Reg. (EU) 1308/2013 and founded the interbranch organization *Tabacco Italia*, which aims to foster a tight and effective coordination between tobacco producers and the tobacco industry. This organization was the result of an interbranch agreement that rules the primary aspects of deliveries, i.e. price, quality, etc. The aim of the present paper is to explore and describe this unique case study with respect to the transaction cost economics theory of hybrids.

Keywords: transaction cost theory, collective arrangement, Common Agricultural Policy, interbranch organization, tobacco

DOI 10.3920/978-90-8686-844-5_15, © Wageningen Academic Publishers 2017

15.1 Introduction

The tobacco sector benefitted from a substantial number of incentives after the Common Agricultural Policy (CAP) of the European Union (EU) was enacted in the early 1960s. For decades, the EU's Common Market Organization (CMO) provided unlimited support for European tobacco production, making this crop the most highly subsidized in the CAP in relation to the area under cultivation. The organization of the common market with respect to raw tobacco was established in 1970, when its production was primarily centred in France and Italy. At that time, approximately 90% of community tobacco production received national financial aid that was intended to facilitate the distribution and/or disposal of the harvest to ensure a fair income for producers (Joossens and Raw, 1991). The goals of the tobacco policy were to establish a common policy throughout the community, expand tobacco production to disadvantaged farming areas, maintain a reasonable income for farmers and adapt production to match those tobacco varieties for which there is demand. As such, these goals implied an additional goal, namely, to reduce tobacco imports.

The key mechanisms of the tobacco policy were norm prices, premiums, export refunds, intervention prices, and maximum guaranteed quantities. Norm prices were prices that the EU wanted the processor to pay the farmer, albeit there was no obligation for the processor to do so. Premiums were those subsidies paid to the processors by the commission to persuade the processors to buy their leaf tobacco from community farmers, and as such, these premiums represented the cornerstone as well as the most expansive part of the entire system. Export refunds were those subsidies paid to processors who sold unwanted varieties of tobacco outside the EU to develop world markets. Intervention prices were the prices, set at 85% of the norm price, guaranteed to farmers or processors for the tobacco they were unable to sell on the open market. Thus, tobacco bought this way was stored by the member states in intervention stores. Finally, maximum guaranteed quantities were quotas for which the premium and the intervention prices were guaranteed. Accordingly, if these quotas were exceeded, both were reduced with the aim to make this variety less attractive economically thereafter.

During its lifetime, several gradual improvements have been implemented with the aim to create a better balance between production and demand (EP, 2009). According to Council Regulation (EC) No. 864/2004, the CAP reform adopted in 2003 was applied to the tobacco sector. The revision process of the tobacco CMO was characterized by the objective to reduce the production of a crop that engendered injury to human health as a response to society's demand. In Italy, 40% of tobacco aid (reference period 2000 to 2002 premium) was decoupled from production and integrated into the single payment scheme (SPS) during a transition period that extended from 2006 to 2009. Hence, in 2010, the tobacco CMO was terminated, and the tobacco market was fully liberalized. From 2010 forward, all subsidies were decoupled, with 50% being transferred to the SPS and 50% being shifted to the restructuring envelope.

Concurrently, the remaining subsidies were gradually reduced by 2013, and only a diminished area-based flat rate survived.

Consequently, in the short term, the cultivation of less profitable tobacco varieties in the EU ceased because the transfer of the tobacco premium into the SPS encouraged producers who were not covering their variable production costs to reconvert to some other land use. As a result, farmers introduced less profitable crops as substitutes for a higher value/higher labour-intensive crop, i.e. tobacco. Thus, this implied that highly intensive crops were replaced by less demanding and more mechanized crops, which cause a substantial increase in unemployment (Manos *et al.*, 2009). With respect to Italy, which, in 2009, was a leader in the European tobacco sector, with 28,564 hectares producing 97,798 tonnes of tobacco, the impact was significant, as five years later, the cultivated area of 18,248 hectares (-36.1%) produced only 53,924 tonnes (-44.8%). This enormous decrease completely reshaped the territorial redistribution of the cultivation of tobacco, as farmers continued cultivating tobacco in only four regions, namely, Umbria, Toscana, Veneto, and Campania. Moreover, the effects of these policies, which were immediately felt in the agricultural production phase, were reflected in the subsequent processing phase, causing a reduction in those sectors with industries active throughout the tobacco processing chain (Arfini *et al.*, 2005). Thus, these changes resulted in a serious issue for the tobacco supply chain as a whole. Therefore, to ensure stable and sufficient deliveries of raw tobacco for the industry during the period 2010 to 2014, a coupled payment (article 68 of Regulation EC 73/2009) was introduced in Italy that allowed farmers, via producer's organizations, to receive direct aid only after signing a cultivation contract with the industry.

More recently, Article 52 of Reg. (EU) 1307/2013 stated that tobacco could not benefit any longer from coupled payments, and as a consequence, it removed all of the justifications for fostering the adoption of contracts between producers and the industry to regulate provisions regarding tobacco (Ciliberti and Frascarelli, 2014). Thus, it follows that the absence of a specific incentive may generate two main issues for the tobacco supply chain in Italy: a further decrease in tobacco production, especially in less suitable and less developed areas, and a greater uncertainty of deliveries towards first processors and manufacturers. This condition could threaten the functioning of the entire tobacco industry in Italy because, without a competitive supply stage that guarantees stable deliveries, all stages thereafter, i.e. processors and manufacturers, could not function effectively. Therefore, to confront such a dangerous situation, shareholders have exploited the new rules of the CAP 2014 to 2020 reform and, more specifically, Reg. (EU) 1308/2013 (CMO). This has resulted in the founding of the interbranch organization *Tabacco Italia* (OIT), which aims to reorganize the entire supply chain by fostering a tight and effective coordination between farmers and the industry due to an interbranch agreement (IA) that oversees the main aspects of deliveries, namely, price, quality, etc. The main purpose of the present paper is to explore and describe, with respect to transaction cost economics (TCE) theory on hybrids, the functioning of the aforementioned interbranch organization (IO) that represents a unique case study for the tobacco sector in Europe and thereby provide useful insights for policymakers, scholars and stakeholders.

15.2 Theoretical framework

Transaction costs theory focuses on the conceptual development of hybrids, particularly the work of Menard (2004). The family of hybrids is located between markets and hierarchies and ranges from contract-based approaches, which are closer to the market, to more integrated structures, which are closer to the hierarchy. Williamson (1991) argues that hybrids generally mix the two types of adaptation found in polar modes of organization, namely, autonomous adaptation, as in markets, and cooperative adaptation, as in hierarchies. Three main characteristics play a key role in hybrids. First, parties pool part of their resources, while keeping their property rights and their associated decision rights distinct, which makes the selection of partners a critical factor. Second, the main mechanism implemented for coordinating parties who maintain separate decision-making rights is contractual. Third, competition persists among partners in hybrid arrangements (Menard and Valceschini, 2005). Thus, it follows that in such arrangements, contracts provide a framework for facilitating the organization of transactions and thereby promote confidence among the parties that because the relationship is valuable to all, it deserves the requisite efforts and investments. Moreover, contracts play a crucial role among coordinating partners, as they specify the number of parties to be included in the arrangement, the duration of the relationship, the requirements concerning quantities and quality standards, and the adaptation clauses and complementary safeguards.

Hybrids tend to develop in highly competitive markets where pooling resources is perceived as a way to manage significant uncertainties as well as a way to survive. This introduces one of the main characteristics of hybrid arrangements, specifically, the degree of uncertainty surrounding the transactions they intend to organize. The literature on hybrids is quite unanimous with respect to the role played by uncertainty in decisions regarding the level at which partners pool resources. In the case of the tobacco sector, for instance, such uncertainty can grow out of a hybrid because of the inputs required, the outputs expected or the transformation process itself. Assuming competitive markets, problems with inputs are often connected to issues of quality, quality control and the risk of free-riding. However, uncertainty may also result from factors exogenous to the arrangement, such as the institutional environment (e.g. the new CAP reform) that influences the preference for a hybrid mode of organization rather than market or integration, as well as from the choice of a specific form of hybrid (Menard, 2004). However, among the determinants of hybrid arrangements, uncertainty is secondary to the existence of specific investments. Indeed, without a minimal degree of mutual dependence in assets, there would be no hybrid arrangement. It follows that the more specific mutual investments are for both parties, as in a highly labour-intensive and highly technological crop such as tobacco, the higher the risks of opportunistic behaviour and the tighter the forms of control implemented. In fact, hybrid organizations develop because of the advantages partners find in linking some of their investments, and by so doing, they establish and accept mutual dependence. In practice, it is the combination of opportunism, which is due to asset specificity, and miscoordination, which is due to consequential uncertainty, that determines the governance characterizing hybrid organizations while taking into account that parties have a strong incentive to choose the arrangement that minimizes their costs of governance. Hence, in the tobacco sector, these types of organizational arrangements are expected to be implemented in a way consistent with strategies oriented towards minimizing transaction costs, particularly those costs depending on uncertainty (due to the new institutional environment) and on the degree of specificity of investments among parties, i.e. farmers, first processors and manufacturers.

Indeed, problems of coordination across the different stages of the supply chain, with high transaction costs as a consequence, always arise with respect to agricultural products. Since this issue is particularly relevant in the agro-food industry, alternative arrangements, grouped under the concept of hybrids, have played an important role (Martino and Pampanini, 2006). Over the last twenty years, there have been changes in society's expectations of the appropriate role of public intervention in agriculture as well as an increasing desire to reduce the levels of financial expenditures. Competition, therefore, is expected to increase among producers and areas, and production is expected to concentrate in areas with comparative advantages for price competition. However, the latter can be avoided through integration, which not only consists of the merging of productive units, but it can also assume collective forms, such as contract setting within value chains, marketing cooperatives that are common in the primary stage of agricultural markets and all forms of hybrid governance structures. The integration strategy engages both collective and public resources, and moreover, it requires market facilities and coordination instruments that correspond to public instruments and collective structures such as IOs (Dervillé and Allaire, 2014). Towards this purpose, European public institutions have provided a new regulatory framework to improve coordination across supply chains by means of legitimizing IOs. These IOs are important examples of collective organizations of the hybrid type. Generally, an IO is defined as a private organization recognized by the State that gathers all segments of an agro-food chain with the objectives to elaborate contractual policies, guarantee equity among partners and allow the enhancement of chain performance. Once decisions have been approved by the IO, all producers and industrial entities must comply with them. These arrangements could reduce opportunistic behaviour, enforce contractual arrangements and improve coordination to better adapt supply to demand in the long term. In addition, whereas the governance mechanisms, on the one hand, can mitigate transaction costs that would otherwise be carried by the central governments, on the other hand, they can also mitigate risks associated with contractual agreements. Royer *et al.* (2016) recognized that collective organizations, such as IOs, aim to: (1) reduce the variability of deliverables; (2) reduce the variability in price; (3) foster innovative investments; (4) reduce the risk of opportunism from downstream partners through contract enforcement; and (5) face various intensities of uncertainty that affect transactions.

Another important feature of IOs is that they are embedded in a legal framework that institutes a centralized and compulsory negotiation between partners by imposing various conditions regarding product distribution and supply management (Martino *et al.*, 2012). This institutional embeddedness provides IOs the legitimacy needed to adopt and implement measures (Royer *et al.*, 2012). Therefore, on the one hand, IOs are organizational hybrids with partners who have different property and decision rights to construct an agreement by which they pool some of these rights and abandon part of their autonomy. On the other hand, these arrangements are also institutional hybrids in that they mix public and private interests in a specific way as they combine self-regulation mechanisms operated by private partners and a legal framework that supervises and enforces these mechanisms. Consequently, IOs can be considered as hybrid arrangements that combine a self-regulation mechanism operated by private partners along the supply chain with a legal framework that determines the conditions by which these mechanisms operate. Ideally, an adequate legal framework brings to the parties a downward shift in the costs of governance, making the choice of a hybrid arrangement more advantageous and institutionally legitimate.

15.3 Objectives and methods

15.3.1 Objectives

Against the theoretical framework, the aims of the present paper are threefold. First, it accurately describes the main features of the OIT and analyses the effects of the extension of the rules (*erga omnes*), which were applied for the first time in the European tobacco sectors. Second, the work represents a first attempt to shed light on the hybrid's nature of the OIT. Third, it aims to evaluate the interaction between the evolution of the CAP regulatory framework (2014-2020) and the adoption of a new (or revisited) form of governance in the Italian primary sector.

Thus, it follows that this is a (preliminary) exploratory study that investigates the impacts of the new CMO regulation on the governance of the tobacco supply chain and describes the functioning and main features of the new collective arrangement represented by the OIT.

15.3.2 Materials and methods

The most suitable method to pursue the aims of this paper is the case study. This technique is particularly suited to the comprehension of real phenomena, even though it cannot provide any quantitative indication on the phenomena analysed nor can it yield any statistical generalizations. Therefore, the study should not be considered as an interview conducted in a classical survey, and the dimensioning of the sample should be considered irrelevant (Yin, 2009). The research protocol involves elaborating on the following research questions with respect to the policy and theoretical framework:

- How has the new institutional environment influenced the governance of the Italian tobacco supply chain after 2013?
- What are the effects of the new form of governance of the Italian tobacco supply chain after 2013?

Both questions reveal the exploratory and descriptive natures of the present paper and clarify the purposes of the case study approach.

After the specification of the research questions, the appropriate unit of analysis must be selected. To this purpose, the interbranch OIT is identified as the unit of analysis, since it is broadly functional to effectively carry out a case study research. It is particularly suitable to this aim, as it is strongly involved in the changes introduced by the new European regulatory framework in the tobacco sector.

Accordingly, a specific design for the case study depends on the number of case studies that are analysed (e.g. single versus multiple case) as well as the number of units of analysis, i.e. a unitary – holistic case study or a multiple – embedded case study (Figure 15.1).

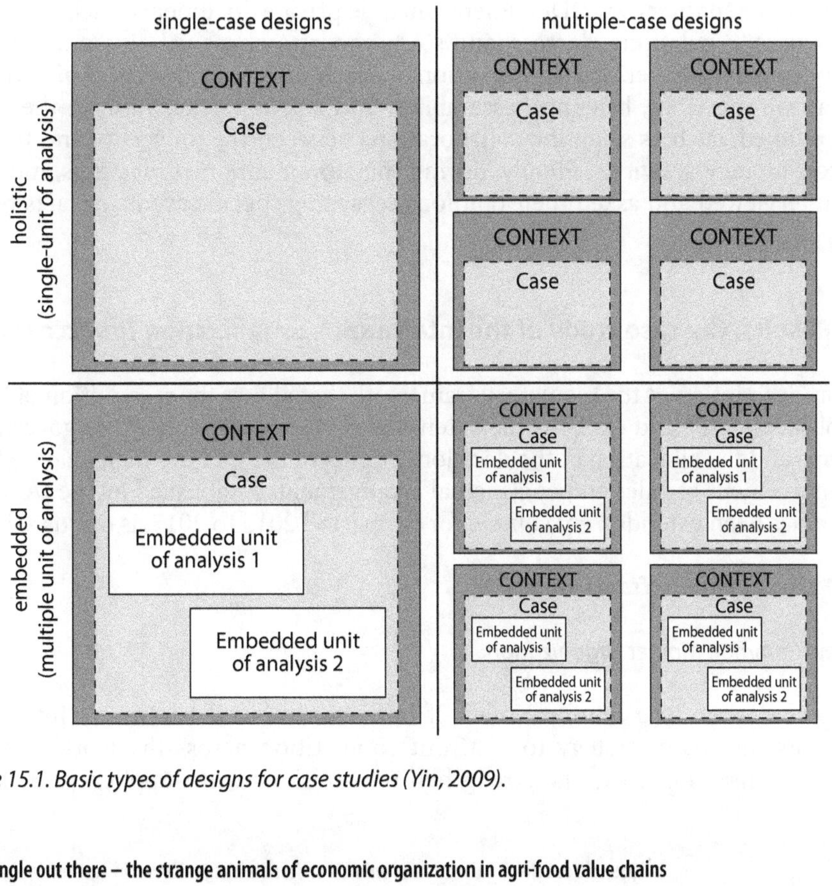

Figure 15.1. Basic types of designs for case studies (Yin, 2009).

Taking into account that the OIT, as a unit of analysis, represents a unique case (Yin, 2009), as it was the first IO in the tobacco sector to benefit from the extension of rules in Europe since the new CMO regulation was enforced, it follows that the holistic single case is the most valid and reliable case study design to address the research questions. Furthermore, because the case study examines only the global nature of an organization (i.e. the OIT), a holistic design is advantageous, considering that no logical subunits can be identified.

With respect to materials, the reliability of the case study research approach stems from the ability to obtain empirical evidence from multiple sources of data. As no single source has a complete advantage over all others, a case study should use as many sources as are relevant to the study (Yin, 2009). If data converge in the same direction, the triangulation of evidence is achieved, and as a consequence, the data gathered are corroborated, and the reliability of the case study is substantiated. The triangulation occurs when various types of evidence support the event or fact constituting the case studies. Indeed, multiple measurements of the same phenomenon lead to a high quality final result.

The information gathering was conducted from June 2015 to February 2016 using three main sources of evidence: documents, multiple (direct and participant) observations and open-ended interviews. Documents include private documents, such as statutes, memoranda, regulations, decrees, and study reports, whose validity and reliability have been carefully verified. Observations consist of site visits of formal activities, such as job meetings, briefings, assemblies, and active participation in the events being studied, such as scientific collaborations between the university and the OIT and consultancy activities. Finally, during the aforementioned meetings, members were interviewed and asked their opinions regarding specific events or facts related to the case study.

15.4 Results: the case study of the interbranch organization *Tabacco Italia*

A short description of the regulatory framework that allows the recognition of IOs in the tobacco sector and explains the extension of rules is provided. The governance, the aims and the allocation of the decisional rights of the OIT are then analysed, and the enforcement of rules and contractual arrangements established by the IA, which in turn has been extended *erga omnes* for the period 2015 to 2017, is discussed.

15.4.1 The regulatory framework

The new Common Market Organization

In a phase where the reinforcement of both vertical and horizontal integration represents the main strategy to confront competition across the world markets, decrease transaction costs and strengthen relationships with suppliers and clients, the stakeholders of the Italian tobacco sector took advantage of the new regulatory framework introduced by the CMO 2014 to 2020 (Table 15.1). Article 157 of the regulation states that member states may recognize IOs that: (1) are composed of representatives of economic activities linked to production and to at least one of the stages of the supply chain; (2) are formed according to the initiative of all or some of the organizations or associations that they constitute; and (3) are in pursuit of a specific aim and take into account the interests of their members and of the consumers, which may include specific objectives as listed in the Regulation[1]. In addition, the new CMO, i.e. Article 164, states that in the tobacco sector, an organization/association could request the extension of rules where it is deemed representative in the economic area or in areas involving a member state, that is, when it accounts for at least two-thirds of the volume of the production of, the trade in, or the processing of the product or products concerned.

Accordingly, the OIT was officially recognized by the national government on the basis of Reg. (EU) 1308/2013, and the Ministry of Agricultural, Food and Forestry Policies (MAFFP) extended its rules *erga omnes* by means of Departmental Decree n.2988 dated 7 August 2015, as the OIT gathered both representatives of producers and first processors, representing 84.8 and 80.3%, respectively, of the tobacco contracted in Italy in 2014 (i.e. 69,166 tonnes). Finally, the OIT also obtained the extension of financial contributions paid by its members as a result of Article 165, which states that:

> individual economic operators or groups which are not members of the organisation but which benefit from those activities shall pay the organisation all or part of the financial contributions paid by its members [...].

[1] Article 162 states that for tobacco sectors, the specific aims may also include at least one of the following objectives: (a) concentrate and coordinate the supply and marketing of the produce of the members; (b) adapt the production and processing jointly to the requirements of the market to improve the product; (c) promote the rationalisation and improvement of production and processing.

Table 15.1. EU regulatory framework for interbranch organizations and extension of rules.

Article	Title
157	Interbranch organizations
158	Recognition of interbranch organizations
162	Interbranch organizations in the olive oil, table olives and tobacco sectors
164	Extension of rules
165	Financial contributions of non-members
168	Contractual relations

15.4.2 Governance and aims

15.4.2.1 Representative bodies and allocation of decisional rights

Members of the OIT are both representatives of producers (*Unione Nazionale Tabacco* – UNITAB and *Organizzazione Nazionale Tabacco Italia* – ONT) and of first processors (*Associazione Professionale Trasformatori Tabacchi Italiani* – APTI) (Figure 15.2). The UNITAB is an association of Italian producers that are also members of the European Association of Tobacco Growers. The ONT, which was founded in 2002 and legally recognized by the MAFFP in 2010, gathers tobacco producer organizations (POs) from Campania, Umbria, and Toscana and represents more than 60% of the Italian production in volume. The APTI represents both Italian first processors and exporters. Its members are the biggest firms in this sector, e.g. multinational enterprises and cooperatives, and it processes approximately 75,000 tonnes of tobacco every year, i.e. 60% of the national volume processed.

The decision-making body of the OIT includes the board of directors (BOD), the president, the members' assembly (MA) and the audit committee. The BOD is equally composed of representatives of both producers and processors, and its members administrate the OIT for a period of three years. The President is alternately a

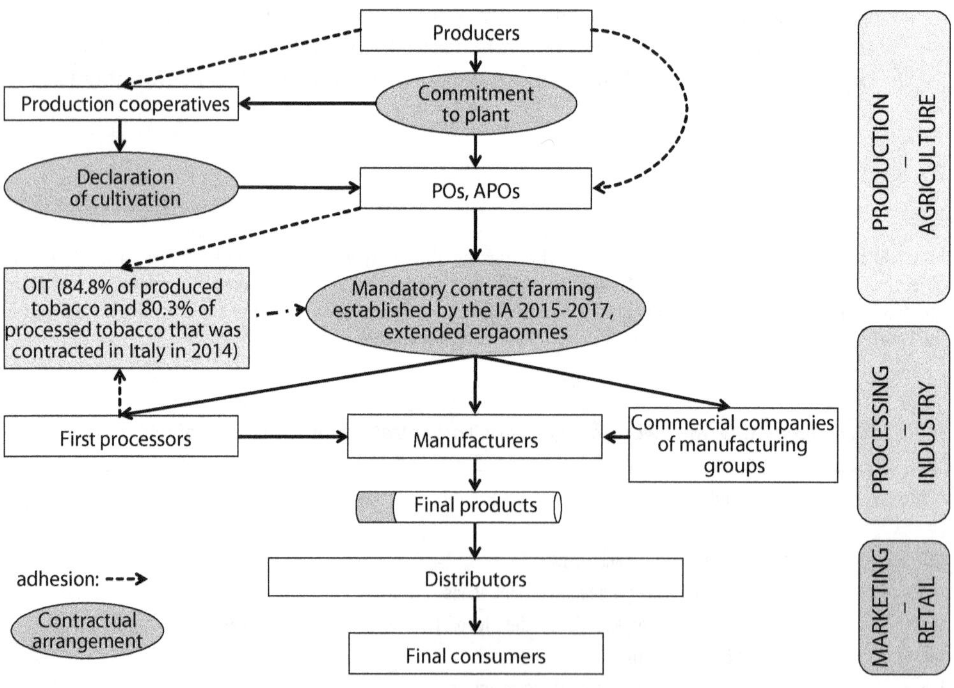

Figure 15.2. The Italian tobacco supply chain and the role of the organization Tabacco Italia.

representative of the farmers or the industry. He promotes the constitution of specific product committees (PCs) and makes decisions about the participation of the OIT in other national, European or international organizations. The MA is composed of two groups that represent, on the one hand, tobacco producers and, on the other hand, processors and traders. PCs participate in the MA but cannot vote. The MA gives directives to the OIT, adopts rules of functioning for the CPs and the OIT, approves the budget and elects the BOD members. Furthermore, according to Article 164 of Reg. (EU) 1308/2013, some decisions adopted by the MA can be extended *erga omnes* to other operators or groups that are not members of the organization.

15.4.2.2 Organization Tabacco Italia aims

According to Article 157 of Reg. (EU) 1308/2013, the OIT pursues specific aims while taking into account the interests of their members. To allow for the proper analysis of OIT functioning, the aims are attributed to the following activity areas:

- organization of supply and market control;
- coordination of supply chain relationships;
- quality of production and competition policy;
- research and development;
- environmental compliance and sustainable production.

Table 15.2 reports the classification of the OIT objectives for the five activity areas previously identified.

15.4.3 The interbranch agreement: enforcement of rules and contractual arrangements

According to the statutory aims, the OIT elaborated an IA for the period 2015 to 2017 that was approved by the MAFFP. The IA focuses on the production and trade conditions that compel the Italian tobacco industry to comply with market requests. The IA consists of specific rules for tobacco producers, their POs or their associations (APOs) and the industrial companies (first processors, manufacturers), and it introduces several possibilities for: (1) marketing coordination; (2) contract framing; (3) qualitative features of tobacco; (4) harvest insurance; (5) production methods; and (6) financial contribution. Table 15.3 describes the primary elements of the IA.

15.4.3.1 Contractual relationships, extension of rules and enforcement

The IA is a collective arrangement that disciplines many aspects to foster the coordination and marketing of raw tobacco in Italy. Accordingly, it defines the main elements of contract farming between producers and processors/manufacturers (Article 2) as well as qualitative requirements for raw tobacco (Article 3). The IA presents a model of the contract valid for the period 2015 to 2017. The contract involves PO and/or APO as well as first processors or manufacturers. Furthermore, it specifies duties for both parties as well as additional information, as reported in Table 15.4.

Table 15.2. Organization Tabacco Italia: objectives and aims for activity areas.

Organization of supply and market control	Concentrate and coordinate supply and marketing
	Coordinate and improve way products are placed on the market
	Draw up standard contracts for sale of agricultural products
	Collaborate with public administrations
Coordination of supply chain relationships	Forecast production potential and record public market prices
	Explore potential export markets
	Promote rationalization and improvement of production and processing
Quality of production and competition policy	Develop methods and instruments to improve product quality
	Establish stricter production and marketing rules
	Fully exploit product potential
	Adapt production and processing to requirements of the market
Research and development	Foster R&D activities by means of research and market studies
	Improve knowledge and transparency of production and the market
	Develop methods and tools that will increase economic efficiency
Environmental compliance and sustainable production	Seek ways to restrict the use of animal-health or plant-protection products and ensure soil and water conservation
	Promote and conduct research into integrated production
	Contribute to the management of by-products and the reduction and management of waste

Table 15.3. Interbranch agreement: main elements.

Marketing coordination	AGEA (Agenzia per le erogazioni in agricoltura, i.e. Italian Agricultural Payments Agency) communicates to the Organization Tabacco Italia (OIT) information concerning cultivated areas and expected production with respect to cultivation contracts
Contract farming	Producers, first processors and manufacturers sign and fulfil contractual conditions established by the interbranch agreement (IA)
Qualitative features of tobacco	Tobacco must fulfil minimum qualitative requirements established by the IA
Harvest insurance	Promotion of the drafting of insurance contracts against weather damage.
Production methods	Producers must comply with guidelines concerning good agricultural practices and products specifications
Financial contribution	Members must make Interbranch fixed financial contribution (established every year by the OIT) to the organization to cover costs directly incurred during production, marketing and R&D activities

Furthermore, according to Article 168 of Reg. (EU) 1308/2013 and Article 62 of the law decree 24 January 2012 (Cresci Italia Decree), the OIT establishes that every delivery of raw tobacco in Italy by a producer to a processor must be covered by a contract among all parties and that the first purchasers (processors/manufacturers) must make an offer for a contract (Ciliberti and Frascarelli, 2013). Furthermore, any

Table 15.4. Interbranch agreement: main elements of contracts for 2015-2017 (own elaboration on Agea Circular n.2015.394).

Item	Decision
Recognition of sellers/purchasers	To sign contracts, parties must be recognized by national authorities
Purchaser's responsibility	Purchasers must communicate the certificate of recognition by 31 March
	Purchasers must transmit the list of associated firms by 25 September
General rules for contracting	The same purchaser may sign more than one farming contract with the same producer organization (PO)/association of producer organizations (APO), but with different farmers
Time	Contracts must be signed by 15 May and PO/APO must send them to AGEA by 20 June
Typology of contracts	1. Contract farming stipulated between the PO/ member by name of a specific number of members with first processors or a manufacturer
	2. Commitment to plant stipulated between the PO/APO member and the PO/APO or between a cooperative member and the cooperative of production
	3. Declaration of cultivation stipulated: (1) between the APO and the PO, when the former signs a 'farming contract' with purchasers, refers to the totality of the commitment to plant; (2) between a cooperative of production (associated with a PO or with a consortium of a cooperative of production) and the PO/consortium, refers to the totality of the commitment to plant
Change of ownership	Not allowed unless the purchaser cannot fulfil the contractual obligations.
Breach of contract	Causes definitive revocation of the recognition that forbids new contract negotiations
Delivery	From 9 January to 15 March for all varietal groups (except for Kentucky and Nostrano del Brenta – 9 January to 31 March)
Payment	Purchasers must pay within 30 days of delivery
	Within 10 work days after payment, APO transfers money to PO
	Within 10 work days of previous money transfer, PO remits money to producers
Administrative controls	Regarding purchasers: controls are concerned with compliance with the time of payment (30 days)
	Regarding PO/APO: controls are concerned with the transfer of money to members
Technical controls	Performed on a sample of 5% of producers involved in contract farming
	Conducted both at the producers' barns/warehouses and at first processors' storage centres

contract or offer for a contract shall be made in writing and shall be made in advance of the delivery. Contracts must include the following elements: (1) price; (2) quantity and quality of the products and the timing of deliveries; (3) duration of the contract; (4) details regarding the payment period; and (5) arrangements for collecting or delivering products. However, it is further noted that all elements of the contracts for the delivery of agricultural products concluded by producers, collectors, processors or distributors shall be freely negotiated among the parties.

Article 9 of the IA requests an extension of the rules and of financial contributions according to Articles 164 and 165 of Reg. (EU) 1308/2013 to all Italian producers, POs,

APOs, processors and manufacturers who are not members of the OIT. As previously noted, the MAFFP allowed for the extension of rules *erga omnes* in August 2015, since it regroups more than two-thirds of the production volume. Table 15.5 presents the rules that have been extended to all actors of the tobacco supply chain in Italy for at least three years, i.e. 2015 to 2017.

Finally, the enforcement of the *erga omnes* is ensured by the Central Inspectorate for Agro-food Products Quality Protection and Fraud Repression of the MAFFP. In the event of non-compliance with the IA, the agents lose, temporarily or definitively, the ability to sign contracts and are assessed an administrative fine not to exceed 10% of the contract value. This penalty system is aimed to dissuade free-riders and hinder the violation of rules. Moreover, in the case where the IA establishes a specific arbitration commission, which is composed of five members (two chosen by producers, two by purchasers and a president chosen by the MA), the aim of the commission is to resolve any controversies between parties and guarantee the enforcement of the agreement.

Table 15.5. Extension of rules for activity areas.

Activity area	Extended rule (item)
Organization supply and market control	Production and market reporting
	Measures to promote and exploit the potential of products
Coordination of supply chain relationships	Development of standard contracts that are compatible with EU rules
	Organization Tabacco Italia (OIT) financial contributions
Quality of production and competition policy	Measures to protect organic farming as well as designations of origin, quality labels and geographical indications
	Stricter production rules than those laid down in the EU or national rules
	Definitions of minimum qualities and minimum standards of packing and presentation
	Use of certified seed and monitoring of product quality
Research and development	Studies to improve quality of products
	Research of methods of cultivation permitting reduced use of plant protection or animal health products and guaranteeing conservation of the soil and conservation or improvement of the environment
	Management of by-products
Environmental compliance and sustainable production	Protection of the environment
	Research to add value to the products through new uses that do not pose a threat to public health
	Plant health or food safety

15.5 Discussion

The present paper highlights the main reasons for the foundation of the OIT and the extension of rules in the tobacco supply chain. Since 2009, the CAP reforms have caused a complete reorganization of the supply chain, but despite the European institutions strongly reducing public aid for tobacco, the production has maintained a central role in the Italian agriculture sector both in terms of value added and employment. Moreover, the rapid decline in the number of inefficient producers, middlemen and first processors from 2009 to 2014 has laid the groundwork for the strengthening of both vertical and horizontal collaborations. Indeed, the liberalization of the tobacco market in the EU has forced farms and industries to establish organizational arrangements that will enhance the competitiveness of the Italian supply chain, improve the quality of raw materials, and deliver a stable supply of products.

These elements note the strong connection between the regulatory framework and the economic results of the production dynamics regarding one of the most funded crops of the European agriculture that has led Italy to create the OIT and resulted in the collaboration of POs, first processors and manufacturers. As previously stated, what makes the OIT an interesting and unique case study is that, since it is highly representative of the Italian tobacco sector, it was allowed to extend *erga omnes* the rules established by the IA, which regulates the main aspects of the deliveries, such as price, quality, etc., as well as contractual arrangements between producers and industrial companies of the Italian tobacco industry.

In this regard, Table 15.6 depicts the influence of the regulatory framework on the determinants of hybrid arrangements with respect to the case study under investigation. Specifically, it highlights: (1) whether and how CMO rules and IA measures interacted with the traditional determinants of transaction costs in the agricultural sector, i.e. asset specificity and uncertainty, according to scholars (Menard and Valceschini, 2005; Royer *et al.*, 2012, 2016); and (2) whether and how the collective arrangements of the IA stimulated resource pooling, which typically characterizes the hybrid form of governance.

Moreover, the contractual arrangement established by the IA represents an organizational solution that fosters resource pooling between farmers and raw tobacco buyers to weaken free-riding activities and manage significant uncertainty in the tobacco world market. Written contracts offer legal protection that acts as an incentive for the extremely high investments characterized by the tobacco supply chain, both from the perspective of supply, i.e. farmers, and demand, i.e. first processors and manufacturing companies.

Moreover, the contractual arrangement entails many conditions that influence the governance of the tobacco supply chain. The commitment to comply with conditions concerning good agricultural practices reduces uncertainty often connected to issues of the quality and the quality control of raw materials while simultaneously requiring

Table 15.6. The influence of the regulatory framework on the determinants of hybrid arrangements: evidence from the Organization Tabacco Italia (OIT) case study.

Common Market Organization (CMO) rules and main interbranch agreement elements / arrangements	Specificity of investments	Contrast to free-riding activities (uncertainty)	Resource pooling
Producers, first processors and manufacturers sign and fulfil contractual conditions (according to Art. 168 CMO) by means of 3 types of written contracts (contract farming, commitment to plant and declaration of cultivation) that include elements such as price, quality, duration, payment period, and delivery	X	X	X
Producers committed to comply with guidelines concerning good agricultural practices and product specifications	X	X	
The same purchaser may sign more than one contract farming with the same producer organization (PO)/association of producer organizations (APO), but with different farmers			X
Change of ownership not allowed unless the purchaser cannot fulfil contractual obligations		X	
Breach of contract causes definitive revocation of the recognition that forbids new contract negotiations		X	X
To sign contracts, both parties must be recognized by national authorities		X	
Extension of rules and financial contributions of non-members (according to Art. 164 and 165 CMO)		X	X
OIT members pay a fixed financial contribution to cover costs directly incurred during production, marketing and R&D activities		X	
Promotion of the drafting of insurance contracts against weather damages	X		
Administrative controls concerning compliance with time of payment and transfer of money to OP/AOP members		X	

specific investments that increase the mutual dependence among assets. Additionally, the official recognition of contractors by national authorities aims to further reduce the risks of opportunistic behaviours, whereas the revocation of such recognition in the event of a contractual breach stimulates resource pooling.

With respect to the extension of rules and financial contributions to non-members, these factors represent an innovative organizational solution, which is embedded in the legal framework of the CMO reform, that enables the improving of coordination and the pooling of resources to more effectively face uncertainties that affect transactions along the tobacco supply chain. With regard to the promotion of insurance against weather damage, such insurance increases the degree of mutual dependence among assets consistent with tighter forms of coordination and governance, such as that established by the IA. Lastly, the administrative controls concerning compliance with on-time payments and transfers of money to the OP/AOP members represent an attempt to enforce contractual clearances to reduce free-riding activities and uncertainty.

The preliminary attempt to analyse the OIT under the lens of the TCE reveals that, even in the absence of specific financial contributions, the OIT may generate efficiency gains and reduce transaction costs by gathering thousands of transactions and eliminating many intermediaries, according to Menard and Klein (2004). At the same time, such a hybrid arrangement maintains the autonomy of producers and industrial companies, thus benefiting more from incentives than what vertical integration would allow while reducing agency problems for processors and distributors. As a consequence, uncertainties over quantity, quality, and price that plague so many agricultural activities, such as tobacco, are reduced, which potentially stimulates investments and lowers transaction costs even more. Accordingly, these arrangements are not automatically detrimental to the final consumers, as is often argued by opponents of these arrangements.

Furthermore, the case study approach applied to the OIT sheds provides information on the institutional embeddedness of the tobacco sector and on the consequences derived from the relationship between the regulatory environment and private initiatives. More specifically, as noted in the extant literature, hybrids, such as the OIT, are characterized by the fact that the two dimensions, namely, private organizational arrangement and institutional embeddedness, are tightly interwoven, which may explain their stability and even their success over time. According to Menard and Valceschini (2005), it follows that both of these dimensions must be considered: the institutional environment, i.e. rules, laws, and policies, and the organizational arrangements embedded in this environment and implemented by agents, i.e. IO members.

In Europe, the agricultural policy established by the EU has long encouraged the formation of networks, some of which were officially promoted and supported. Currently, despite the 'old' tools of the CMO, e.g. quotas, import duties, and export refunds that have been dismantled and/or deactivated, policymakers strive to maintain market stability and improve agricultural competitiveness by means of promoting more vertical and horizontal integration and coordination (Ciliberti and Frascarelli, 2014, 2015). Jolink and Niesten (2012) reported that several studies demonstrate that rules and regulations of the institutional environment influence the formation of hybrids. Such rules mostly concern European or national efforts to introduce various industries to competition that conversely causes an increase in hybrid organizations. Therefore, the general trend is that the liberalization process leads to the creation of quasi-markets, in which efforts to mimic private markets are combined with the provisions of public services.

The OIT case study indicates that the institutional environment influences the choice of a specific mode of hybrid and its characteristics, as a result of the rapid and unpredictable changes over the last several decades in the environment of organizations and that these changes have led to a substantial increase in the number of hybrid collaborations among organizations. Conversely, hybrid formation is surely affected by the consequences of large environmental uncertainties. Indeed, while a strong inducement to form hybrids results from the rapid and unpredictable changes in a firm's environment, a larger environmental uncertainty increases the complexities involved in agreeing on price, quality and volume in alliance contracts. In conclusion, what emerges is that the OIT is a private organization with institutionally embedded functions and that it aims to improve communication activities along the tobacco supply chain, reduce information asymmetries, increase trust levels in the relationships between producers and the tobacco industry and elaborate common growth strategies. It follows that the formation of hybrids, such as the OIT, has been affected by, on the one hand, rules and regulations in the (European) institutional environment that have imposed competition in industries and, on the other hand, by the large (and growing) uncertainty due to the gaps left in the same institutional environment as a result of the dismantling of the protectionist and distortive CAP aids in effect until 2009.

15.6 Concluding remarks

Whereas the CAP reform 2014-2020 abolished all forms of coupled support to the tobacco sector beginning 1 January 2015, during the period 2010 to 2014, farmers received Article 68 payments (Reg. EU 73/2009) if they signed a contractual arrangement with the first processors. Without such a financial incentive, the risk was that contractual relationships would be abandoned and more uncertainty would be introduced. To address the issue of a dangerous deregulation in the Italian tobacco sector that could increase transaction costs along the tobacco supply chain, stakeholders, e.g. representatives of both the agricultural and the industrial world, seized the opportunity offered by the new CMO and undertook an innovative path towards the diffusion of collective arrangements in the Italian agricultural sector due to Regulation (EU) 1308/2013, i.e. CMO 2014-2020.

The present paper reveals that OIT represents, as of now, an interesting and unique case study of how, in the absence of any form of financial contributions, the interaction between the regulatory framework and the institutional environment affects the organizational arrangements in the Italian agro-food sector to increase the efficiency and profitability of the tobacco sector. Indeed, as a reaction to the elimination of a financial incentive that aimed to promote contracts along the tobacco supply chain, this chain initially faced a period of strong and rapid decline in Italy in terms of both production and employment. However, soon afterwards, it promptly reacted to the increasing uncertainty due to the liberalization of the European market by taking advantage of the three policy tools established by the new CMO to promote coordination among farmers and first processors/manufacturers.

The case study approach was used in this preliminary explorative study to analyse the features of the OIT with respect to the TCE theory of hybrids. In this regard, the main characteristics, such as governance, aims, and allocation of decision rights, the enforcements of rules and the contractual arrangements along the supply chain IA, were investigated. This study revealed that the organizational arrangements may reduce transaction costs by means of limiting uncertainty caused by opportunistic behaviour and influenced by the changing regulatory framework and that they may foster specific investments in the tobacco sector.

Finally, special attention was devoted to the close relationship between the private and public dimensions, i.e. the form of the organization and institutional environment, as enabling such a collective arrangement could ensure the stable supply of raw materials for the industrial companies, i.e. first processors and manufacturers, and guarantee a better quality tobacco, which would then stabilize markets and increase the competitiveness and profitability of the entire supply chain.

15.6.1 Limitations and future developments

The present paper has certain limitations to be considered. First, because it addresses a recent topic, it is not feasible to allow any generalization of the results by means of quantitative methods. Thus, the case study approach is the most suitable in that it allows for a preliminary exploration of the OIT case with respect to the theory on hybrids. Notwithstanding, as the holistic single case is often considered less compelling than the multiple cases, the study is considered less robust; however, it must be considered that, by definition, an unusual case, such as the one described in the present paper, is likely to involve only a single case by definition, since the rationale for multiple case designs cannot be satisfied. It follows that the contribution of this study should be considered as preliminary, and therefore, it is recommended that a quantitative research or multiple case study approach with the OIT members as the unit of analysis be conducted after the IA enforced by the OIT produces tangible effects on the Italian tobacco supply chain. Such a study would be extremely useful because it would provide insights regarding the role of this hybrid promoted by the EU institutions and with respect to its effectiveness to achieve a more competitive European agro-food sector.

References

Arfini, F., Donati, M. and Menozzi, D., 2005. Analysis of the socio-economic impact of the tobacco CMO Reform on Italian tobacco sector. Paper prepared for presentation at the 11st Congress of the EAAE 'The future of Rural Europe in the Global Agri-food system'. August 24-27, 2005. Copenhagen, Denmark.

Ciliberti, S. and Frascarelli, A., 2013. L'obbligo dei contratti di cessione dei prodotti agricoli e alimentari: una valutazione degli effetti dell'articolo 62 della legge n. 27/2012. Economia Agro-alimentare 25: 37-61.

Ciliberti, S. and Frascarelli, A., 2014. Direct payments 2014-2020: a qualitative method for evaluating resource allocation scenarios in Italy. Politica Agricola Internazionale 4: 7-22.

Ciliberti, S. and Frascarelli, A., 2015. A critical assessment of the implementation of CAP 2014-2020 direct payments in Italy. Bio-based and Applied Economics 4(3): 261-277.

Dervillé, M. and Allaire, G., 2014. Change of competition regime and regional innovative capacities: evidence from dairy restructuring in France. Food Policy 49: 347-360.

European Parliament (EP), 2009. Alternative and sustainable production for tobacco cultivated areas in the European Union. EP, Brussels, Belgium. Available at: http://tinyurl.com/yakogfpy.

Jolink, A. and Niesten, E., 2012. Recent qualitative advances on hybrid organizations: taking stock, looking ahead. Scandinavian Journal of Management 28: 149-161.

Joosens, L. and Raw, M., 1991. Tobacco and the European common agricultural policy. British Journal of Addiction 86: 1191-1202.

Manos, B., Bournaris, T., Papathanasiou, J. and Chatzinikolau, P., 2009. Evaluation of tobacco cultivation under the common agricultural policy (CAP). Journal of Policy Modeling 31: 225-238.

Martino, G. and Pampanini, R., 2006. Natura economica delle organizzazioni dei produttori agricoli: un'interpretazione alla luce della teoria neoistituzionale. Economia & Diritto Agroalimentare 11(1-2): 129-145.

Ménard, C., 2004. The economics of hybrid organizations. Journal of Institutional and Theoretical Economics 160: 345-376.

Ménard, C. and Klein, P.G., 2004. Organizational issues in the agrifood sector: toward a comparative approach. American Journal of Agricultural Economics 86(3): 750-755.

Ménard, C. and Valceschini, E., 2005. New institutions for governing the agri-food industry. European Review of Agricultural Economics 32(3): 421-440.

Martino, G., Pampanini, R. and Morbidelli, F., 2012. Integration policy in the agri-food chains: theory and empirical evidences. Agricultural Economics 58(9): 409-424.

Royer, A., Menard, C. and Gouin, D.M., 2012. Marketing boards as hybrid governance: a study of the Canadian hog industry. Selected paper for presentation at the IAAE Triennal Conference. August 18-24, 2012. Foz do Iguaçu, Brazil.

Royer, A., Ménard, C. and Gouin, D.M., 2016. Reassessing marketing boards as hybrid arrangements: evidence from Canadian experiences. Agricultural Economics 47: 1-12.

Williamson, O., 1991. Comparative economic organization: the analysis of discrete structural alternatives. Administrative Science Quarterly 36(2): 269-296.

Yin, R.K., 2009. Case study research: design and methods. Sage Publications, Thousand Oaks, CA, USA.

16. Interventionist policies and market failures of European wine regulations

P. Corsinovi[1] and D. Gaeta[2]*

[1]Center of Economics, Hochschule Geisenheim University, Von-Lade-Straße 1, 65366 Geisenheim, Germany; [2]Department of Business Administration, University of Verona, Via Cantarane 24, 37129 Verona, Italy; paola.corsinovi@hs-gm.de

Abstract

Over the year, EU wine policy changed significantly its structure from a policy based on and protection of domestic markets, to a policy that aims to stimulate the quality production and competitiveness on the international scene. The authors, seeks to explain the implementation of the EU wine policies applied to production management with the aim to understand what determines interventionist policies and their (possible) market failures. These 'markets failures', according to the authors, are shown follow two examples: gain of competitiveness for third countries wine producers respect those the EU, due to an increase of EU constraints, and price volatility, especially for bulk wine. This analysis is based on an in-depth analysis of EU wine policies, the financial expenditures and the market relationship of EU wine sector. First, authors have analysed the main interventionist policies adopted in the new single Common Market Organization (Reg.1308/2013; European Commission, 2013) using the financial allocations (2013-2018) in the EU Member States. Second, focus the effect of wine regulations in terms of market and competition. After 47 years of implementations of these regulations, have these reforms allowed the growth of the firms' income and rise the development of demand for the EU wines? EU policy seems to be arrived late to understand the increasing threat posed by the competitive growth of third countries. Many doubt remains instead on the timing and the costs of these regulations, and therefore effectiveness, compared to the benefits.

Keywords: CMO, EU, policy, supply, wine

16.1 Introduction

The Common Market Organization (CMO), of the wine sector is the legal and regulatory basis of the European wine market, covering everything from vineyards to wine production, and in terms of its scope it is one of the most complex structures within the Common Agricultural Policy (CAP). This is due to the complexity and heterogeneity of the policymakers, the institutions and the lobby organizations involved. Following, at least in part, the objectives of the CAP (Article 40 TFEU), the wine policy has introduced a number of instruments designed to address the problems of income stability, controlling supply and to improve competitiveness. Over the years, the EU domestic context and international scenes have (probably) played an important role in the design of wine strategies and interventions, since their inception in 1962. Historically, the EU has often been accused of having policies to achieve protectionist goals: from distortive interventions in the 1970s and 1980s to less distortive payments in the 1990s (after the MacSharry reform), to decoupled payments in the 2000s (after the Fischler Reform), to target payments from 2010 until 2013 (Anania and Pupo d'Andrea, 2015).

The strong arm of the CAP was based on direct support measures implemented through price intervention, a high level of customs protection and by setting a guaranteed average price. Thus generating imbalances in the domestic context between agriculture in the north of Europe and the Mediterranean and between the policies for supporting the income of agriculture (I pillar) than those focused on the rural development programme (RDP) (II pillar).

The complex difficulties and contradictions of an unpredictable market, accentuated issues relating to surplus production and farm abandonment, created market imbalances and distorted income of the wine market (Corsinovi and Gaeta, 2017).

Despite the fact that the overarching aim of all the agricultural and wine reforms had been to control supply and reduce spending, the gap between surplus and demand had become unsustainable and the measures adopted had increase the budgetary costs of the EU and they have not had the desired effects. The result was the increase of the EU cost for support of this system and, the CAP is still, the most integrated of all EU policies and accounts for the largest share of the EU budget – 38% by 2013. This is equal to 363 billion euros of the 960 billion total (EU) (50 billion euros/year): a significant reduction compared to the 417 billion euros (2007-2014).

Agricultural policy literature (Alston and Hurt, 1990; Alston and James, 2002; Bureau and Mahé, 2015; De Gorter and Tsur, 1991; Gardner, 1987; Gardner and Rausser, 2002) has demonstrated that when the market is uncertain, prices are fluctuating and income is unstable, adopting certain agricultural policies can attenuate the negative effects to different extents. Although neoclassical economic theory shows that in the long term liberalist policies are preferable to interventionist policies, the reality of the public support facet of agricultural policy demonstrates the exact opposite. There are different types of impact that public intervention can have on the agricultural sector. It can affect production and consumption structures through rules, standards and institutions that regulate agricultural production, trade and even consumption across the world. Public intervention regulations have taken several different forms and have helped to attenuate volatility on the market. They can also affect income through levies, transfers and controls on profit and income, as well as the price of goods through levies, subsidies and of course price control itself.

The wine sector can be considered itself a textbook case. Many economists have been addressing wine policies and their effects (such as Anderson, 2004, 2014; Anderson and Jensen, 2016; Colman, 2010; Corsinovi and Gaeta, 2017; Dal Bianco *et al.*, 2015; Malassis, 1959; Meloni and Swinnen, 2013; Pomarici and Sardone, 2001, 2009; Spahni, 1988).

This analysis seeks to explain the implementation of the wine policies applied to production management in EU with the aim to understand what determines interventionist policies and their market failures. These 'markets failures', according to the authors, are shown follow two examples: gain of competitiveness for third countries wine producers respect those the EU, due to an increase of EU constraints, and price volatility, especially for bulk wine.

Authors have analysed the main interventionist policies adopted in the new single CMO (Reg.1308/2013; European Commission, 2013) using the financial schedule (2013-2018).

Second, focusing on the effect of wine regulations in terms of market and competition. Price volatility, taking also as example, represent a signal that an EU policy needs to be rethinking. Price variations become problematic when they are large and cannot be anticipated and, as a result, create a level of uncertainty that increases risks for producers, traders, consumers and governments and may lead to suboptimal decisions and represents an 'open door' to extrude new producer with a better cost/income standard. After 47 years of implementations of these regulations, have these reforms allowed the growth of the firms' income and rise the development of demand for the EU wines? Many doubt remains instead on the timing and the costs of these policies, and therefore effectiveness, compared to the benefits for the sector.

16.2 The single Common Market Organization as regards the wine sector: horizon 2014-2018

Under the framework of the CAP in 1962, the CMO was created in order to enable a gradual convergence of prices and the elimination of customs barriers, with the aim of establishing a single market for products with one common customs tariff for the rest of the world.

In 1970 two formal CMO wine regulations relating 'table wine' were published, with the aim of market intervention designed to regulate the wine overproduction or surplus by a rules concerning product control, planting rights; and regulation for quality wine. These two policies kept their autonomy for 29 years, until 1999. Meloni and Swinnen (2013) provide interesting analyses of the political economy mechanism through the historical origins of the EU quality policies. The most important wine regulations being adopted in 1987 (Reg. 822/1987) and in 1999 (Reg. 1493/1999). After less than ten years the new reform entered into force on 2009 (Reg. 479/2008). From 2007 agriculture products are governed by Single CMO (Reg.1234/2007), which replaced all 21 CMOs created between 1962 and 1971. Wine has been included in the single CMO since 2009, after the 2008 reform (Reg. 479/2008).

The new single CMO is actually part of the CAP reform (2014-2020) and it was approved in December 2013.

Only with the 2008 and 2013 reforms, wine policy changed significantly from a policy based on subsidising production and the protection of domestic markets from non-European producers, to a policy that aims to stimulate quality production and the competitiveness of the wine sector on the international scene (Corsinovi and Gaeta, 2017).

There were a multitude of political reasons behind how and why the CMO was modified and how and why the reform process was accelerated. However, Regulation 1493/99 had failed to highlight that the structure behind both the supply and production sectors in Europe was increasingly inadequate in the face of the new consolidated market situation. The idea of a global market dominated by highly competitive large international groups with the concentration of organized distribution and free from the constraints facing Community businesses posed a real threat.

On one hand, the high concentration of the wines from third countries in the USA wine market has become highly competitive for the EU wineries. The growth of wine exports from third countries, which reached 20 million hl in volume in 2015; only 20 years before, in 1995, the export was about 11 million hectolitres (mhl). The EU wine market is still dominated by Spain, France and Italy as the main EU exporting countries, and Germany and UK are the leading importers[1]. Since 2000, the international scenario has been dominated by the USA in terms of volume and value. The EU wine flow into the US market accounts for 3 billion euros, and is a market that seems to be growing (OIV, 2016).

On the other hand, on the demand and consumers' side, globalisation has had a profound impact. Social and religious culture, diet issues and alcohol policies have contributed to major changes in the geography of consumption: with a reduction in

[1] http://tinyurl.com/yasrcuy3.

the traditional EU wine countries versus outside the EU (Corsinovi and Gaeta, 2017; EC, 2006a,b).

The scenario of the EU and the rest of the world's consumption showed two situations: on one side, domestic consumption trends showed a decline in the main EU wine-producing countries (France -43%; Italy -36% and -45% in Spain in the last fifteen years from 2000-2015) with only a few positive trends in Germany 0.8% and the UK with 28%. On the other side, there was a remarkable growth in wine consumption outside the EU (USA 32%; Canada 49%, China 49%, Russia, 53%; Australia 27% from 2000-2015) (Gaeta and Corsinovi, 2014; OIV, 2014, 2016).

While the EU debated how to resolve the high budgetary costs and the consequences of overproduction; the international competition was becoming ever stiffer (in 2005 accumulated wine stocks represented the equivalent of one year of production and the structural surplus was estimated at approximately 14,5 million hl, equivalent to 8.5% of the total production). The EU wine producers were finding themselves at a disadvantage compared to those from third countries, who were often represented by a few restrictive production and markets rules, large multinationals and massive marketing operations.

The problem of overproduction and expenditure have been addressed by means of a policy of limiting production potential, with a 'quota' on new plantations, an incentive for the permanent abandonment of production through a grubbing-up premium (from 2009-2011). In the same time have been set up a gradual withdrawal of distillation subsidies until 2012 (phasing out measures) for crisis distillation, potable alcohol distillation and the use of concentrated grape must measures.

In the history of wine policy, were spent about 782 million of Euro for distillation and 130 million for aid must. This was 73 and 12% of the total budget CMO. These values 'fell' to 270 and 106 million between 2008 and 2011.

At the same time, one of the greatest fears for the European Commission was the 'attack' of new wine players or 'new world wine' on EU wines market (EC, 2006a,b). The measure of promotion in third countries was established with market oriented in order to recover the market share from EU wines (Corsinovi and Gaeta, 2016; Gaeta and Corsinovi, 2014). Regulation 1308/2013 within the Single CMO represents the last wine reform (European Commission, 2013).

According to the financial scheme of the EU and Reg. 1308/2013, the wine budget for the period 2014-2018 is still almost 1 billion per year (6,240 billion in total). Budget to be shared among the EU wine producing countries. From 2008 resources were distributed via national envelope (National Programme Support, NPS).

The regulation was based on measures that can be achieved through national support programmes managed directly by the Member States according to their strategies. Table 16.1 highlights the financial plan for each Member States.

From 2008 with Reform 479/2008 the CMO's financial resources are distributed via national envelopes. The budgetary limits for National Support Programs is 1,248,913 million of € in 2014 and 1,105,054 million of € in 2015-16-17 and 2018.

Looking at the distribution of funds for 2014-2018, we can see that the majority of CMO funds are found in historical viticulture countries, such as Spain, France, and Italy. Looking at small-scale viticulture countries (this term includes small viticulture areas as Cyprus, Malta, Croatia, Austria, UK, Luxembourg and Lithuania), Austria is the most important country in terms of budgetary support, while Malta and UK have no CMO support beyond 2014. Romania and Hungary are the main viticulture countries in Eastern Europe (the support is distributed among Bulgaria, Hungary, Czech Republic, Romania, Slovakia and Slovenia).

Table 16.1. Budgetary plan (2014-2018, €1000) (based on European Commission, 2013).

	2014	2015	2016	2017	2018
Historic viticulture Member States					
Spain	353,081	210,332	210,332	210,332	210,332
France	280,545	280,545	280,545	280,545	280,545
Italy	336,997	336,997	336,997	336,997	336,997
Greece	23,963	23,963	23,963	23,963	23,963
Portugal	65,208	65,208	65,208	65,208	65,208
Germany	38,895	38,895	38,895	38,895	38,895
Small-scale viticulture					
Cyprus	4,646	4,646	4,646	4,646	4,646
Malta	402	-	-	-	-
Croatia	11,885	11,885	11,885	11,885	11,885
Austria	13,688	13,668	13,668	13,668	13,668
United Kingdom	120	-	-	-	-
Luxembourg	588	-	-	-	-
Lithuania	45	45	45	45	45
Viticulture in eastern Europe					
Bulgaria	26,762	26,762	26,762	26,762	26,762
Hungary	29,103	29,103	29,103	29,103	29,103
Czech Republic	5,155	5,155	5,155	5,155	5,155
Romania	47,700	47,100	47,100	47,100	47,100
Slovakia	5,085	5,085	5,085	5,085	5,085
Slovenia	5,045	5,045	5,045	5,045	5,045
Total	1,248,913	1,105,054	1,105,054	1,105,054	1,105,054

The regulation provides for more streamlined national programs with eight eligible measures rather than eleven of the 2008 wine policy: (1) promotion in third countries; (2) restructuring and reconversion; (3) green harvesting; (4) mutual funds; (5) harvest insurance; (6) investments; (7) innovation; and (8) by product distillation: all measures are programmed for 2014-2018.

The new programs no longer contain measures such as: potable alcohol distillation, crisis distillation, funds for grape must and for the use of concentrated grape must that had dominated the scene for more than forty years achieving the goal to 'stabilize markets and ensure a fair standard of living for the agricultural community concerned' (Preamble, Reg. 822/87) by finding some balance between wine production and market outlets, and provides for price support (Spahni, 1988).

The scenario for 2014-2018 marks officially the new orientation of the EU wine policies focusing on the competitiveness and quality of wines. It is currently marked to restructuring and to vineyards reconversion. It still represents the largest share of the future wine policy (45.31% of the budget), followed by the measures implemented with the 2008 wine reform such as promotion and investments (which account for respectively 21.19 and 21.05%) (Figure 16.1).

The financial distribution of each EU countries for national support measures is shown in the following Table 16.2. Member States have decided the distribution based on the expenditure incurred in the previous national program (CMO of 2008).

Restructuring and vineyards reconvention represent the largest share of the future wine policy followed by the investments and promotion. The aim was to improve

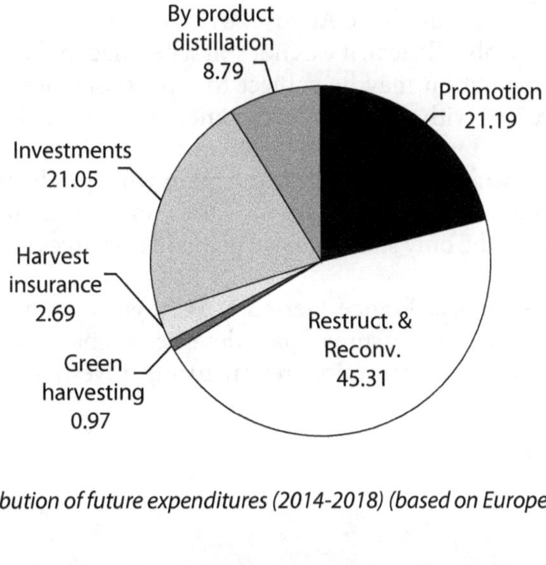

Figure 16.1. The distribution of future expenditures (2014-2018) (based on European Commission, 2013).

Table 16.2. National support programs (2014-2018, €1000) (based on EU Commission, Wine Unit statistics: http://tinyurl.com/y9kmyv2q).

Measures	2014	2015	2016	2017	2018	EU 28 total 2014-2018
Promotion	227,216	231,716	232,716	234,589	234,589	1,160,826
Restructuring and reconvention	537,968	486,605	488,418	485,190	484,613	2,482,793
Green harvesting	10,600	10,600	10,600	10,600	10,600	53,000
Mutual funds	not provided					not provided
Harvest insurance	29,503	29,506	29,503	29,503	29,503	147,515
Investments	181,869	244,732	241,919	242,221	242,798	1,153,358
Innovation	not provided					not provided
By product distillation	96,300	96,300	96,300	96,300	96,300	481,498

the quality of wines and by adapting their vineyards to market demand (varietal changes) through paying compensation for the loss of revenue while a vineyard is being adapted, and as a contribution to the costs of restructuring and conversion.

Among the instruments introduced to support the competitiveness, promotion in third countries, represent an interesting opportunity for the competitiveness. The main beneficiaries are the Appellation of Origin (within the Consortia organization) as well as the temporary associations of businesses and the producer groupings.

The resources were used by firms working in the United States, Canada, China, and Japan to strengthen the markets. The majority of projects concentrated on joint participation in trade show events, contact with journalists from the countries of origin, and training activities for operators in the sector. No amounts seem to be related to innovation measure, since August 2013 (the date of communication of the expenditure by the Member States), it was not still scheduled in the new programming. Probably its implementation may be subject to a postponement due to alignment problems (demarcation) with the rural development programs of the member states.

Table 16.3 shows the strategies adopted by the main wine producing countries. Italy has planned to al locate resources for six measures on eight totals. Italy among the main wine countries is the only states to have planned resources for green harvesting.

Other Member States such as France, Germany, Portugal, have planned to distribute resources thought main four measures. Spain deserves a separate comment: this State has addressed resources into promotion, restructuring, investments and by-products distillation.

Table 16.3. Financial expenditures (2014-2018, €1000) (based on EU Commission, Wine Unit statistics: http://tinyurl.com/y9kmyv2q).

Member States	Promotion	Restructuring and reconvention	Green harvesting	Mutual fund	Harvest insurance	Investments	Innovation	By product distillation
Germany	8,700	83,490	not provided	not provided	8,765	93,520	not provided	not provided
Spain	250,000	418,660	-		-	224,000	-	159,000
France	250,000	502,705	-		-	450,000		200,000
Italy	509,985	700,000	50,000		100,000	225,000		100,000
Portugal	49,998	226,048	not provided		34,998	-		14,998
Total EU 28 MS	1,160,824	2,482,793	53,000	not provided	147,513	1,153,538	not provided	481,498

16.3 The effect of Common Market Organization wine regulations in terms of market and competition

As mentioned above, for many decades, the European wine market has suffered from issues relating to its recurrent overproduction, although the extent of this market imbalance has grown and shrunk over time. The EU has always considered control over vineyards a necessity in order to avoid possible production increases. Previous reforms were characterized for the most part by intervention aimed at controlling surface production in order to reduce surpluses wine quantities and guaranteeing the income stability of producers. EU action to rebalance the market has probably even exceeded its own expectations in terms of creating a progressive contraction of supply, particularly in the main surplus-producing countries.

A focus on the effects on the evolution of the area under vine (1000 ha), after the aids for abandonment and grubbing up, and intervention for supply control with the planting right system, shows that there has been a gradual and continuous reduction in surface area.

After a wave of plantings authorised within the EU by the new CMO at the start of the century (going from 15 member states in 2000 to 27 in 2012 and 28 in 2015), the economic situation led to a trend to replant being globally pushed back or slowed down in some vineyards and to plan grubbing up in others.

In the twenty-five years from 1990, the total EU vine production area fell by 19% (from 4,168 million ha between 1986 and 1990 and 3,362 million ha). The downward trend visible on Figure 16.2 from the mid-1980s to the mid-1990s can, in part, be attributed to the continued block on new plantings, due to both EU provisions and individual Member State choices, as well as to the incentives for distillation and private storage measures.

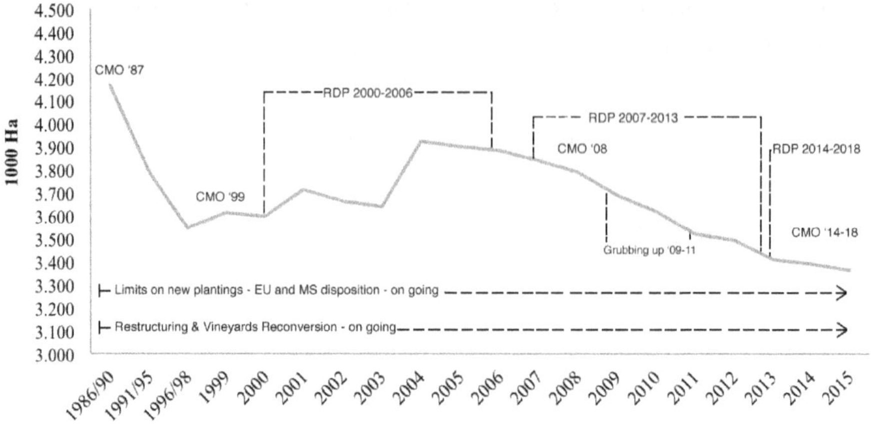

Figure 16.2. Evolution of EU surface area under vine in the main wine countries (1000 ha) (based on Corsinovi and Gaeta, 2017).

The effect of policies on surface area is characterised by a temporal displacement between the grubbing up of vines due to land abandonment, and grubbing up due to the restructuring and reconversion measure first implemented by the CMO in 1980, but which saw its greatest phase of development during the 1999 reform (Gaeta and Corsinovi, 2014). Examining the reasons and the causes of the reduction in EU vineyards raises a great many hypotheses: firstly, it is clear that while on the one hand the EU policy incentivised the abandonment, on the other the ever more insistent intervention – the restructuring and conversion measure that has the objective of increasing the competitiveness of wine producers – had a significant impact on large areas of vineyards across Europe (Gaeta and Corsinovi, 2014). Secondly, the increase in area starting from 2003 was probably linked to the introduction of the 2000-2006 programming period for the RDP, which characterised the second pillar of the CAP with the introduction of measures linked to support the development of farms (Corsinovi and Gaeta, 2017). An example of this in the first RDP programming period is the measure to help young farmers with their first farm investments. The necessary restructuring of old vineyards was done on the basis of increasing production quality, and introducing all the innovative processes offered by modern technology.

Looking at the implementation in the main wine producing countries, France specifically focused its plan for plantings on the years around the turn of the century. For the most part, France concentrated its grubbing-up plan on the same time period, reducing quantity by 28% and systematically reducing surface area under vine in areas where there was the highest risk of overproduction. This drop appears to have been controlled and spread over the last decade, whereas the gradual reduction in Spain and Italy has been followed by accelerated reductions in recent years. The Portuguese wine sector on the other hand saw relative stability until 2009 when there was a reduction of almost 9,000 ha. This is most probably linked to the grubbing-up scheme, but more particularly to the financial crisis, which saw Portugal rank highly among the member states where the tax burden increased between 2010 and 2011(Gaeta and Corsinovi, 2014).

Focusing on the production side, the EU is still the world's leading wine producer, accounting for almost half the global vine-growing area and about 60% of production by volume. Italy, France, and Spain are the EU's largest wine-producing countries, representing 80% of total output, followed by Germany, Portugal, Romania, Greece, Austria, and Hungary. An analysis of the evolution of European production shows clear trends and phenomena. Figure 16.3 shows the historical production (official production 1000 hectolitres) from 1980 to 2015 in the mainly EU wine member states: Italy, France, Spain, Germany, and Portugal. Looking at Figure 16.3 we can see that production peaked around 1980, particularly in Italy and France with more than 80 mhl. Spain saw a more modest change in production levels up to 1990 when it reached 40 mhl, but figures fell between 1991 and 1995 (Gaeta and Corsinovi, 2014). Over the last six years, total production levels fell by 10.22 thousand hectolitres. The year has been marked by significant climatic hazards, which are at the root of these developments, particularly in Europe. All main EU wine-producing countries have suffered drops in production levels, albeit to differing extents. In 2015, EU wine production is likely to reach 165.8 mhl: Italy has become the biggest producer in the world (49.5 mhl), followed by France (47.5 mhl) and Spain (37.2 mhl). Global wine production has reached 274 mhl.

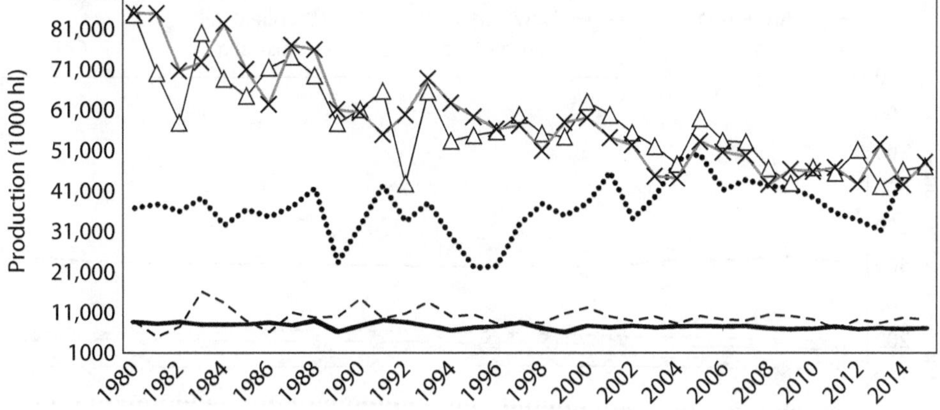

Figure 16.3. The main wine producer countries: historical analysis (based on Eurostat database: http://tinyurl.com/yasrcuy3).

Despite multiple policy interventions to attempt to control them since 1962 that wine has continued to rise. Between 1981 and 2012, the level of European stock grew steadily with peaks in the marketing years 2001-2002 and 2005-2006. In 2002, stock levels were exactly the same as they had been 11 years earlier in 1989 and between 2004 and 2006 stocks progressively increased to 20 mhl. From 2009-2013, the stocks progressively decreased probably as a consequence of grubbing up (2009-2011), and the abolition of distillation measures from 2012. However, distillation measures adopted in the past effectively resulted in transferring the structural surplus issue from the wine sector to the alcohol sector all at a huge cost (Gaeta and Corsinovi, 2014).

In the time that the EU was trying to solve its internal tensions (increasing costs of distillation measures, grubbing up and restructuring measures), the popularity of the wines from third countries, has increasingly become a threat to the member states and the EU has found itself floored, unprepared and playing catch up.

The international situation, shown in the Figure 16.4 and 16.5 shows the dynamics in the main wine producing third countries. Argentina, Chile, Australia, and China represent the first major sign of change in wine geography. China is one of the newcomers to the global wine scene and over the past 15 years it has seen its surface increase significantly. This country is the main driver of the growth of the world area under vines and thus confirmed its position attained in 2014 as the country with the second largest vineyard surface area worldwide (830,000 ha), even if the data include the production of table grapes, which is very popular in China (OIV, 2014, 2016). While vineyards in the United States, New Zealand, South Africa, have remained

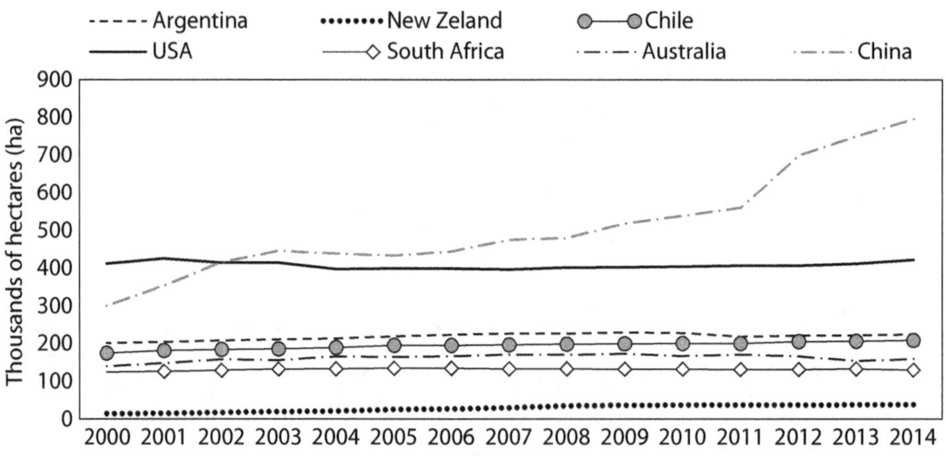

Figure 16.4. Evolution of surface area under vine in the main world wine countries (1000 ha). (based on Eurostat database (http://tinyurl.com/yasrcuy3) and OIV, 2014).

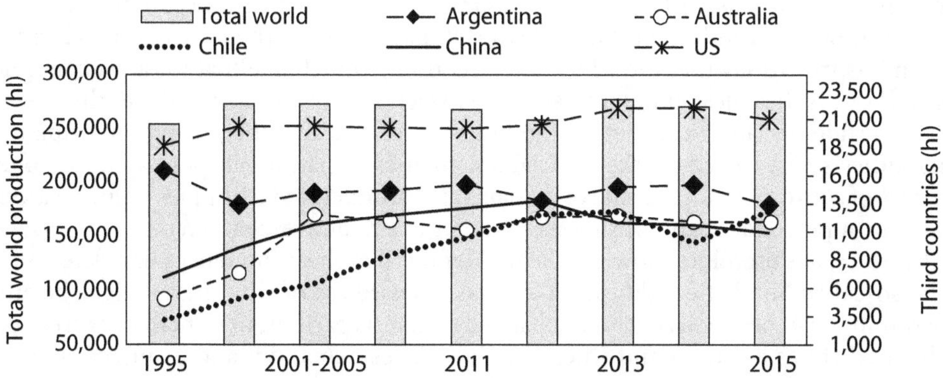

Figure 16.5. The EU's competitors: evolution of production (based on OIV, 2014).

stable, there have been reductions in surface area in Argentina, Australia, Brazil and South Africa.

The EU wine supply on the international market at the end of the 1990s was at a significant disadvantage compared to third countries, which were supported by a liberalist wine production economy with much lower administrative costs and sectorial conditions more inclined to economies of scale. International competition has been booming since the end of the 1990s, although its roots run deeper, and EU policies were very slow to realise the growing threat posed by this. On the production side, between 1990 and 2010 (Figure 16.5), non-EU countries such as Chile, South Africa, Australia and China saw their now famous supply increase. The US was the first confirmed wine producers outside the EU (22.1 mhl) that have seen some reductions since 2011-2012, followed by Argentina (13.4 mhl), Chile (12.9 mhl), Australia (11.9 mhl) and China (11 mhl).

As shown, around the turn of the century more and more 'new world' wines available on the import market, including in the EU (EC, 2006b,c,D). In the years following this, European businesses learnt how to defend, and then regain, their market positions. However, this came at a cost. Large investments on the market let to a marked and in some cases dangerous drop in profitability and competitiveness respect to the farmers in Third countries which are favoured by liberalistic policies (without production restrictions and subsidies).

16.4 What has failed: the politics, the market or both of them?

At the root of the political intervention lies the conviction that the market economy alone is cannot to guarantee either social efficiency or equality. Regardless of the nature of, or the reasons for, a market failure, the State can intervene, but the question here is how much and when should it stop? If markets left to their own devices can fail then it is true to say that the wine sector economy (much like that for any economic good) obeys the rules on the interaction between supply and demand and the effects that certain variables can determine both prices and quantity, using an academic hypothesis of a closed market. Variables, such as variations in per-capita income, prices of substitutes, on the side of demand and production techniques or innovation, on the supply side, can considerably alter the predictive character of any model. Therefore, it is useful to know and clearly define all of these influences that determine the so-called 'market condition'. In any case, regardless of what these might be and assuming that these will remain unchanged or ceteris paribus, in its perfect expression the market has its ethics and when supply satisfies demand it and its prices are said to be in balance. Although it lends itself to many interpretations, the term 'market balance' is defined by equality between the quantity demanded and the quantity supplied $D(p)=S(p)$. Under these conditions, no one has any incentive in changing the situation and the independent decisions of individuals with conflicting interests become compatible (Blanchard, 2006; Talamona, 2001). Sometimes, the balance is altered by internal or external (e.g. hail affecting a vineyard) factors and, demand and supply are not immediately fully adjusted except that prices change. This kind of situation can cause several price and quantity fluctuations and public intervention can act to mitigate the negative effects of a market left to its own devices.

For many years, one of the main problems often encountered in the wine sector relates to the choice and management methods for wine in cellars or present on the market for restocking. When examining the production of supply of wine we need to identify the function that indicates the quantity of output that the businesses are capable of producing corresponding to the various price levels and variations in the amount of stock on the market which at least in the short term can influence the overall quantity of supply. At any price above the equilibrium price, the quantity supplied exceeds the quantity demanded resulting in a surplus or stocks. At any price below the equilibrium price, the quantity demanded exceeds the quantity supplied resulting in a shortage. Only at the intersection of demand and supply are the quantities demanded and supplied equal. Stock variations may be a valid instrument for interpreting trade in bulk wine and the relationship between price fluctuations and changing stock.

Generally, stock variation could generate speculation which in turn can have both stabilizing and destabilizing effects on price fluctuations on the DO wine market. If for example, as shown in Figure 16.6, demand moved from D1 to D2 the equilibrium would also move from a to b and this would cause a price variation from p1 to p2 (falling). Faced with such a situation, wine businesses (particularly those with large volumes and dominant market positions) can decide to wait to sell their produce, thinking that the crisis is only temporary and that prices will return to previous level. With prices so low, consumers may be encouraged to purchase more. This once again moves demand from D1 to D3 and the equilibrium moves to point c equal to price p3 (Sloman, 2004; Blanchard, 2006).

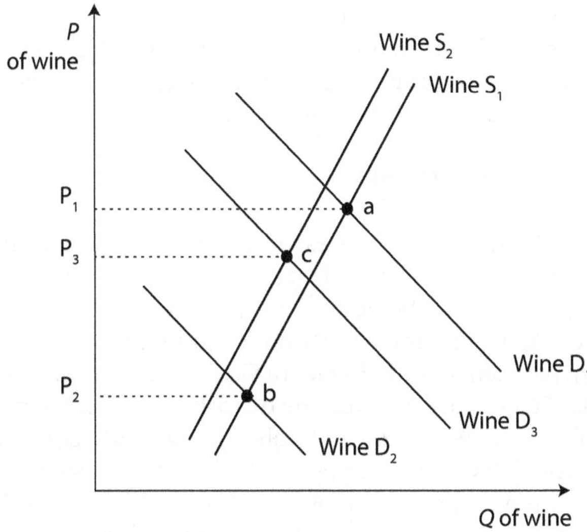

Figure 16.6. Supply control through speculation (based on Sloman, 2004).

This is an example of possible speculation with stabilizing effects. The wine producer knows that after harvest there will be an increasing of supply and if all the produce comes on to the market at the same time, the price will drop. Wine producers may, therefore, keep some of their wine or even buy up competitors' stocks once there is a surplus and the price is low so that they can sell it on once supply diminishes and prices rise. This action can often create price fluctuations which can have real destabilizing effects on the market in the short term, but which eventually lead to price stabilization in the medium term. Thus, a seemingly negative phenomenon can have a calming effect on the market (however, it can be said that this is speculation with a positive effect). In other words, the concentration of large masses of wine in a few manufacturers and bottlers in contact with the large distribution element acts as a 'ceiling price'. On the one hand provides an outlet to the small wine producers (although penalizing of low prices), from the other side eliminates the wine mass market that would otherwise generate more collective bid price decreases.

However, when the price of the good being produced falls, production also tends to gradually decline until it reaches a final equilibrium position. This does not happen instantaneously, but over a certain period of time (the opposite is true if the price rises instead of falling).

In order to respond to this variation in price, the producers may decide that instead of reducing grape production, they will reduce the amount of supply on the market in terms of bottled wine and increase sales of bulk wine or stocks; slowly adapting supply until it reach a new balance point.

As shown in Figure 16.7, the case of European wine prices is unique and at the same time interesting. From the figure, it is clear that the wine sector has not been immune to the phenomena of volatility and price bubbles. Sudden price increase, have been followed by fluctuations not only on an annual basis, but also month on month. Figure 16.7 covers only the historic prices for bulk red wine (€/vol/hl) for the main EU producer countries from 1986 to 2013.

Although on the one hand price increases were sustained by expectations, which in the short-medium term were proven correct, thereby encouraging various forms of investment and acquisition, on the other continued agricultural policy intervention in the form of direct support (from distillation to aid for private storage) and indirect support (single farm payment (within I Pillar of CAP), mutual funds, agri-environment support under Pillar II, etc.) accentuated the distortions. In fact, aiming to intervene to correct market failings, the agricultural policy has actually ended up contributing to market failure through its income support, which aimed to balance the distribution of wealth and its reaction to the issue of overproduction. Price volatility is also a sign that a policy rethink is needed. Even if volatility is typical for the agricultural sector, due the characteristic rigidity of its supply and demand curve, the seasonal nature of its produce, the production's vulnerability to uncontrollable variables (e.g. climate, frost, or unexpected events) and the perishable nature of its produce in some cases, public intervention has not managed to have the desired effect. As shown, between 1996 and 2013 under three wine CMOs Reform, the price of bulk wine continued to fluctuate.

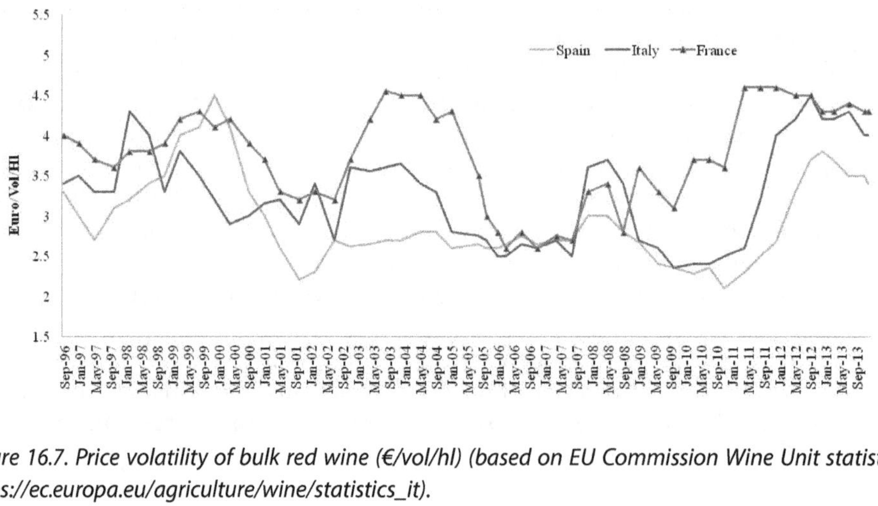

Figure 16.7. Price volatility of bulk red wine (€/vol/hl) (based on EU Commission Wine Unit statistics: https://ec.europa.eu/agriculture/wine/statistics_it).

16.5 Concluding remarks

Wine has become one of the most globalised products in the world. Even if the globalisation phenomenon is not new for the wine sector (Anderson, 2004), its influence over the years has changed significantly. The EU is still the first wine producing area with around 60% of world production, but Europe's share of the world's vineyards has declined from 63% in 2000 to 54% in 2014, the effect of the EU policy to control the production such as the permanent abandonment premiums, ended in 2011 (OIV, 2016). In the meantime, the share of all other world wine regions is increasing, in particular in Asia, which now accounts for 25% of the world's vineyards (OIV, 2016). While continuing on the path of reform which started in the early '90s, it was only with the 2008 and 2013 reforms that wine policy changed significantly from a policy based on subsidising the production and protection of domestic markets from non-European producers, to a policy that aims to stimulate the quality production and competitiveness of the wine sector on the international scene. Only, since the beginning of 2008 the wine policy focuses on competition (promotion in third countries) and interventions that are more structural in nature (like the supports to investments, innovation, restructuring, etc.). The CMO promotion measure (from 2008) was probably the best solution that the EU could imagine. Even if the ability to dispense the budget and the typical Eurocratic bureaucracy have made this policy difficult to use, the interest of wine farmers is increasing daily – as it should be as a result of this measure. The results of public allocation and measures seem to be identified in Rausser definitions about the PERTs and PESTs policy (Rausser, 1982). The policies that are generated to correct market inefficiencies and failure that have the aims to improve a hypothetical collective welfare system, PERTs policy, finish in configuration as PESTs intervention made for private interest of the decision makers (Rausser, 1982). It is also true to say that the European legislator created significant complications for the sector. In many cases, the legislator also granted the Member States the ability to implement important simplifications and it was national bureaucracy, which resulted in these not being introduced. Often when this kind of possibility is granted, people are concerned with how institutions and civil servant positions will continue to exist once there is no longer any reason for them and it is always nicer to be able to pass the buck (Corsinovi and Gaeta, 2017). Previous reforms were characterized for the most part by intervention aimed at controlling production in order to reduce surpluses. The EU holy mission of market equilibrium between supply and demand, based only on quotas or production, was strongly controlled, creating a bureaucracy over cost of production.

As shown in these analyses, restructuring and reconversion measures has become the cornerstone of the EU wine policies. It was probably the best and most efficient policies ever made, both by helping vine growers to rebuild their quality vineyards, control the supply and readapt the grape varieties to consumer trends. A return to a protectionist regulation is obvious in the EU's wine policy where the 2008 decision to liberalise the vineyard planting rights system was overturned in 2013 and a new set of regulations (such as the authorisations system) was implemented. Indeed, the world wine market has been over-regulated, and the effective level of protectionism has probably not changed at all (Dal Bianco *et al.*, 2016; Foster and Spencer, 2002).

EU policy seems to be arrived late to understand the increasing threat posed by the competitive growth of third countries. The future of EU financing and its impact on the CAP, continues to animate the political scenario. Today, the EU is called to the design the new strategies post 2020. The future of the wine policies and the Common Agriculture policies post 2020 is already debated in various instances. Many doubt still remains instead on the timing and the costs of these regulations, and therefore effectiveness, compared to the benefits.

References

Alston, J.M. and Hurd, B.H., 1990. Some neglected social costs of government spending in farm programs. American Journal of Agricultural Economics 72(1): 149-156.

Alston, J.M. and James, J.S., 2002. The incidence of agriculture policy. In: Gardner, B.L. and Rausser, G.C. (eds.) Handbook of Agricultural Economics. Vol. 2B. Elsevier, New York, NY, USA.

Anania, G. and Pupo D'Andrea, M.R., 2015. The 2013 Reform of the Common Agricultural Policy. In: Swinnen, J.F.M. (ed.) The political economy of the 2014-2020 common agricultural policy an imperfect storm. Rowman and Littlefield International Ltd., Centre for European Policy Studies, London, UK, pp. 33-84.

Anderson, K., 2004. The world's wine markets: globalization at work. Edward Elgar, Cheltenham, UK.

Anderson, K., 2014. Excise taxes on wines, beers and spirits: an updated international comparison. American Association of Wine Economists, Working Paper No. 170.

Anderson, K. and Jensen, H., 2016. How much does the European Union assist its wine producers? Journal of Wine Economics 11(2): 289-305.

Blanchard, O., 2006, Macroeconomics. Pearson Prentice-Hal, Englewood Cliffs, NJ, USA.

Bureau, J.C. and Mahé, L.P., 2015. Was the CAP Reform a success? In: Swinnen, J.F.M. (ed.) The political economy of the 2014-2020 Common Agricultural Policy an imperfect storm. Rowman and Littlefield International Ltd., Centre for European Policy Studies, London, UK, pp. 87-132.

Colman, T., 2010. Wine politics: how governments, environmentalists, mobsters, and critics influence the wines we drink. University of California Press, Berkeley, CA, USA.

Corsinovi, P., and Gaeta, D., 2016. The EU wine policy orientations through the budget expenditure analysis (1970-2015). Annual American Association of Wine Economists (AAWE) Conference Bordeaux (France), June 21-25, 2016.

Corsinovi, P. and Gaeta, D., 2017. The European wine policies and their consequences on the global wine trade. Food Economy 19(1): 59-88.

Dal Bianco, A., Boatto, V., Caracciolo, F. and Santeramo, F.G., 2015. Tariffs and non tariff frictions in the world wine trade. European Review of Agricultural Economics 43(1): 31-57.

De Gorter, H. and Tsur, Y., 1991. Explaining price policy bias in agriculture: the calculus of support maximizing politicians. American Journal of Agricultural Economics 73(4): 1244-1254.

European Commission (EC), 2006a. Wine – common market organisation. Available at: http://tinyurl.com/yap3am72.

European Commission (EC), 2006b. Towards a sustainable European wine sector. Available at: http://tinyurl.com/y9r4svxc.

European Commission, 2006c. Communication from the Commission to the Council and the European Parliament – Towards a Sustainable European Wine Sector. Available at: http://tinyurl.com/y8erk3hm.

European Commission, 2006d. Impact assessment, annex to the communication from the Commission to the Council and the European Parliament 'Towards a Sustainable European Wine'. Available at: http://tinyurl.com/y7ql8hr5.

European Commission, 2013. Regulation (EU) No 1308/2013 of the European Parliament and of the Council of 17 December 2013 establishing a common organisation of the markets in agricultural products and repealing Council Regulations (EEC) No 922/72, (EEC) No 234/79, (EC) No 1037/2001 and (EC) No 1234/2007. Official Journal of the European Union L 347/671. Available at: http://tinyurl.com/y7kwqjzl.

Foster, M. and Spencer, D., 2002. World wine market: barriers to increasing trade. ABARE Research Report 02.6. ABARE report for the Grape and Wine Research and Development Corporation. Australian Bureau of Agricultural and Resources Economics, Canberra, Australia.

Gaeta, D. and Corsinovi, P., 2014. Economics, governance, and politics in the wine market. Palgrave Macmillan, New York, NY, USA.

Gardner, B.L. and Rausser, G., 2002. Handbook of agricultural economics. Vol. 1A-1B, 2A and 2B. Elsevier, New York, NY, USA.

Gardner, B.L., 1987. Causes of U.S. farm commodity programs. Journal of Political Economy 95(2): 290-310.

Malassis, L., 1959. La politique agricole de 1946 à 1958. La recherche d'une politique agricole. In: Bulletin de la Société Française Économie Rurale (eds.) Économie rurale. No. 39-40. L'économie agricole Française 1938-1958. Socieàteà Française D'eàconomie Rurale, Paris, France, pp. 233-251.

Meloni, G. and Swinnen, J.F.M., 2013. The political economy of European wine regulations. Journal of Wine Economics 8(3): 244-284.

Organisation Internationale de la Vigne et du Vin (OIV), 2014. The wine market: evolution and trends. Available at: http://tinyurl.com/yda6mlrc.

Organisation Internationale de la Vigne et du Vin (OIV), 2016. State of the vitiviniculture world market. Available at: http://tinyurl.com/y7gt8zlc.

Pomarici, E. and Sardone, R., 2001. *Il Settore Vitivinicolo in Italia. Strutture Produttive, Mercati e Competitività alla Luce della Nuova Organizzazione Comune di Mercato.* INEA, Rome, Italy.

Pomarici, E. and Sardone, R., 2009. L'OCM Vino. La Difficile Transazione Verso una Strategia di Comparto. INEA, Rome, Italy.

Rausser, G.C., 1982. Political economic markets: PERTs and PESTs in food agriculture. American Journal of Agricultural Economics 64(5): 821-833.

Sloman, J., 2004. Essential of economics. Pearson Education, Harlow, UK.

Spahni, P., 1988. The common wine policy and price stabilization. Gower Publishing, Farnham, UK.

Talamona, M., 2001. Riflessi sull'economia. Etas Libri, Milan, Italy.

17. Labelling strategies in the agri-food context: the case study of RSPO certification

C. Riganelli[*] *and A. Marchini*

Department of Agricultural, Food and Environmental Science, University of Perugia, Piazza dell'Università, 06123 Perugia, Italy: chiara.riganelli@gmail.com

Abstract

The research takes into account a new empirical application related to palm oil issue: the case study of environmental certification RSPO. Starting from the mandatory introduction of palm oil among the list of ingredients, as established by Regulation No 1169/2011 of European Parliament and of the Council, the study aims to analyse the strategic firms' behaviour and the consumers' reaction in this matter. We attempt to determine the market effects of labelling program, taking into account the case of RSPO certification. The question we aim to answer is: what is the consumer reaction to voluntary certification of firms about palm oil issue? Results show a positive impact of this new type of certification. Policy and managerial implication are emerged, in order to understand the dynamic of a new mandatory information disclosure in the market.

Keywords: palm oil, certification, labelling strategy, regulation

DOI 10.3920/978-90-8686-844-5_17, © Wageningen Academic Publishers 2017

17.1 Introduction

The vertical information flow can be considered a key issue in a competitive and global food system, especially when new policy dynamics about consumer information occur. Since quality, in its different perceptions, is a difficult issue to measure and to observe, firms voluntarily move to light on their decisions, especially through a certification of their procedures. This condition is particularly interesting when environmental topics are taken into consideration. They have a specific interest among consumers, which represent their environmental consciousness and the green image of firms.

The research takes into account the definition of a current problem statement. On the 25th of October 2011 the European Union established a Regulation regarding the provision of food information to consumers (Regulation No 1169/2011 of the European Parliament and of the Council). According to the Article 1, this Regulation provides the basis for the assurance of a high level of consumer protection in relation to food information. In particular, the main interest focuses on the mandatory wording of the origin of vegetable oil used in the product preparation, starting from December 2014. This event had led to many debates, both on the consumer and firm side. On the one hand, public opinion has raised many concerns about the huge diffusion of the ingredient in the most part of baked products. The sustainability of oil palm plantation was immediately one of the main issues to debate, with regard to the strong and negative environmental impact of plantation in specific areas devoted. So, this phenomenon has lead them to pay greater attention in consumers' purchasing choices.

On the other hand, firms face up to new 'informational dynamics' able to influence their reputation. Tackling against a power general public opinion about this environmental concern, some firm seek Roundtable on Sustainable Palm Oil (RSPO) certification. This is a voluntary certification, and works as an assurance to the consumer about the sustainable standard of palm oil. Through certification process it is possible to insert the RSPO trademark in the packaging.

The research focuses on this 'information dynamic' related to firms behaviours about palm oil issue, investigating the entire population of Italian firms in the NACE classification 'manufacture of rusks and biscuits; manufacture of preserved pastry goods and cakes'. In this study we attempt to determine the market effects of labelling program, taking into account the case of RSPO certification, through a new empirical application. We take into consideration a way through which an increase on the provision of information to consumers is likely to impact the behaviour of firms, and consequently the reaction of consumers. The study aims to analyse the impact of this certification in their outcome, considering the effect in the product information and quality, as expression of consumer reaction. From these evidences, and considering the main literature about voluntarily information disclosure, we aim to answer to the question: what is the consumer reaction to voluntary disclosure of firms about palm oil?

The paper is structured as follows. First, we present a review of literature that considers impacts of environmental quality disclosure on firms behaviour and consumers preferences. This is then followed by a description of the particular palm oil case study, through the analysis of the RSPO certification and the hypothesis that we are going to test. The description of the research methodology and the presentation of our findings follow. These are then discussed and conclusion drawn.

17.2 Theoretical framework

17.2.1 Environmental disclosure

Since the Akerlof paper (1970), many scholars were interesting in the way to model how scarce quality signalling due to information asymmetry between buyer and seller caused a market failure in the agri-food context (Hennessy, 1996; Hobbs, 2003; Holleran *et al.*, 1999).

The consumers' attention grows both in terms of sustainability and health hazard, also because of recent food scandals (Grunert and Wills, 2007; Henson and Northen, 2000; Loader and Hobbs, 1999). Indeed, the consumption habits and lifestyles are increasingly focused both on nutrition's care and environmental consciousness. For this reason, producers are increasingly lead to improve the quality and the quantity of information provided to consumers.

Specific attention was been reserved to environmental issues, about which many scholars try to model the dynamics (Khanna, 2001; Lyon, 2013; Lyon and Maxwell, 2007; Melnyk *et al.*, 2003). Following Khanna (2001), environmental regulation has traditionally taken the form of mandatory regulation. In the recent years we have seen the creation of a demand for environmental self-regulated firms' products. This trend allows specific signals, both for firms and consumers: for the formers, an environmental protection mission declares a particular attention of their production to sustainability. Consumers, otherwise, can signal their preferences for environmental friendly products.

Following Dranove and Jin (2010), we can define an information disclosure as an effort to measure and report verifiable product quality information. A particular way through which disclosure can be effective and have an important role in the consumer choice: a policy label programs. During the last years product labelling has become a policy tool and many scholars have try to model the dynamics (Blackman, 2008; Caswell and Mojduszka, 1996; Lyon and Maxwell, 2007; Teisl and Roe, 1998).

While a policy label program can be a mandatory information provision, firms can be aligned to new policy dynamics through a self-labelling program. As a matter of fact, the literature of information disclosure shows how and why mandated information disclosure policies can induce changes in firms' organization behaviour. This is what is called 'labelling strategy'. In terms of environmental issue, label can take the form of an eco-label, that are voluntarily obtained 'seals of approval' certified by public or private organizations or self-declarations (Khanna, 2001; Van Ravenswaay and Blend, 1999). Following Delmas and Grant (2014), the objective of eco-labels is to reduce information asymmetry between the producer of green products and consumers by providing credible information related to the environmental attributes of product and to signal that the product is superior in this regard to a non-labelled product (Crespi and Marett, 2005).

What is interesting to evaluate, through a new empirical case, is the effectiveness consumer reaction about environmental quality signal, especially when a new information policy turned out (Garvin, 1984; Ittner and Larcker, 1997; Marchini and Riganelli, 2015; Marinelli *et al.*, 2015; Powell, 1995).

17.2.2 Voluntary labelling strategies about environmental issues

There is a wide body of literature trying to evaluate the effectiveness of voluntary actions and their consequences in terms of qualitative and quantitative outcomes (Blackman, 2008; Khanna, 2001; Khanna and Damon, 1999). Particularly about environmental issues, voluntary agreements have a huge diffusion. Environmental labels can be considered as a firm's effort to answer to specific needs of consumers, and environmental labels allows the design of market forces in support of a cleaner environment (Lyon, 2013).

Following Khanna (2001), firms' incentive for environmental self-regulation can be stated to be arising for several reasons: pre-empting the threat of mandatory regulation, better relations with government and with stakeholders which include consumers, suppliers, communities and other firms in the industry.

Consumers and investors have the potential to influence firms' behaviour through actions taken in their self-interest. Indeed, one of the most important effect of a voluntary disclosure is the positive effect on the public opinion and the improvement of relations with customers and the community, increasing the degree of trust and fidelity (Caswell and Padberg, 1992; Ippolito and Mathios, 1990; Jin and Leslie, 2003; Mazurek, 1998; Storey *et al.*, 1999). Green products are considered credence good, so that consumers cannot ascertain the environmental and sustainable impact during purchase or use. So, firms' participation in environmental voluntary programs provides an increase of environmentally friendly reputation and the attempt to convert credence goods in experience goods (Khanna, 2001).

Considering that demand by consumers have been the major motivation to introduce environmental standards (Bjørner *et al.*, 2004; McEachern and McClean, 2002), in this research we aims to analyse the reaction of consumers, read into public opinion key, when a voluntary disclosure about environmental issue is done by firms. So, we enunciate the follow hypothesis:

H: A voluntary environmental disclosure leads to a positive reaction of consumers demand.

17.3 Empirical analysis

17.3.1 The case of palm oil

In the recent years many efforts in the European Union policy were centred on the traceability along the vertical information flow, in terms of transparency and assurance. Recently the EU established a Regulation regarding the provision of food information to consumers (Regulation no 1169/2011 of the European Parliament and of the Council). The Regulation is rightly made

> in order to achieve a high level of health protection for consumers and to guarantee their right to information, it should be ensured that consumers are appropriately informed as regards the food they consume. Consumers' choices can be influenced by, *inter alia*, health, economic, environmental, social and ethical considerations. […] New mandatory food information requirements should however only be established if and where necessary, in accordance with the principles of subsidiarity, proportionality and sustainability.

Particular interest was reserved to Annex VII, point 8, part A, focused on the mandatory wording about the origin of vegetable oil used in the product process. This indication was mandatory since December 2014, and particular attention was reserved to products containing palm oil. This news has had a great impact on the baked products' sector, which turned out to be one of the biggest user if palm oil. This fact is due to the particular characteristics of the oil: it is able to resist at high temperature, is solid at ambient temperature and is liquid during the cooking. These and others characteristics make this oil technologically very efficient in production phases.

However, immediately critics arise on the sustainability of this oil. An important environmental issue turns out. In 2011, palm oil made up 32.8% of the global total vegetable consumption, and roughly three quarters go into production of foodstuffs. Total vegetable oil demand is predicted to double between 2010 and 2050, from 120 to 240 million tons annually (World Bank, 2011). This phenomenon leads to important environmental issues due to answer at a growing demand by industries. In particular, Malaysia and Indonesia are the world's largest suppliers of palm oil, accounting approximately 85% of the total production. This phenomenon leads to increase in land use devoting to the palm oil plantations, included forests. Deforestation leads to several problems, both in terms of global warming and biodiversity.

17.3.2 Roundtable on Sustainable Palm Oil certification

What is the firms' reaction about this palm oil issues? Tackling against a power general public opinion, some firms seek RSPO certification ('RSPO Rules on Market Communication and Claims, 2015'). This may be considered as voluntary action in order to disclose the quality of product, maintaining the fidelity of clients and enhance the brand reputation. RSPO is a voluntary certification, and works as an assurance to the consumer about the sustainable standard of palm oil. The RSPO is a global, multi-stakeholder initiative, formed in 2004, including participants coming from many different industries and from many countries, palm oil producers or users. The members come from many different backgrounds, including plantation companies, manufacturers, NGOs, financial institutions and retailers of palm oil products, environmental and social NGOs, from country that produce and/or use palm oil ingredient. RSPO have three models to promote sustainable palm oil and their commitments: Identity Preserved, Segregated, Mass Balance and Book and Claim. The first three are controlled through the RSPO Supply Chain Certification Standard and relate to the physical control of oil palm along the complete supply chain. The forth one is a certificate trading scheme that starts with the production of certified and sustainable oil palm from the certified plantations.

The main goal is 'to promote the growth and use of sustainable palm oil through cooperation within the supply chain and open dialogue between its stakeholders'. A crucial point of the RSPO certification regards the passage of information about the use of sustainable palm oil products, which must be done in an accurate and clear manner that can be easily understood by stakeholders receiving this information.

The RSPO certification is made by a third-part verification and certification by an accredited Certification Body. The latter certifies sustainable palm oil production, as started in the RSPO Certification System. Growers are assessed for certification once every 5 years and, if certified, are assessed for compliance annually.

The certification, each of the four models, allows use of RSPO trademark on the label, in order to permit consumers to discern firms that participate to the voluntary program. The RSPO registered mark consisting of a rounded palm top accompanied by the letters 'RSPO', the TM mark and accompanied by a valid license number.

In light of this evidence, in the research we may consider certification as a strategic choice of firms in order to enhance their quality and communicate these improvements to consumers. What we consider interesting to investigate is the impacts of this voluntary certification in the consumers' demand. Reformulating the previous hypothesis, now we have:

H: RSPO certification leads to a positive impact on consumers demand.

17.3.3 Data

The subject of this research are Italian firms in the 'manufacture of rusks and biscuits; manufacture of preserved pastry goods and cakes' sector (as defined by the 4-digit NACE code). We decided to select this class of firms because of the great interest of public opinion about palm oil consumption in these kind of products. The data are extracted from AIDA database[1], in which we have the name, the contacts and financial data of Italian companies. We have collected financial data over the last available two years, 2013-2014.[2] These data are matched with information about certification, collected through the database on the RSPO and GreenPalm websites. So, in this way we can collect information about typology and the date of certification over the years considered.

Regarding the information about the use of palm oil among ingredient, do not exist any database containing this feature. A possibility could have been to check the packaging, or website dedicated to palm oil issues. However, a more fast and precise collection of this simple information is through telephone interviews, contacting the quality office of firms and asking for this information. The interviews are made during the first week of March 2016. During the call we ask about use of palm oil among their ingredient.

Considering only active firms, we have a database of 259 firms.[3] An interesting issue concerning the use of palm oil: a part of sample (37% of firms) never used the vegetable oil, because of the respecting of traditional taste and processing, or for firm's policy and mission.

Palm oil certification is a time variant dummy variables, explaining whether and when certification of ingredient is done. We built the variables considering all four types of RSPO certification. This is the independent variable of our analysis that we aim to investigate. As dependent variables we choose the one that best represent the hypothesis that we aim to test. Our data include firms' revenue during the two years considered. Following the study of Jin and Leslie (2003), revenue data alone is insufficient to estimate a demand function. However revenue data alone is sufficient to verify if the demand is responsive to the new information disclosure.

To overcome model misspecification, the study also controls for several variables that influence the revenue and performance. Firm size, measured in class by ISTAT

[1] Bureau Van Dijk – 2015.

[2] Although the absence of 2015 limits a part of significance of the analysis, however we consider interesting this preliminary analysis, taking into account the mandatory wording related to vegetable oils by the end of 2014.

[3] Only for 6 firms we cannot reach information, because they do not answer to the questions. A possibility could be to obtain the information through a research in the supermarkets. However, this a small number, so proceeding without these did not introduce important bias in the analysis.

classification[4], controlled for scale of economies, visibility and reputation on firms to the public opinion (Henriques and Sadorsky, 1996; McWilliams and Siegel, 2000). Firm age was also included as a control variable, in order to control for firm life cycle stage, which may affects profitability and managerial priority (Hanks *et al.*, 1993). Finally, geographical locations of firms are taken into account.

17.3.4 Methodology and results

This section introduces the methodological approach adopted for the empirical analysis, used to test the hypothesis previously proposed. The purpose of the research is to test the nature of the relationships between our interest variable, RSPO certification, and the dependent variable revenue, controlling for some specific characteristics.

Before proceeding to analyse data, it is important to consider the issue related to palm oil use. As previously mentioned, some firm never use palm oil among ingredients, although potentially they could use it. It is important investigate ifthe observable selection factor leads to a bias in the estimation of the outcome equation. For this reason, an Heckit model was employed to test the independence of the selection and the regression equations (Greene, 2011). Testing for revenue regression equation, we can accept the null hypothesis of independence of selection and regression equation, and we can proceed only with firms that use palm oil ingredient. Anyway, this makes analysis more appropriate considering that we want test for certification of ingredient use.

The data collection allows a multivariate panel data analysis (Baltagi, 2005; Greene, 2011), of a strongly balanced panel with 164 firms and 2 observations per firm, so 328 observations in total. Age and Revenue were measured by the logarithm of variables. Furthermore, in order to eliminate possible outliers, all continuous variables are winsorized at the cut points of 1 and 99 (Cox, 1998). Stata Software 12 was used to perform statistical analysis. Table 17.1 shows the descriptive statistics of variables considered in our analysis.

From descriptive statistics we can observe how revenue variable vary among certifiers and not-certifiers. From Table 17.2 we can see bigger mean values of revenue in the group of certifiers than in the not-certifiers one. This evidence suggests a propensity to certify among large firms.

To analyse the effect of our interest variables, palm oil certification, we explicit the relationship of the hypothesis. So, we estimate the following equation, obtained by firm *i* in the year *t*.

[4] ISTAT ranks firms by n. of employees and turnover in Micro (less than 10 employees and less than 2000 euros of turnover), Small (employees between 10 and 50 and turnover between 2,000 and 10,000 euros) and Medium-Large (employees more than 50 and turnover more than 10,000 euros). In the analysis we consider the same classification, but we take into account only the information about the number of employees, in order to avoid endogeneity issue.

Table 17.1. Descriptive statistics.

	Variables	Description	n	Mean	SD	1	2	3	4	5
1	Revenue	Log revenue from sales and service	302	8.117	1.931	1.000				
2	Palm oil certification	Dummy variable coded 1 in the year a firm become certified and thereafter	328	0.122	0.328	0.386	1.000			
3	Size	Class size: micro, small and medium-large	328	1.918	0.735	0.789	0.390	1.000		
4	Age	Log age of firm	322	2.423	0.309	0.534	0.141	0.439	1.000	
5	Area	Geographic area categorised in North, Centre and South-Islands	328	1.805	0.904	-0.432	-0.205	-0.342	-0.266	1.000

Table 17.2. Mean values of revenue of palm oil certification.

	2013 revenue (€ millions)	2014 revenue (€ millions)
Palm oil certified	64.517	42.805
Palm oil not-certified	8.602	6.916

$$REV_{it} = \alpha_i + \beta\, CERT_{it} + \sum_k \gamma_k X_{it} + \delta_4 year_t + \varepsilon_{it}$$

REV_{it} denote the outcome variable, the log Revenue. *CERT* is the variable that represent whether or not palm oil certification was undertaken. X_{it} represents the set of time-variant independent control variables, $year_t$ are the full set of year dummies and ε_{it} is the error term. As show in the equation, we include firm fixed-effects α_i. We also include robust estimation of standard errors, clustered for geographical areas, in order to avoid possible serial correlation and to taking into consideration the violation of normality assumption and presence of outliers.

Following Cameron and Trivedi (2009), fixed effect model allows for unobserved individual heterogeneity that may be correlated with the regressors. Such unobserved heterogeneity leads to omitted variables bias, and in practice it can be difficult to obtain a valid instrument. Data from a short panel, with as few as two periods, offers an alternative way to proceed if unobserved individual-specific effects are assumed to be time invariant. In particular, in our analysis Hausman Test show significance, so fixed effect is considerate. Results are presented in Table 17.3.

Table 17.3. Panel data analysis.[1]

Model variables	Fixed effects revenue[2]
Palm oil certification	0.119** [0.025]
Year dummy – 2014	Yes
Size: small vs micro	0.982* [0.301]
Size: medium-large vs micro	0.890 [0.363]
Age	1.138** [0.146]
Constant	4.572*** [0.393]
Observations	302
Overall R^2	0.380[3]

[1] There are robust standard errors in brackets. Regression include city random-effects (i.e. we cluster the standard errors by geographical area with Huber-White standard errors).

[2] ***P<0.01, **P<0.05, *P<0.10.

[3] Reported R-squares is overall firm measure.

From the results, we can answer to the previous hypothesis:

H: RSPO certification leads to a positive impact on consumers demand.

As we can see in Table 17.3, there is a positive impact for our interesting variable related to palm oil certification. As previously mentioned, for the model Hausman Test shows a significant P-value, which rejects the consistency of random effect estimator. So, we consider results from fixed effect estimators. Considering the entity of the impact, and controlling for some variables, belonging to the group of palm oil certifiers allows an increase in revenue of 14% with regard to those that not belonging it.

17.4 Conclusions

Considering the previous studies about environmental information disclosures, this paper contributes to underline the dynamics taking into account a new economic phenomenon regarding voluntary environmental certifications. The research focused on palm oil issue in the NACE sector 10.72.00, evaluating whether participating in a voluntary certification program has an impact in the consumers' reactions and preferences.

Starting from the literature, it is commonly known that an eco-labelling strategy provides advantages in terms of public opinion. Furthermore, there is a growing trend for policymakers to issue policies about labelling information in order to guide behaviour of market participants. Considering these evidences, empirical studies in this phenomenon have a great interest among economists, in order to determine the market effects of a new labelling program, the reaction of firms and the consumers' response. This research aims to contribute to improve the empirical evidence in this field of study, carry out a new phenomenon investigation as further evidence about disclosure information dynamics about environment, but also their impact in the market.

Panel data analysis considers two year, 2013 and 2014, is was implemented. The results show a positive and significant impact of palm oil certification, with an increase in revenue of 14% for firm certified with regard to those ones that do not have RSPO certification. So, we can conclude that there is a consumers' demand expression for environmentally friendly products. RSPO certification claim on products' pack is a clear signal of sustainable palm oil use and the consumers express their demand for these kinds of products. Anyway, there is an important issue to consider in this result. The palm oil concern is a really new matter, taking place only in the recent years. In particular, a health hazard related to palm oil consumption has emerged in the recent months. Adding the 2015 data can complete our analysis. Anyway, considering the years available, that correspond to the year of policy change (2014) and the previous one (2013), we can interpret our results at a very beginning palm oil concerning.

Our results have interesting implication for managers and policymakers. For the former, the study helps to understand the strategic activities about palm oil information disclosures in order to respond at consumers' demand and improve firm's performance. More in general, policymakers can see the impact of a new labelling regulation, and the effective reaction of firms toward this new issue.

The research is in a very first step about the study of palm oil phenomenon, which is quite complex and affects a lot of study areas. Firstly, there is an important health issue to take into consideration. In order to tack against a power general public opinion about this last aspect, many firms start to reformulate their ingredients composition, considering others oil as substitutes of palm oil. For this reason can be interesting analyse the impact of this kind of choice in the consumers demand.

Furthermore, the study takes into consideration only a specific industry using the ingredient, the 10.72.00 Nace classification. However, considering the great interest and the huge use in this sector, this analysis can be considered as a first approach in the study of palm oil issue. Can be interesting an examination of phenomenon also in a qualitative way, through a firm case study. Analysing a firm through an internal point of view allowing a detailed investigation of the dynamics related to palm oil. As a matter of fact, a qualitative analysis through interviews with organizational actors involved in decision making about voluntary actions and disclosures allow for significant theory building. To conclude, additional studies are needed in order to complete this first tempt, but also in order to evaluate the global dynamic of this new and complex phenomenon.

Acknowledgements

The authors thank the University of Siena for providing access to the AIDA database. A different version of the work was presented at the 153 EAAE Seminar, held in Gaeta in 2016.

References

Akerlof, G., 1970. The market for lemons: quality uncertainty and the market mechanism. Quarterly Journal of Economics 84: 488-500.

Baltagi, B.H., 2005. Econometric analysis of panel data, 3rd edition. John Wiley and Sons Ltd., Oxford, UK.

Bjørner, T.B., Hansen, L.L.G. and Russell, C.S., 2004. Environmental labelling and consumer choice: an empirical analysis of the effect of the Nordic Swan. Journal of Environmental Economics and Management 47: 411-434.

Blackman, A., 2008. Can voluntary environmental regulation work in developing countries? Lesson from case studies. Policy Studies Journal 36(1): 119-141.

Cameron, A.C. and Trivedi, P.K., 2009. Microeconometrics using stata. Stata Press, College Station, TX, USA.

Caswell, J.A. and Mojduszka, E.M., 1996. Using informational labeling to influence the market for quality in food products. American Journal of Agricultural Economics 75(5): 1248-1253.

Caswell, J.A. and Padberg, D.I., 1992. Toward a more comprehensive theory of food labels. American Journal of Agriculture Economics 74: 460-468.

Cox, N., 1998. WINSOR: Stata module to Winsorized a variable. Statistical Software Components, Boston College Department of Economics, Chestnut Hill, MA, USA.

Crespi, J.M. and Marett, S., 2005. Eco-labelling economics: is public involvement necessary? In: Krarup, S. and Russell, C.S. (eds.) Environment, information and consumer behavior. Edward Elgar Publishing, Cheltenham, UK, pp. 93-110.

Delmas, M.A. and Grant, L.E., 2014. Eco-labeling strategies: the eco-premium puzzle in the wine industry. Business and Society 53(1): 6-44.

Dranove, D. and Jin, G.Z., 2010. Quality disclosure and certification: theory and practice. Journal of Economic Literature 48(4): 935-963.

Garvin, D.A., 1984. What does 'product quality' really mean?. MIT Sloan Management Review 26(1): 25-43.

Greene, W.H., 2011. Econometric analysis, 7th edition. Prentice Hall, Upper Saddle River, NJ, USA.

Grunert, K. and Wills, J.M., 2007. A review of European research on consumer response to nutrition information on food labels. Journal of Public Health 15: 385-399.

Hanks, S., Watson, C., Hansen E. and Chandler, G., 1993. Tightening the life cycle construct. Entrepreneurship Theory and Practice 18: 5-30.

Hennessy, D.A., 1996. Information asymmetry as a reason for food industry vertical integration. American Journal of Agriculture Economics 78(4): 1034-1043.

Henriques, I. and Sadorsky, P., 1996. The determinants of environmentally responsive firm: an empirical approach. Journal of Environmental Economics and Management 30: 381-395.

Henson, S. and Northen, J., 2000. Consumer assessment of the safety of beef at the point of purchase: a pan-European study. Journal of Agricultural Economics 51(1): 90-105.

Hobbs, J.E., 2003. Information, incentives and institutions in the agri-food sector. Canadian Journal of Agriculture Economics 51(3): 413-429.

Holleran, E., Bredahl, M.E. and Zaibet, L., 1999. Private incentives for adopting food safety and quality assurance. Food Policy 24(6): 669-683.

Ippolito, P.M. and Mathios, A.D., 1990. Information, advertising and health choices: a study of cereal market. RAND Journal of Economics 21(3): 459-480.

Ittner, C.D. and Larcker, D.F., 1997. Quality strategy, strategic control system, and organizational performance. Accounting, Organizations and Society 22(3-4): 293-314.

Jin, Z.G. and Leslie, P., 2003. The effect of information on product quality: evidence from restaurant hygiene grade cards. Quarterly Journal of Economics 118(2): 409-451.

Khanna, M., 2001. Non-mandatory approaches to environmental protection. Journal of Economic Surveys 15(3): 291-324.

Khanna, M. and Damon, L.A., 1999. EPA's voluntary 33/50 program: impact on toxic releases and economic performance of firms. Journal of Environmental Economics Management 37: 1-25.

Loader, R. and Hobbs, J.E., 1999. Strategic response to food safety legislation. Food Policy 24: 685-706.

Lyon, T.P., 2013. The pros and cons of voluntary approaches to environmental regulation. Reflections on Responsible Regulation Conference. March 1-2, 2013. Tulane University, New Orleans, LA, USA.

Lyon, T.P. and Maxwell, J.W., 2007. Environmental public voluntary programs reconsidered. Policy Studies Journal 35(4): 723-750.

Marchini, A. and Riganelli, C., 2015. The quality management in the olive oil SMEs: an analysis in the southern Italy. Quality – Access to Success 16(146): 97-102.

Marinelli, N., Simeone, M. and Scarpato, D., 2015. Does quality really matter? Variables that drive postmodern consumer choices. Nutrition and Food Science 45(2): 255-269.

Mazurek, J., 1998. The use of unilateral agreements in the United States: the responsible care initiative. ENV/EPOC/GEEI(98)25/Final. Organization for Economic Cooperation and Development, Paris, France.

McEachern, M.G. and McClean, P., 2002. Organic purchasing motivation and attitudes: are they ethical. International Journal of Consumers Studies 26(2): 85-92.

McWilliams, A. and Siegel, D., 2000. Corporate social responsibility and financial performance: correlation or misspecification? Strategic Management Journal 21: 603-609.

Melnyk, S.A., Sroufe, R.P. and Calantone, R., 2003. Assessing the impact of environmental management systems on corporate and environmental performance. Journal of Operations Management 21(3): 329-351.

Powell, T.C., 1995. Total quality management as competitive advantage: a review and empirical study. Strategic Management Journal 16(1): 15-37.

Storey, M., Boyd, G., Dowd, J., 1999. Voluntary agreements with industry. In: Carraro, C. and Leveque, F. (eds.) Voluntary approaches in environmental policy. Kluwer Academic Publishing, Dordrecht, the Netherlands.

Teisl, M.F. and Roe, B., 1998. The economics of labeling: an overview of issues for health and environmental disclosure. Agricultural and Resource Economics Review 27(2): 140-150.

Van Ravenswaay, E.O. and Blend, J., 1999. Using ecolabeling to encourage adoption of innovative environmental technologies in agriculture. In: Casey, F., Schmitz, A., Swinton, S. and Zilberman, D. (eds.) Flexible incentives for the adoption of environmental technologies in agriculture. Springer, Dordrecht, the Netherlands.

World Bank, 2011. The World Bank Group Framework and IFC Strategy for Engagement in the Palm Oil Sector. IFC International Finance Corporation. Available at: http://tinyurl.com/y8kcxkbx.

18. Short food supply chains and shared value on the multifunctional farm: an analysis of determinants

C. Nazzaro, G. Marotta and M. Stanco*

Department of Law, Economics, Management and Quantitative Methods, University of Sannio, Via Delle Puglie 82, 82100 Benevento, Italy; cnazzaro@unisannio.it

Abstract

New models of value creation identify short food supply chain as a competitive strategy of the multifunctional farm, in which citizen-consumers demands are expressed. Society's repeated requests have imposed changes in cultural practices and in the economic players' behaviours, defining new challenges particularly for the agriculture sector that is called to meet consumers' expectations. In this scenario, the European agricultural model promotes diversified and multifunctional activities, and the farm, through the short chain, attracts the citizen-consumer, enhancing the public goods created. The short chain becomes a tool enabling the citizen-consumer to directly satisfy his immaterial needs and the preferred approach to the farm (using goods and services) and the territory. In this perspective, the short chain represents a competitive strategy that opens new opportunities for the repositioning of the multifunctional farm and a tool to distribute value. In this paper, the short food supply chain, in particular, direct sales on the farm, is interpreted as an approach that allows the internalization of multifunctionality in the market and the creation of shared value between producer and citizen-consumer. More specifically, the analysis aims at evaluating value creation generated through direct sale only considering the component that benefits the farmer. The objective of the paper is to investigate the impact of specific variables on the creation and the sharing of value between the producer and the citizen-consumer in the short chain, giving rise, in each case, to a premium price or a minus price. The results identify the circumstances underlying the sharing value and the variables able to influence the success of this type of governance in multifunctional farms. The results of the study provide useful implications for policy makers.

Keywords: wine farms, direct sales, public goods, value creation

18.1 Introduction

The last decades were characterized by meaningful changes of the social, economic and institutional scenarios and of the behaviour of the players of the economic system. These changes, which have redefined the role agriculture has in the post-modern society, represent the consequence of many market failures that affected both the economic model and (food) markets.

Since the mid '80s, the unsustainability of the economic and industrial development model and the crisis of the Fordian and urban-centric model have generated negative externalities, social and territorial imbalances and depletion of natural resources, leading to economic marginality and environmental impoverishment. Between the end of the '90s and the new millennium, new awareness and social instances, mainly in the environmental, nutrition and health fields, have determined an evolution of the purchase and consumption model towards a need for greater responsibility. Citizen-consumers express new requests of healthiness, safety and quality of food products, focusing on their origins, functionality and value/identity and assigning a greater relevance to the social costs linked to the production cycle (Mancini *et al.*, 2015). Sensitivities and social needs are changing, and citizen-consumers are becoming increasingly aware of the relation between nutrition and health for the person's wellbeing (Marotta *et al.*, 2014). Moreover, there is a request for a stronger link with the territory and a need (for greater safety of the products consumed) to recover a direct relation with farmers and place of origin of the products, as well as with rural culture in general. In short, there is a demand for a quality of life that enables the rural territories to be not only a place of production but also of fruition and consumption of goods and services.

The new consumer, conscious of the non-neutrality of his purchases, is careful about the social, ethical and political aspects of the products consumed (Leonini and Sassatelli, 2008; Mancini *et al.* 2012). Critical and responsible towards himself, towards others and the environment (Bovone and Mora, 2007), the post-modern citizen-consumer has been the object of increasing attention in recent literature, and a wide debate over the issue has developed that gives rise to different definitions, some of which relate to the environmental and ethical-social dimensions. The first dimension is the object of definitions that leverage on the green consumer (Carrigan *et al.*, 2004), who buys products and services that have a positive impact on the environment (François-Lecompte and Roberts, 2006), while the second dimension is the focus of definitions that evaluate purchase and consumption based on moral conventions and personal values (Crane and Matten, 2004) oriented towards what society recognizes as 'good' (Smith, 1990).

A consumer's behaviour can be defined as conscious and responsible when the attention for the social quality of a good and/or service becomes a determining factor in buying choices, i.e. when considerations about social and ethical costs associated with the good and/or service in the short and mid-long terms prevail in determining the decision to purchase. In particular, the attention for social quality emerges when the critical and responsible citizen-consumer is reflexive (DuPuis, 2000; Vermeir and Verbeke, 2006) and, when making purchasing decisions, increasingly refers to variables such as origin of the product, production techniques, environmental externalities linked to the production model and social aspects. In this way, through their buying decisions, the consumers express the need to re-discover and preserve identity, values, traditions, cultures and territories (Cembalo *et al.*, 2010), preferring products that have 'other' attributes compared to their immediate value in use (Long and Murray, 2013). This reveals a contractual power of consumers who, with their choices, are able to punish firms (Zamagni, 2004).

In this context, society's renewed demands have imposed changes in cultural practices and in economic players' behaviours, defining new challenges, particularly for agriculture. For this reason, with the reform of the Common Agricultural Policy (CAP), the European Union has tried to give concrete answers through the promotion and support of a multifunctional and diversified agricultural model that values the social functions of agriculture (production of public goods). This European agricultural model adds diversified and multifunctional activities to the traditional ones and, through the short chain, attracts the citizen-consumer to the farm to valorise the public goods created (Marotta and Nazzaro, 2011). The short chain becomes a tool enabling the citizen-consumer to directly satisfy his immaterial needs ('alternative' values embedded in the product, Brunori and Bartolini, 2013) and becomes the preferred approach to the farm (using goods and services) and the territory. In this perspective, the short food supply chain becomes a tool not only to enhance the farm's multifunctionality but also to distribute the value that itself helps to create.

The aim of the paper is to develop an analysis of the short chain, in particular of the direct sale on the farm, as a strategy that internalizes multifunctionality in the market and creates shared value between producer and citizen-consumer. More specifically, the analysis evaluates value creation generated through direct sale only in terms of the component that benefits the farmer.

In this perspective, after presentation of the theoretical background of the multifunctional farm, a model of the short food supply chain is developed as an instrument to create shared value, whose determining variables are investigated through empirical testing.

18.2 Short food supply chain in the multifunctional farm: theoretical background

18.2.1 Multifunctional and multi-value farm

The changing processes described have valorised the social role of multifunctional agriculture in response to the new citizen-consumers' needs and have led to an innovative perspective on value creation in farms.

In the European agricultural model, the multifunctional farm becomes a space of consumption and use of a set of intangible values (health and wellbeing, environment and landscape, biodiversity, traditions and rural cultures, etc.). In this way, multifunctionality directs the farm, although still implicitly, towards social responsibility because it not only responds to market dynamics but also tends to satisfy various immaterial requirements of society linked to the production of public goods, which create inclusive wellness.

This new interpretation of the farm functions has developed a wide theoretical debate examining the topic of multifunctional agriculture from different points of view. In particular, recent studies (Marotta and Nazzaro, 2011) have highlighted how in the secondary functions of the farm, multifunctionality (production of positive externalities), influenced directly by the citizen-consumers' demands, becomes a competitive strategy for value creation. In the theoretical model of value portfolio (Marotta and Nazzaro, 2011), multifunctionality is interpreted as the ability to produce localized public goods (Mollard, 2002), monetized through diversification in the activities of the short chain. Multifunctionality has encouraged a new entrepreneurial vision with respect to environment and food safety issues, internalized in productive activities as resources/opportunities, in the context of a general new focus of the farm not only on consumers but also on society.

In this direction, the modern multifunctional farm is configured as a complex reality that produces market and non-market outputs. According to this scheme, the productive diversification/differentiation of the farm, approaching the citizen-consumers who choose an aware and responsible purchasing behaviour, practicable preferably through the short chain, allows them to live experiences of direct contact with the farmers and benefit from the public goods jointly with the products and services offered. In this way, the short chain, more specifically direct sales, increases the consumers' willingness to pay (premium price for the tradable goods and services purchased, as they incorporate the value of intangible public goods present in the business environment; Marotta and Nazzaro, 2011). For this reason, the short chain becomes the tool that allows the farm to internalize the public goods in the market (marketability of multifunctionality) and keep more value (shared value) for the producer and for the citizen-consumer. In these terms, the most positive effects of this organizational form of market are emphasized in the literature (Brown and Miller, 2008; Hinrichs, 2003; Marsden *et al.*, 2000; Mundler and Laughrea, 2016), in particular at the level of adoption of sustainable agricultural practices, farmers' skills development and job creation. These effects create value for the enterprise, the territory and the local community (citizen-consumers).

In the framework suggested in this paper, the short food supply chain (direct sale) is interpreted as an instrument to valorise multifunctionality, create new market spaces and share value. In these terms, the short food supply chain becomes a competitive strategy that opens new opportunities for the repositioning of the multifunctional farm.

18.2.2 Short food supply chain as a strategy to create shared value: conceptual framework

The multifunctional farm transforms itself from a simple production space to 'market space' through diversification strategies that draw the citizen-consumer closer to the farm (direct sales, agri-tourism, farmer's market, GAS, etc.). In this space, through the short chain, the citizen-consumer becomes the protagonist of his own purchasing choices because it is able to establish a direct relationship with both the producer and the places of production, acquiring a set of information (quality, safety, seasonality of products, processing methods, etc.) that reduces the information asymmetries along the traditional agri-food chain. In some cases, the channels of the short chain allow the citizen-consumer to go further in becoming a co-decision maker in the farm production choices: the strong connection of trust between the farmer and the citizen-consumer allows the latter to orient production decisions of the enterprise on the basis of individual needs and preferences. Hence, the short chain, through direct exchange, generates different value chains, shared by the farmer and the citizen-consumer:

- The farmer will enjoy a premium price obtained from the difference between the selling price at the wholesale and the selling price on the farm, which internalizes the value of localized public goods in the market.
- The citizen-consumer, in turn, will benefit of a minus price, obtained from the difference between the selling price on the farm and the price applied for the same product in retail (Figure 18.1).

The direct sale (short chain) thus produces a convergence of interests between two virtuous behaviours, the farmer's (multifunctionality) and the citizen-consumer's (responsibility), that determines win-win commercial transactions, generating shared value between the two protagonists of the exchange, in the form of a premium price (farmer) and a minus price (citizen-consumer). However, depending on the cases, the value created through the short chain will sometimes benefit the producer and other times the citizen-consumer, based on the price established for the direct sale. The higher the price, and the more in line with the retail price, the greater the portion of premium price for the producer. Vice versa, the lower the price, and the more in line with the wholesale price, the higher the portion of the minus price of the citizen-consumer.

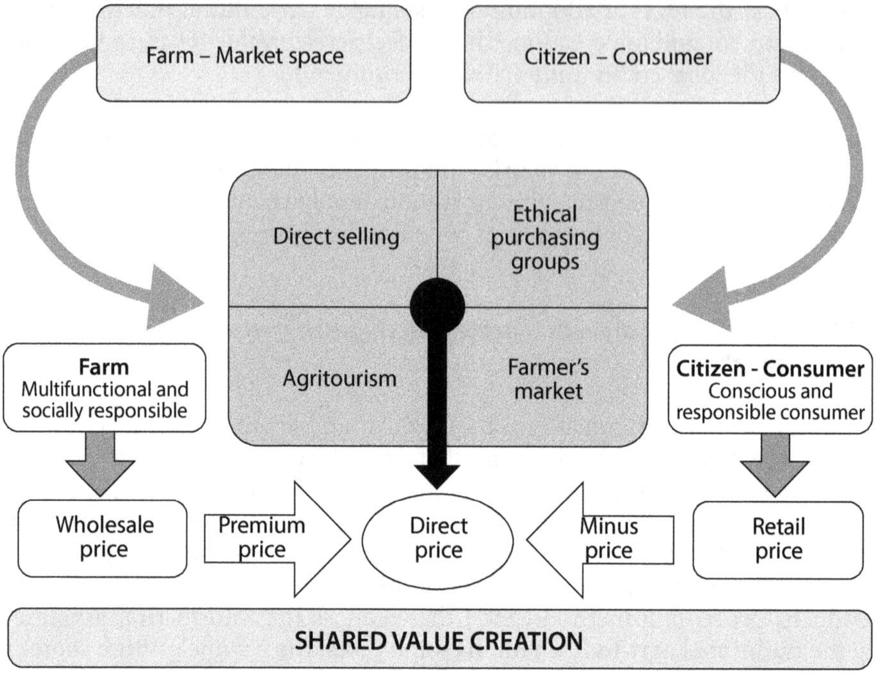

Figure 18.1. Short food supply chain and shared value.

In this model, the direct sale represents for the farmer the instrument through which multifunctionality finds a market recognition represented by the premium price on products sold directly on the farm.

In the light of the above, we verified empirically which variables influence the sharing of value creation (definition of the direct sale price) through the short chain. In particular, we investigated the weight of the variables of human capital, multifunctionality and social responsibility in the short chain performance to verify how direct sales create shared value.

18.3 Empirical evidence

18.3.1 Research design and methodology

The aim of this empirical analysis is to investigate the impact of certain variables on the creation and sharing of value between the producer and the citizen-consumer in the short chain, giving rise in each case to a premium price or a minus price.

The variables considered for the analysis relate to three areas of interest within the internal farm resources: human capital and relational networks, intangible assets with multifunctionality (production and use of localized public goods) and orientation towards social responsibility (care of social aspects).

Specifically, the variables human capital and relational networks are key factors in the definition of the short chain strategy that, intercepting the new and articulated consumption demand, may enable the farm to reposition itself towards new markets. As evidenced by the wide literature on the subject (Albert, 1989; Child, 1975; Freeman, 1986; Herdt and Mandac, 1981; Kalirajan, 1990; Ulrich, 1997; Ulrich and Brockbank, 2005), human capital, and in particular entrepreneurship, represents a strategic variable to direct the organizational assets of the enterprise towards virtuous results and to favour the success of the short chain experiences, particularly their diffusion at a regional level (Marotta *et al.*, 2013).

The factors considered as determining the quality of human capital are:
- age (<40; 40-50; 51-60; >60);
- level of education (primary school, junior high school and high school, bachelor, postgraduate master).

The quality of human capital also determines the farm propensity to relational networks, i.e. construction and participation in a network and adherence to consortia and cooperatives (co-operation). This last aspect cannot be ignored, particularly if referring to small, family and uncompetitive businesses such as those under investigation.

The attention to the business network, i.e. the inter-enterprise relational networks for the exchange of information, shared rules and associations of resources, contributes to enhancing the human and social capital, increasing the competitiveness of farms and creating shared value. In fact, Vrontis *et al.* (2011) stress the need for networking even with competing local brands, in the form of (what they call) a 'collective action', towards international market penetration, brand recognition, country-of-origin image management and international competitive 'flocking' of SMEs.

This factor has been investigated through a dummy variable:
- participation in enterprise networks.

Human capital and the social capital are therefore determinant variables in production diversification strategies and in approaching successful short chain pathways that create value for the farm. In fact, such variables determine the capability of the enterprise to respond to the citizen-consumer's new environmental, ethical and social requests, integrating such aspects in the production and commercial strategies, adopting multifunctional and socially responsible practices.

Multifunctionality has been analysed as production of public goods within the farm context, i.e. protection of the landscape and biodiversity conservation, maintenance of rural traditions and farm atmosphere.

The orientation towards the farm's social responsibility, the third variable observed, depends on the degree of the entrepreneur's consciousness about environmental and social issues and on the investments in the implementation of multidimensional sustainable processes, environmental and food security (protection of natural resources and biodiversity and the right to health), territoriality (inclusive local development) and ethics (inclusion and rights of the individual). For this investigation, the orientation towards corporate social responsibility has been analysed through the sole variable of 'care of the social aspects' (in terms of contribution to local employment, initiatives for the local community and general activities of social inclusion).

To evaluate the level of the variables of multifunctionality and social aspects in the specific enterprise context, three classes have been used, namely, low, medium and high. Through the definition of specific ranges that account for the minimum and maximum values expressed by the entrepreneurs, such classes allowed detection of the farm's behaviour.

The level of production of localized public goods has been detected through the use of the following classes and ranges:
- low, for values lower than 30;
- medium, for values between 31 and 45;
- high, for values over 46.

To evaluate the care of the social aspects, the ranges were defined as follows:
- low, for values lower than 3;
- medium, for values between 4 and 6;
- high, for values greater than 7.

For the purpose of this analysis, a questionnaire was prepared to conduct a direct survey of a sample of 100 wine enterprises that have undertaken multifunctionality trails and short chain through direct selling on the farm. These enterprises operate in the territorial system of the 'Telese Valley' of Sannio (Southern Italy), which is particularly suited to quality wine growing. Enterprises were selected through focus groups with the sector operators (trade organizations). Of the 100 wineries sampled, 50 are associated with the consortium 'PDO Sannio', and 50 do not belong to any kind of consortium or enterprise network. The two groups were chosen to develop a comparative analysis. The questionnaire analysed the main socio-economic and structural aspects of the surveyed farms that were functional to the objectives of the study and was articulated in seven parts:
- Section 1. Human Capital (entrepreneur's age, gender, education level, years of experience, any generational transitions of the enterprise and members of the family involved in the business).

- Section 2. Physical and financial characteristics of the enterprise (size expressed by the ratio of current UAA/total UAA; number of employees; number and type of business diversification activities; distance from the city centre; size of woodlands; turnover class; membership in enterprise networks, associations, consortia, etc.).
- Section 3. Access to policies (access to funds of the RDP 2007/2013; use of LEADER funds; use of national funds).
- Section 4. Techniques of wine production and promotion (adoption of biological and/or low-impact techniques; presence of sulphites or preservatives; types of wine products and presence of quality brands; number of bottles of wine produced annually; typology of actions aimed at the promotion and care of the corporate image; price of the bottles of wine sold to the wholesaler; selling price of the bottles of wine sold on the farm; number of bottles sold through direct sales).
- Section 5. Farm atmosphere and traditions (diversification activities exclusively related to wine-making; presence in the enterprise of activities to enhance rural traditions and preserve cultural identity).
- Section 6. Localized public goods (level of maintenance and preservation of the rural landscape of the farm; level of protection of groundwater and maintenance of water balance; possible measures to reduce water needs or recovery of rainwater for irrigation purposes; level of conservation of biodiversity and natural habitat; production of energy from renewable sources).
- Section 7. Economic and social relations with the local community and inclusion (contribution to local employment; initiatives in favour of local communities; non-agricultural activities within the farm; social activities).

The questionnaire involved wineries exclusively; therefore, the survey directly detected only the premium price, i.e. the creation of value generated through direct sales only in terms of the component that would benefit the farmer. The minus price was calculated indirectly as the difference between the value created in the enterprise through direct sales and the value retained by the farmer.

18.3.2 Results and discussion

The analysis of the collected data through a direct survey highlighted the strong relationship among the different variables investigated and the creation of shared value in the farms that practise the short chain.

Specifically, it was possible to observe the strategic role of internal farm resources. Human capital represents one of the main drivers explaining the different performances of the enterprises that practise the short chain. For the purpose of the survey it was decided to consider the two dimensions of analysis related to human capital that are considered by the literature as the most important: age and educational level.

The analysis revealed that young entrepreneurs show a more dynamic behaviour; they are open to the introduction of innovative business processes and have a better capability to provide answers to the questions of the new conscious and responsible consumer. Together with age, educational level acts in a positive way: the higher a farmer's knowledge and skills, the more efficient the farmer's business management and capability to intercept new markets, as exposed in the literature (Kalirajan, 1990; Lockheed *et al.*, 1980). Level of education is often correlated with age and with other elements such as experience (Freeman, 1986) and the entrepreneur's gender. Some studies (Mathijs and Vranken, 2001) have focused on the interaction between the educational level and the gender of the entrepreneur, showing that enterprises managed by women appear more efficient and able to use internal resources. In particular, women show better performance in the implementation of multifunctional activities and a better capability to access credit.

The specific analysis revealed that, also in the wine sector, there is a positive relation between entrepreneur's age and the premium price, as entrepreneurs under 40 hold a higher share of the value created at 45.6% (Figure 18.2). Positive performance in this direction is also achieved by entrepreneurs aged between 41 and 50, who are able to keep a great part of the value created, equal to 44.2%. This percentage decreases significantly, reaching a value of 38.7% in the form of a premium price, if we consider entrepreneurs over 60.

Vice versa, the minus price[1], understood as the value obtained by the citizen-consumer in terms of amount of money saved by buying directly on the farm instead of in retail, is proportional to the increase of the agricultural entrepreneur's age, starting from a percentage of 54.4% with entrepreneurs under 40 and reaching a maximum quote of 61.3% for those over 60. For this reason, entrepreneurs in old age do not appear very dynamic in capturing the challenges of the new markets and in approaching new commercial strategies and are not able to create value through the short chain. This confirms the strategic role of human capital in the repositioning strategies of the farm.

[1] In this survey, the average price of retail has been estimated on a representative sample of 50 stores located throughout the regional territory.

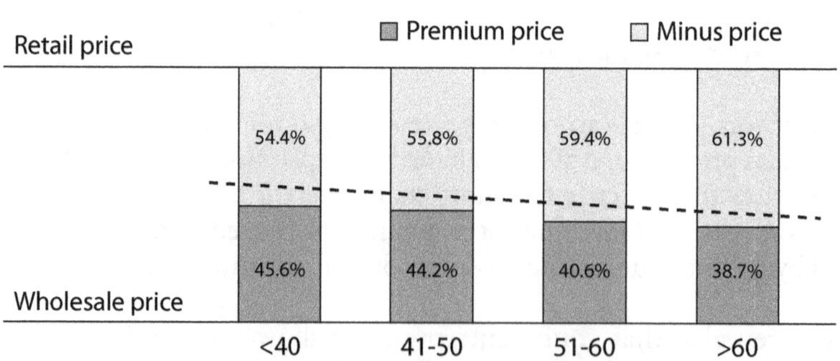

Figure 18.2. Premium price and minus price in wine farms by age groups.

In general, young producers, thanks to their greater individual autonomy and higher desire for fulfilment when compared with the older producers, are more open to change and interested in experimental activities of productive diversification that increase the enterprise's appeal and in undertaking new commercial strategies that can capture unexplored markets and increase incomes. The strong propensity to implement innovative processes of farm diversification may also be encouraged by the younger entrepreneurs' greater capability to access community policies. Through the survey, producers were asked to state whether they had benefitted from EU funds (RDP 2007/2013) and to specify which measures they had adopted. The analysis revealed that, in addition to the measures of Axis 1, entrepreneurs under 40 benefited from the measures of Axis 3 of the RDP, oriented towards improving the quality of life in the rural areas and developing activities of diversification. Finally, the level of dynamism and the propensity towards innovation could be correlated with the enterprise's image and reputation. In particular, younger managers tend to be more risk-oriented and more associated with policies of corporate growth compared to older managers (Child, 1975), while older producers tend not to introduce radical innovations because of the associated risks. Younger producers with a higher educational level who invest in these entrepreneurial activities and aim to improve their reputation and create their own market niche appear to be more dynamic in the implementation of product, process and commercial innovations. From the data, it is possible to observe how the premium price changes from 29.2% for entrepreneurs with just a primary licence, to 52.8% for those who hold a bachelor's degree, and finally to 53.9% for those holding a post-graduate degree (Figure 18.3).

The survey shows, therefore, that there is a close correlation between age, educational level and premium price. The data shown in Figures 18.2 and 18.3 show that the premium price appears to be higher for entrepreneurs of young age and with a high level of education. As highlighted in the theoretical debate, the empirical analysis confirms

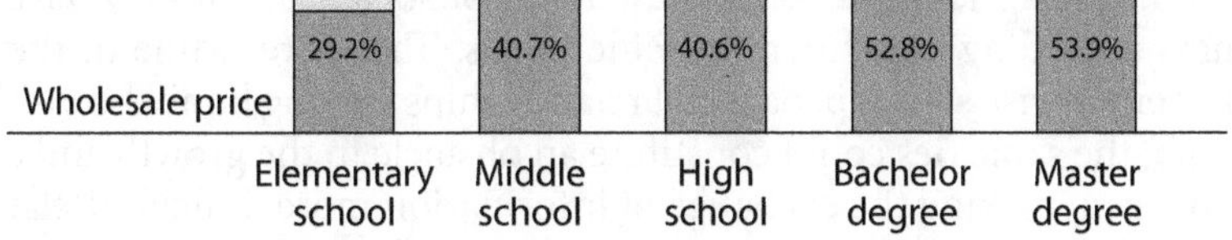

Figure 18.3. Premium price and minus price in wine farms by educational level.

that younger and more educated entrepreneurs have greater bargaining power and, therefore, are able to create and retain more value through the short chain channels.

The age, knowledge and skills of entrepreneurs are the strategic factors for the success of short food supply chains in terms of both implementation of these paths and organizational strategies on the farm and have a significant impact on profit growth and on the premium and minus price formation and distribution. As shown in the literature (Lucas, 1988; Romer, 1990), the stock of human capital significantly affects the development of new ideas and innovations, representing one of the main drivers for value creation. The more educated entrepreneurs have a clearer awareness of innovations and new marketing practices: this increases their ability to quickly understand the new consumer's demands and their competitive advantage in key markets.

Conversely, low levels of education and advanced age determine a lower bargaining power for an entrepreneur in commercial relations with wholesalers and in direct relations with citizen-consumers. The farmer is able to get a lower price for the wholesale of products and to retain a lower premium price for direct sale on the farm. Value creation (premium price) through the short chain improves if, in addition to human capital, social capital is implemented through membership/participation in consortia networks and/or enterprise networks. Furthermore, the quality of human capital defines the enterprise's capability to form networks through the way in which economic players interact and organize themselves to generate growth and economic development (Grootaert, 1998) within the local context in which they operate.

The willingness to join together and create networks or interpersonal relations to share information, expertise, resources and values is closely related to the quality of human capital. In other words, a good quality of human capital also favours the creation of business networks. The spread of knowledge, skills and experience transmitted from the most qualified human resources helps to increase the quality of the territorial capital. The analysis shows that the quality of human capital is a crucial factor for the access to enterprise networks. In fact, it seems that the higher the level of education, the greater the entrepreneurs' willingness to join business networks.

Based on the two different dimensions of social capital recognized in the literature, bonding social capital and bridging social capital (Dasgupta, 1999), formal links (between family members, political participation and civic consciousness and voluntary organizations) and informal links (relationships with friends and acquaintances) are created. In the agricultural field, these types of social capital may have different effectiveness according to farmers' specific needs. Therefore, while in the phase of agricultural enterprise start-up, parental relationships (strong bonds) could be useful to the farmer, the same ties could constitute an obstacle in the growth and expansion phase of the firm, making the exchange of information more difficult. Relations with acquaintances (weak ties) could be more useful to allow access to information and strategic resources necessary to improve business performance (Granovetter, 1973).

In light of these reflections, the business network, understood as the set of inter-enterprise relational networks for the exchange of information, sharing of practices, standards and resources, appears as strictly related to the economic growth of businesses. This occurs especially in a context such as that of the wine sector investigated, where the small size of the enterprises constitutes a limit for the full development of their potential. In fact, Prashantham and Dhanaraj (2010) and Kontinen and Ojala (2011) emphasize that networking can help overcome the limitation associated with the availability of resources and small dimensions by supporting the enterprises in developing new strategies and products and moving them towards internationalization.

Membership of the business network, then, in addition to becoming a corporate asset able to improve competitive advantage, can be a strategic factor for the success of the short chain and for the distribution of the value created in favour of the producer.

Empirical results show that wineries that adhere to the network are able to retain more value from short chain channels, particularly direct sales, with a premium price percentage of 47.7%, while farms that do not adhere to any consortium/cooperative are able to retain only 36% of the value created (Figure 18.4).

The advantages of the adhesion to a network of wine enterprises are therefore meaningful in terms of value creation. The results show that sharing information, resources and commercial practices can allow farms to overcome the limits imposed by material and immaterial internal resources. Moreover, by taking advantage of the synergies with other wine operators, networks allow firms to increase the level of innovativeness and define new competitive strategies useful to keep a greater value through the short chain, with positive effects on incomes. The business network could allow companies to reduce costs and introduce sustainable innovations, contributing to creating internal economies for the farms.

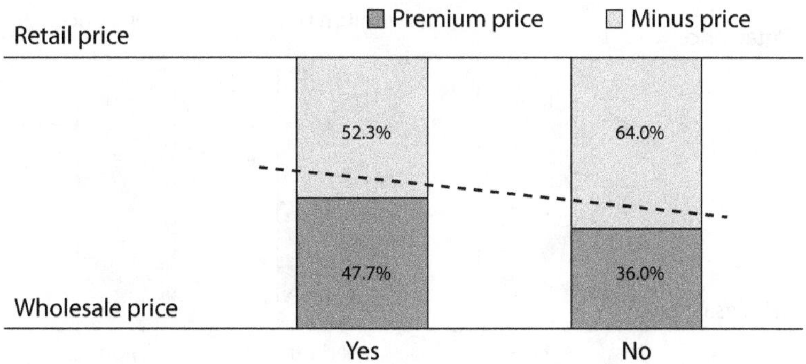

Figure 18.4. Premium price and minus price in wine farms by adhesion to networks.

The analysis reveals, therefore, that human capital and social capital represent determining variables for shared value creation, and their quality determines the enterprise's capability to intercept the new instances of the society and produce non-market outputs (i.e. public goods). The positive externalities (multifunctionality) generated by the farm give the products added value. The short chain attracts the citizen-consumer who will benefit from business goods and services and localized public goods and will express a greater willingness to pay for them. Thus, the short chain helps to transform the farm from a simple production area to a market space (purchase and use). The survey confirmed that there is a positive correlation between premium price and production of high-quality localized public goods (care of the farm landscape, presence of biodiversity, rural traditions and farm atmosphere). The higher the level of public goods produced on the farm, the greater the consumers-citizens' willingness to pay for the use of such goods. In fact, the empirical analysis showed that entrepreneurs that maximize the production of positive externalities hold a maximum percentage of value equal to 65% of the value created, while farms that neglect these aspects retain, on average, a premium price of 38.6% of the value generated through direct sales (Figure 18.5).

The increased focus of the multifunctional wineries (with a performance that in some cases goes beyond what is required by policy and by law) on environmental and socio-cultural public goods implicitly represents their orientation towards socially responsible production models that implement multidimensional sustainability business processes but also inclusion and local welfare.

In this paper, we wanted to consider just the variable of the care of social aspects, understood as the farm's attention to the local welfare, measured in terms of contribution to local employment, initiatives in favour of the territorial community and, more generally, in favour of social inclusion.

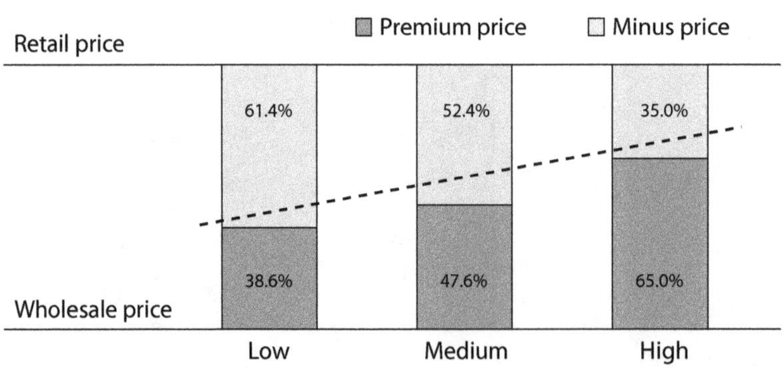

Figure 18.5. Premium price and minus price in wine farms by farm public goods.

The care of these aspects enhances the enterprise's reputation and improves its market position thanks to a greater appreciation of its products by the modern citizen-consumer. The latter expresses a greater willingness to pay a premium price to purchase products that attribute particular importance to social aspects. In fact, in cases where business activities with a high impact on the local welfare are very significant, entrepreneurs are able to benefit from a premium price equal to 51.1% of the value created through direct sales. A premium price of 44.9% is held by farms that show less interesting performance in the social sphere. Finally, there are wineries that show a lack of attention for social aspects, for which the premium price is equal to 39% of the value created; in that case, the majority of the value will benefit consumers in the form of a minus price (Figure 18.6).

In sum, the empirical evidence demonstrates that the multifunctionality and orientation towards social responsibility represent real competitive levers for value creation. The costs that the farm bears for environmental investments and the pursuit of ethical and social values result in an increase of intangible enterprise assets that enhance reputation and market positioning and consequently profits. This is because the farm products enjoyed through a short chain are perceived by aware and responsible citizen-consumers as more sustainable and ethical and qualitatively better than those of competing farms.

18.4 Final remarks and policy implications

The results of this study identify strategic variables that have a major impact on shared value creation within the strategies of short food supply chain and provide useful information for practitioners and policy makers. The empirical analysis confirms the conceptual framework: direct sales can be an instrument of commercial exploitation of localized public goods (multifunctionality), allowing the enterprise to reach new markets, reposition itself and share the value created.

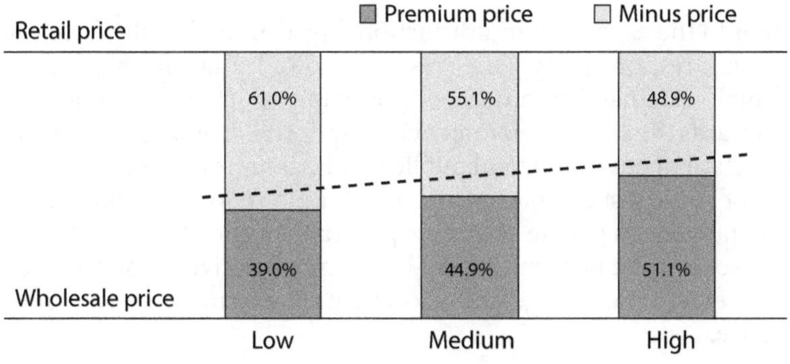

Figure 18.6. Premium price and minus price in wine farms by social aspects.

This research result is quite innovative in the theoretical debate and promises interesting implications for policy. First, the analysis shows that the strategic capacity of the short chain, as part of a renewed and diversified multifunctional farm model, depends on the farm's internal resources. Specifically, the quality of entrepreneurial capital (corporate human capital) is the key cross-cutting variable, which largely determines the effectiveness of short chain practices and the quality and value of the other variables (networking, multifunctionality and social responsibility) in creating value.

The survey finds a close correlation between entrepreneurial skills and premium price. This relationship becomes virtuous, with a premium price that turns out to be higher for entrepreneurs of young age and with a high level of education. This is confirmed by the literature, as noted above, which shows that age and level of education are determining factors of the quality and dynamism of human capital. They allow an improvement in the bargaining power of farmers, who are thus able to retain more value through the short chain channels.

However, human capital acts directly on social capital: the higher the quality of human capital, the greater the ability to create networks. The survey also demonstrated how networking represents an important internal resource in the creation of value through the short chain.

Furthermore, the empirical evidence has confirmed the existence of a positive correlation between premium price and production of business public goods. The higher the level of public goods produced on the farm, the greater consumers-citizens' willingness to pay for their use. The analysis showed that entrepreneurs oriented towards maximizing the production of positive externalities hold a maximum percentage of value (more than half of the total value created). Conversely, enterprises that neglect these aspects can hold, on average, a lower premium price (about a quarter of the entire value generated through direct sales). In essence, the market recognition for intangible assets is commensurate with the maximization of capacity of such assets by the entrepreneur.

The attention to the aspects of multifunctionality also implies the farm's orientation towards social responsibility. The results showed that farmers' awareness of environmental sustainability processes (investments in multifunctional practices) orientates towards the same corporate social responsibility issues. The farmer, engaged in the production of environmental public goods, assumes a greater awareness of the importance of social issues in terms of contribution to local employment initiatives in favour of indigenous people and more generally of social inclusion activities. The focus on these aspects, particularly on high impact activities on the local welfare, improves the enterprise's image and allows entrepreneurs to benefit from a higher premium price.

Ultimately, the premium price created through the short chain is linked to the entrepreneurial skills, which impact the creation of networks, quality of intangible assets and orientation towards social responsibility aspects of the business.

In light of these results, it is possible to note some policy implications. Policies should aim first for the enhancement of human resources and second for the strengthening of intangible corporate assets and social responsibility practices. With reference to the first aspect, sector policies should favour generational turnover through a well-trained human capital (training geared to facilitate the acquisition of organizational and strategic skills along the short chain), with managerial skills and capabilities that open new business opportunities and create solid networks.

For the strengthening of intangible corporate assets measures should be applied to enable enterprises to improve the quality and quantity of public goods produced on the farm, as such goods are strategic elements in promoting new ways to use the property and business services and thus enhance the willingness to pay.

Finally, rural development policy must use appropriate instruments to support the social responsibility practices of farms, specifically social inclusion, through increased employment opportunities offered to the local community, which also contributes to local development.

In conclusion, the analysis showed that the short food supply chain can be a commercial exploitation of the multifunctional lever and can strengthen enterprise and territorial reputation and create shared value if it works the human capital that determines the quality of networks, of the assets of multifunctionality and of the social responsibility of the farm.

References

Albert, F.J., 1989. Les ressources humaines: a tot strategique. L'Harmattan, Paris, France, 75 pp.

Bovone, L. and Mora, E., 2007. La spesa responsabile. Il consumo biologico e solidale. Donzelli Editore, Rome, Italy.

Brown, C. and Miller, S., 2008. The impacts of local markets: a review of research on framers markets and Community Supported Agriculture (CSA). American Journal of Agricultural Economics 90(5): 1298-1302.

Brunori, G. and Bartolini, F., 2013. La filiera corta: le opportunità offerte dalla nuova PAC. Agriregionieuropa 9: 35.

Carrigan, M., Szmigin, I. and Wright, J., 2004. Shopping for a better world? An interpretive study of the potential for ethical consumption within the older market. Journal of Consumer Marketing 21(6): 401-417.

Cembalo, L., Migliore, G. and Schifani, G., 2010. Modelli di consumo e nuovo paradigma della sostenibilità: i Gruppi di Acquisto Solidali in Sicilia. Rivista di Economia Agraria 65: 4.

Child, J., 1975. Managerial and organization factors associated with company performance, part II: a contingency analysis. Journal of Management Studies 12: 1-2.

Crane, A. and Matten, D., 2004. Business ethics. A European perspective. Managing corporate citizenship and sustainability in the age of globalization. Oxford University Press, Oxford, UK.

Dasgupta, P., 1999. Economic progress and the idea of social capital. In: Dasgupta, P.E. and Serageldin, I. (eds.) Social capital. A multifaceted perspective. World Bank, Washington, DC, USA.

DuPuis, E.M., 2000. Not in my body: BGH and the rise of organic milk. Agriculture and Human Values 17: 285-295.

Francois-Lecompte, A. and Roberts, J.A., 2006. Developing a measure of socially responsible consumption in Franc. Marketing Management Journal 16(2): 50-66.

Freeman, R., 1986. Demand for education. In: Ashenfelter, O. and Layard, R. (eds.) Handbook of labor economics. North-Holland, Amsterdam, the Netherlands.

Granovetter, M., 1973. The strength of weak ties. American Journal of Sociology 78(6): 1360-1380.

Grootaert, C., 1998. Social capital, household welfare and poverty in Indonesia. Local level institutions study. Environmentally and Socially Sustainable Development Network, World Bank, Washington, DC, USA.

Herdt, R.W. and Mandac, A.M., 1981. Modern technology and economic efficiency of Philippine rice farmers. Economic Development and Cultural Change 29: 375-399.

Hinrichs, C.C., 2003. The practice and politics of food system localization. Journal of Rural Studies 19(1): 33-45.

Kalirajan, K.P., 1990. On measuring economic efficiency. Journal of Applied Econometrics 5: 75-85.

Kontinen, T. and Ojala, A., 2011. Network ties in the international opportunity recognition of family SMEs. International Business Review 20(4): 440-453.

Leonini, L. and Sassatelli, R., 2008. Il consumo critico. Significati, pratiche, reti. Editori Laterza, Bari, Italy.

Lockheed, M.E., Jamison, D.T. and Lau, L.J., 1980. Farmer education and farm efficiency: a survey. Economic Development and Cultural Change 29(1): 37-76.

Long, M.A. and Murray, D.L., 2013. Ethical consumption, value convergence/divergence and community development. Journal of Agricultural and Environmental Ethics 26: 351-375.

Lucas Jr., R.E., 1988. On the mechanics of economic development. Journal of Monetary Economics 22: 3-42.

Mancini, P., Marotta, G., Nazzaro, C. and Simonetti, B., 2012. A statistical approach to evaluate the effects of obesity on health-care expenditure. Electronic Journal of Applied Statistical Analysis 5(3): 418-424.

Mancini, P., Marotta, G., Nazzaro, C. and Simonetti, B., 2015. Consumer behaviour, obesity and social costs. The case of Italy. International Journal of Business and Society 16(2): 295-324.

Marotta, G. and Nazzaro, C., 2011. Verso un nuovo paradigma per la creazione di valore nell'impresa agricola multifunzionale. II caso della filiera zootecnica. Economia Agro-alimentare 1-2: 215-250.

Marotta, G., Nazzaro, C. and Simeone, M., 2013. Capitale umano e capitale sociale nell'agricoltura multifunzionale: un'analisi delle esperienze di filiera corta nella Campania interna. Economia Agro-Alimentare 3: 149-173.

Marotta, G., Simeone, M. and Nazzaro, C., 2014. Product reformulation in the food system to improve food safety. Evaluation of policy interventions. Appetite 74: 107-115.

Marsden, T., Banks, J. and Bristow, G., 2000. Food supply chain approaches: exploring their role in rural development. Sociologia Ruralis 40(4): 424-438.

Mathijs, E. and Vranken, L., 2001. Human capital, gender and organisation in transition agriculture: measuring and explaining the technical efficiency of Bulgarian and Hungarian farms. Post-Communist Economies 13(2): 171-181.

Mollard, A., 2002. Multifonctionnalite externalite et territories. Les Cahiers de la Multifonctionnalite 1: 28-54.

Mundler, P. and Laughrea, S., 2016. The contributions of short food supply chains to territorial development: a study of three Quebec territories. Journal of Rural Studies 45: 218-229.

Prashantham, S. and Dhanaraj, C., 2010. The dynamic influence of social capital on the international growth of new ventures. Journal of Management Studies 47(6): 967-994.

Romer, P.M., 1990. Endogenous technological change. Journal of Political Economy 98(5): 71-102.

Smith, N.C., 1990. Morality and the market: consumer pressure for corporate accountability. Routledge, London, UK.

Ulrich, D., 1997. Human resources champions-the next agenda for adding value and delivering results. Harvard Business Case School Press, Boston, MA, UK.

Ulrich, D. and Brockbank, W., 2005. The HR value proposition. Harvard Business Case School Press, Boston, MA, UK.

Vermeir, I. and Verbeke, W., 2006. Sustainable food consumption: exploring the consumer attitude-behavioral intention GAP. Journal of Agricultural and Environmental Ethics 19(2): 169-194.

Vrontis, D., Thrassou, A. and Czinkota, M.R., 2011. Wine marketing: a framework for consumer-centred planning. Journal of Brand Management 18(4-5): 245-263.

Zamagni, S., 2004. *L'ancoraggio etico della responsabilità sociale d'impresa e la critica alla RSI.* AICCON Working Paper No.1. AICCON, Bologna, Italy.

19. Policy and governance of payments for ecosystem services in agri-food value chains

F. Marangon and S. Troiano[*]

Department of Economics and Statistics, University of Udine, via Tomadini 30/a, 33100 Udine, Italy; troiano@uniud.it

Abstract

The paper examines the role of markets for ecosystem services deriving from agri-environmental and rural landscape conservation in the agri-food chain. In spite of their importance, sustainable management of agricultural biodiversity and ecosystem services conservation are only partially financed by public institutions. This paper illustrates an innovative economic tool, Payment for Ecosystem Services (PES), used as a financing mechanism to achieve positive environmental outcomes by supporting environmental friendly behaviour in agri-food value chains. Our framework suggests that PES can act as a tool able to support sustainable management of agri-food value chains. In spite of financial crises, demand for ecosystem services is growing, nevertheless their supply need to be supported by innovative incentive schemes. PES could play a fundamental role in increasing ecosystem services provision. The areas for future research that seem to be the most promising include following aspects. First, we look forward to understanding the role of public intervention. Second, several insights can be gained from investigating PES implementation in agri-food value chains. The study suggests that PES schemes have a potential role to play with regards to ecosystem services conservation in agri-food value chains. The paper provides an original contribution by highlighting the role of an innovative economic instrument for financing agricultural biodiversity conservation and supporting land users for the positive externalities in agri-food value-chains.

Keywords: agricultural biodiversity, Payments for Ecosystem Services, incentive schemes, agri-food system

19.1 Introduction

According to the Convention on Biological Diversity (CBD), biodiversity means:

> The variability among living organisms from all sources including, inter alia, terrestrial, marine, and other aquatic ecosystems and the ecological complexes of which they are a part: this includes diversity within species, between species and of ecosystems (UN, 1992).

There are strong linkages between biodiversity and human well-being: biodiversity provides several ecosystem services that, besides other factors, are able to influence human well-being and are in turn affected by changes in it (Millennium Ecosystem Assessment, 2005). Ecosystem services (also called nature's services) are benefits provided to people by natural resources and healthy functioning ecological systems that contribute to making human life both possible and worth living (FAO, 2009).

Biodiversity is fundamental to agricultural production and food security. As stated by the Conference of the Parties to the CBD (2000), agricultural biodiversity includes all components of biological diversity of relevance to food and agriculture, and all components of biological diversity that constitute the agricultural ecosystems (agro-ecosystems).

Agricultural activity is a prevalent form of land management (FAO, 2009). Agricultural ecosystems are both the main providers of ecosystem services (FAO, 2007; Gutman and Davidson, 2007) and dependent on ecosystem services as they support production functions. Some dimensions of agricultural biodiversity are directly managed as main parts of production for supplying food, fresh water, fibre and fuel (provisioning services or ecosystem-goods), while others are not directly intended for production. In fact, agricultural biodiversity also performs ecosystem services such as biophysical processes that control climate, floods, diseases, air and water quality conservation, pollination and maintenance of soil fertility and biota (regulation services or ecosystem outcomes). Ecosystem services contribute to nutrient cycling, formation of soil, photosynthesis, etc. (supporting services or ecosystem-functions), but also to further aspects of human well-being such as the recreational, aesthetic or spiritual benefits produced by an ecosystem (cultural services or ecosystem-benefits).

The Millennium Ecosystem Assessment (2005) found that ecosystems have experienced a consistent change during the last fifty years and several ecosystem services have been declining, in particular regulating services. The improvement or degradation of ecosystems depends on farmers' actions. If they properly manage agricultural lands they can provide several ecosystem services (Palm *et al.*, 2014). Nevertheless, modern agriculture such as the intensive ones could reduce agricultural biodiversity and the provision of its ecosystem services (Krishna *et al.*, 2013). Therefore, the type, quality, and quantity of services provided by an ecosystem can be affected by the farmers' decisions. As in some cases the interests of the farmer who takes decisions is not aligned with the benefits of the beneficiaries, the result can consist of the decline of the availability of ecosystem services (Gomez-Baggethun *et al.*, 2010). In fact, the private value created by farmers from maintaining agricultural biodiversity and ecosystem services provision is often less than the benefits generated by accounting for public good values. While they are generally successful at providing ecosystem services with a market (i.e. provisioning services), they are not able to maintain the level of regulating, supporting and cultural services provision due to high costs and, for some services, the absence of prices (they are free services). These types of services, in fact, often display public good characteristics, which means that they may be enjoyed by any number of people without affecting other peoples' enjoyment (non-excludable and non-rivalrous). Moreover, public goods markets do not exist. The result is a sub-optimal level of service provision that requires external (governmental) intervention.

According to Troiano and Marangon (2011), the creation of appropriate incentive-based economic instruments and the diffusion of targeted information seem to be able to help farmers to improve their role in conserving and managing ecosystem services by reducing the potential negative impacts deriving from their activity and improving their earnings. As regards information, for example, economic valuation of ecosystem services can help farmers and landholders deal with the effects of market failures by measuring the costs to society of their activities in terms of lost economic benefits. These costs can then be imposed, in various ways, on those who are responsible ('Polluter Pays' Principle – PPP) (OECD, 1972), or can be used to determine the value of further actions to enhance ecosystem services provision, such as market creation.

If there is a growing realisation that biodiversity is an enormously valuable infrastructure and its ecosystem services are essential to maintain liveable conditions and for sustainable socio-economic development, economic valuation is able to get the attention of people also by assigning monetary values. Giving an economic value to biodiversity and its benefits should avoid them being ignored and improve appreciation, conservation and provision of ecosystem services (TEEB, 2010).

In this paper we discuss the opportunity to enhance agricultural biodiversity conservation, to maintain the provision of its ecosystem services and to support the sustainable development of the agri-food value chain by using an incentive-based instrument, Payment for Ecosystem Services. We highlight the role of public intervention to avoid gaps and solve questions in order to provide more effective incentive intervention.

19.2 Policy approaches and instruments for agricultural biodiversity conservation and sustainable use

In order to conserve and support the sustainable management of agricultural biodiversity, it is possible to use different approaches and instruments (Table 19.1) (OECD, 2010).

Table 19.1. Policy approaches and tools for biodiversity conservation (OECD, 2010).

Regulatory (command and control) approaches	Economic instruments	Information and other instruments
restrictions or prohibitions on use (e.g. trade in endangered species and CITES)	price-based instruments: • taxes • charges/fees • subsidies	eco-labelling and certification
restrictions or prohibitions to access (e.g. protected areas, legislated buffer zones along waterways)	liability instruments: • non-compliance fines • criminal indictment • performance bonds	voluntary agreements
quality or quantity standards, often enforcing the use of specific technologies (e.g. commercial fishing net specifications)	removal or reform of perverse subsidies market creation and assignment of well-defined property rights payments for ecosystem services	

According to the regulatory tools, both restrictions and prohibitions on use or access and quality or quantity standards can play a vital role in biodiversity conservation.

Nevertheless, in some contexts it seems that market-based instruments (Stephenson, 2012) could better bring all stakeholders along for reducing the human impact on biodiversity and ecosystem services.

These instruments are increasingly included by governments in their policy toolkits that aim to provide behaviour-related incentives and create contexts where decisions reflect more fully the costs or benefits they create since they have the ability to correct market-failures in a cost-effective way (European Commission, 2007; OECD, 2010).

The use of these tools to protect biodiversity is increasingly accepted as it seems to be able to integrate conservation and sustainable use into the decision-making process of economic agents. Moreover, according to a dynamic efficiency assumption, these instruments prove useful in reaching goals for environmental resources conservation and sustainable management in the long term as they give an incentive to develop innovations in order to reduce negative effects on these resources (European Commission, 2007). Reliable updated information for ecosystem services and payment schemes on a number of areas are collected by Ecosystem Marketplace (2016).

Such instruments are diverse and include Payment for Ecosystem Services (PES), which is an innovative market-based tool used to influence peoples' behaviour. PES is: (1) an incentive scheme able to achieve positive environmental outcomes by supporting environmentally friendly behaviour (Wunder, 2005, 2015); and (2) a financing mechanism, especially in a situation of lack of enforcement of environmental legislation. Its use helps to integrate economic, social and environmental aspects in decision making processes according to sustainable development principles ('triple bottom line').

Although PES is a 'multi-facetted term with many diverse definitions coexisting' (Schomers and Matzdorf, 2013, p. 16), an influential definition is due to Wunder (2015, p. 241) focusing on market transactions and construing PES as 'a voluntary transaction between service users and service providers that are conditional on agreed rules of natural resource management for generating offsite services'.

The mechanism is similar to any service: the provider should be paid for supplying that. In fact, payments are made by the beneficiaries of the service in question (i.e. individuals, communities, businesses or government acting on behalf of various parties) to those who provide the service (Wunder *et al.*, 2008). Therefore, with these schemes there is an opportunity to put a price on un-priced services deriving from ecosystems and, in doing so, it is possible to bring them into the economy. Consequently, we can be able to quantity benefits deriving from ecosystem services and to compare them with the costs of their provision. However, according to Fischer and Brown (2014), it is fundamental to avoid any incongruent expectations of sustainable finance from this tool in the absence of conditionality. In fact, investment is based only upon an ecosystem service return.

The conceptual basis of this market-based instrument can be found within neoclassical environmental economics and the basic idea of PES is linked to the Coase theorem that is based on the assumption that, given certain conditions, the problems of external effects can be overcome through private negotiation directly between the affected parties regardless of the initial allocation of property rights (Engel *et al.*, 2008). The result will lead to improved economic efficiency (Pascual *et al.*, 2010). Nonetheless, high transaction costs, power imbalances, or poorly defined property rights can impede the implementation of a Coasean solution.

In some contexts this incentive-based solution has been widening to certain types of government interventions that can be counted as PES-like mechanisms that are more in line with Pigouvian environmental taxation and subsidisation to correct negative and support positive externalities (Troiano and Marangon, 2011). Examples of these Pigouvian concept of PES are a number of governmental payment programs that introduced agri-environmental payments, which provide payments to farmers choosing to implement conservation efforts that improve the environment and/or maintain the countryside on a voluntary basis (Schomers and Matzdorf, 2013).

While in Coasean type PES the beneficiary directly pays the provider of the service with private money on a purely voluntary basis (private negotiation), in the Pigouvian type PES the government intervenes in order either to pay itself or make others pay on behalf of the direct beneficiaries to spur ecosystem services provision (Sattler and Matzdorf, 2013). Both have pros and cons. For example, according to Engel *et al.*

(2008) Coasean PES approaches seem to be efficient, as the actors with the most information about the value of the ecosystem service are directly involved, have a clear incentive to ensure that the mechanism is functioning well, can observe directly whether the service is being delivered, and have the ability to re-negotiate or terminate the agreement if needed. Coasean PES schemes frequently pay land stewards for the provision of ES that are characterized as club goods, with beneficiaries of such ES existing only at local scales that can therefore be directly identified. In the Pigouvian PES scheme there is not the directness of transfer, i.e. the buyers are not the direct users. This PES scheme rather focus on the provision of public goods, and beneficiaries cannot be excluded at all or at reasonable costs. However, this scheme usually keep seller and buyer transaction costs reasonably low.

Among the publications that describe a PES case study in detail, the Pigouvian concept is by far the most dominant approach (Schomers and Matzdorf, 2013). Nevertheless, the most innovative scheme is the Coasean type PES. According to this type of PES and to Wunder's definition (Wunder, 2005, 2015), PES schemes aim to address market failure by developing private market negotiations and providing direct compensation to individuals or communities who manage landscape and environmental resources and influence the provision of ecosystem services. In particular, PES compensation aims to support the additional costs associated with maintaining or enhancing those ecosystem services.

Compared with regulatory approaches and according to market-based instruments, PES schemes have potentially larger cost-effectiveness gains (Alix-Garcia *et al.*, 2003; Engel *et al.*, 2008) as command and control approaches tend to require the same level of conservation and impose uniform restrictions across farmers or landholders. Moreover, PES seem to be more flexible because participation is voluntary and more efficient 'as the actors with the most information about the value of the service are directly involved, have a clear incentive to ensure that the mechanism is functioning well, can observe directly whether the service is being delivered and have the ability to re-negotiate (or terminate) the agreement if needed' (Engel *et al.*, 2008, p. 666). It follows that actors with relatively higher marginal costs of conservation will tend to conserve less than those with lower costs. Furthermore, the lack of success of indirect mechanisms proposed during the 1980s and 1990s to engage the development community and private sector in environmental resources conservation suggests that new approaches for conservation and sustainable use financing might best be both market-based and direct, such as the PES approach.

PES can be used alone or in conjunction with other policy instruments. In fact, it can be part of a policy mix: it can incentivise enhancement in providing ecosystem services over and above that required by existing regulations (OECD, 2010).

19.3 Payments for ecosystem services in favour of agricultural biodiversity conservation

As stated above, agricultural biodiversity includes all the components of biological diversity of relevance to food and agriculture, and those that constitute the agro-ecosystem (e.g. the variety and variability of animals, plants and micro-organisms, at the genetic, species and ecosystem levels, which sustain the functions, structure and processes of the agroecosystem).

Garbach *et al.* (2014) considered biodiversity as a foundation for all ecosystem services, and stated that species richness – measured as the number of species in a given area – is associated with enhanced ecosystem services. Similarly, Scherr and McNeely (2008) affirmed that conservation of biodiversity (genes, species and ecosystems) is considered to be an ethical imperative as it also supports ecosystem services provision. Ottaviani and El-Hage Scialabba (2011) pointed out that the ecosystem services occurring in agricultural biodiversity compartment regard in particular pollination, seed dispersal, pest control, preservation of detritivorous food chains, preservation of genetic diversity.

As stated above, agricultural activity is the main provider of ecosystem services (FAO, 2007) as farmers manage agricultural ecosystems. From decisions about their activity depends the amount and the quality of ecosystem services provided. Moreover, they can deliver or not several ecosystem services depending again on their decisions (Page *et al.*, 2015).

On the one hand they are able to successfully furnish services with a market (e.g. agritourism, teaching farms, social farms, etc.), on the other hand they often fail in maintaining the level of ecosystem services for which a market does not exist. In fact, the provision of some ecosystem services, such as biodiversity protection, has the character of a public good. Consequently, although these ecosystem services benefit society, they tend to be available at no charge and without any reward for farmers who take care of landscape and environmental resources and continuously provide ecosystem services.

It seems that the use of both monetary incentives and information could improve the role of agriculture in environmental conservation and ecosystem services provision. An opportunity could be found in lessons learned from existing innovative PES schemes. From the well-known example of Vittel PES in north-eastern France (Perrot-Maitre, 2006) where a private water bottling company paid local farmers to avoid nitrate contamination due to agricultural intensification to the Danube PES scheme created to maintain aesthetic and biodiversity values (FAO, 2013; WWF, 2014), there are several ideas in order to develop these tools.

Although PES could be applied to a wide range of ecosystem services, it has been largely limited to applications in the context of forest ecosystems, carbon sequestration, wild biodiversity conservation and water management (Engel *et al.*, 2008; Landell-Mills and Porras, 2002; Muradian *et al.*, 2010; Pettenella *et al.*, 2012; Wunder *et al.*, 2008).

Nevertheless, due to the rapid and unprecedented loss of biodiversity in rural areas across the world (FAO, 2007, 2009) there could be several opportunities for PES schemes to conserve threatened agricultural biodiversity and its ecosystem services provision.

Biodiversity conservation in rural areas has the characteristic of a public good as it benefits society but tends to be available at no charge because existing markets fail to value its ecosystem services. Due to this situation, the unsustainable use of biodiversity occurs. One of the key challenges is to devise PES schemes that are able to address the externalities deriving from biodiversity and its ecosystem services in rural areas, improving the provision of these services derived from agricultural activities conducted in a sustainable manner (Narloch *et al.*, 2011).

The chance to create biodiversity markets could firstly provide an opportunity to secure the provision of benefits deriving from biodiversity ecosystem services by paying the efforts of providers in sustainably managing and preserving biological processes as well as habitat and species. Secondly, payments are incentives offered to farmers to adopt more sustainable practices as these practices may lead to higher production costs or lower production.

In order to work, PES must be a win solution for both buyers/users and sellers/farmers. In fact, first of all, it has to be positive from a buyer's perspective as the payments should be less expensive than those associated with any alternative instruments of providing the ecosystem service. Moreover, a PES scheme has to be positive also from a seller's point of view as the level of payment received should at least cover the costs of furnishing the service and, if needed, also of lost agricultural production.

There could be several agents involved in PES programs (OECD, 2010): beneficiaries of ecosystem services who are willing to pay for their conservation or enhancement ('buyers') and farmers or biodiversity resource managers who are able to supply the services ('sellers') are principal groups. Often, there are also some agents joining previous protagonists in order to ease PES design and implementation ('intermediaries') and 'knowledge providers' like resource management experts, valuation specialists, land use planners, regulators and business and legal advisors who can facilitate PES development.

It is important to note that the way that buyers and sellers can be configured in scheme development can vary (Smith *et al.*, 2013). In fact, we can have a 'one-to-one scheme' where a company enters into a contract with a single major land-owner to provide biodiversity conservation or a 'one-to-many' scheme where a company makes arrangements via a broker to pay many farm businesses for biodiversity-sensitive management practices in a key rural area. Moreover, a 'many-to-one' PES scheme could be developed where multiple buyers together invest in the enhancement and conservation of biodiversity, and a 'many-to-many' PES scheme could be created where government pays farmers for biodiversity-friendly land management practices on behalf of the wider public.

19.4 Payments for ecosystem services and agri-food value chain sustainable management

Although PES schemes are an innovative, still emerging tool that rapidly changes and spreads out across countries and institutions, Ecosystem Marketplace (2013, 2016) has tried to map the main PES schemes considering their main sub-categories and key characteristics. The result is a matrix that describes innovative markets and market-like instruments for ecosystem services, which could be of particular interest for agri-food value chain development.

This matrix considers ecological commodities and puts attention also on bundled services. Payments for bundled services makes all or a combination of ecological commodities certain, i.e. carbon, water and biodiversity services. This group also includes payments in which the ecosystem service payments are built into the price of a product, for example for certified agricultural production.

Taking into account the variety of components of agricultural biological diversity, the range of actors involved in maintaining it, and according to a broad definition of PES, such as the definition proposed by FAO (2007) comprehending the green premium price of a product, an interesting opportunity to support agricultural biodiversity seems to come for the agri-food value chain from combining PES with the eco-certification of agricultural products (Robinson and Keenan, 2010) and labelling schemes (Lockie, 2013). These PES could be considered the new generation of PES schemes, which could combine community-based initiatives and certification schemes (Ottaviani and El-Hage Scialabba, 2011) to support agricultural biodiversity. In detail, PES schemes that seeks to combine elements of PES tout court with product certification may allow one to take into account the widest range of ecosystem services attributable to a specific environmental or landscape context with special biodiversity features and its benefits (Ghazoul *et al.*, 2011; Huberman and Shepherd, 2010). The market driver in this case is certification and labelling, which may be both voluntary and compliant.

Further opportunities could also come from the combination of labels for local food and preservation of genetic diversity with PES. We refer to the case of certification referring only to those products that support high levels of biodiversity or that contribute to maintaining a specific landscape and local biodiversity (Arsil *et al.*, 2014; Troiano and Marangon, 2011) involving several stakeholders and actors along the value chain. The certification should allow considering the widest range of ecosystem services attributable to a landscape or local biodiversity that a product is helping to conserve, allowing one to capture the aspects that defy quantification (Huberman and Shepherd, 2010). In this context, the idea of expanding the scope of the PES by creating 'landscape labels' (Ghazoul *et al.*, 2009) designed to label all goods and services originating from a particular area should promote the inclusion also of those ecosystem services that, presenting difficulties for identification and quantification (e.g. cultural services), arouse less interest. This combination of PES schemes with certification seem to be useful as it should be able to share its benefits to all stakeholders contributing to a successful certification process and to contribute to biodiversity conservation. Moreover this combination should make available an easily accessible tool (a label) at local, national, and international level by which relevant information (e.g. including traceability of the products' environmental attributes) can be verified. In addition, this tool should also be inclusive in that the use of a label is not restricted to a particular product, but associated with the wider context, where the provision of a number of ecosystem services provision has been guaranteed along the value chain (e.g. reducing environmental impacts). For example, the evaluation of environmental issues relevant to the international marketplace has highlighted the importance of greenhouse gas (GHG) emissions, water, waste, non-renewable inputs and biodiversity.

However, it is essential to be aware of the possibility that the presence of a multitude of labels could cause user confusion, resulting in negative effects, both in economic terms and in landscape and environmental aspects. Nonetheless, a number of studies showed that communication could play a role in changing consumer attitude toward environmental friendly products (Lombardi *et al.*, 2016).

These particular PES schemes could provide both a basis for premium pricing and more stable agri-food supply chain relationships for certified producers (Lockie, 2013). Several markets may be created as a lot of agricultural and agro-industrial products could potentially be certified (organic, biodiversity-friendly, etc.). Moreover, it seems that the current market size can be increased according to the estimated 15% current rate of growth (Ecosystem Marketplace, 2013). The opportunities to develop these PES schemes are consistent both in developing countries and in developed states. In fact, while markets for certified and labelled agricultural products are still small in developing countries, in several developed countries local demand is rapidly increasing and it is not always possible to provide for local supply. Consequently, there could be opportunities to develop exports from both developing countries and local production.

Through premium pricing it is possible to add value to certified agricultural products. Moreover, production of these products may generate employment and income while improving food security and the ability to adapt to climate change.

While opportunities to develop these PES schemes really exist, often countries face many challenges in fully participating in certified markets from both a production perspective (i.e. high compliance costs) and a demand point of view (i.e. in certain developing contexts, building sufficient consumer interest and creating a consistent demand could be a problem for PES creation) (Page and Bellotti, 2015). A broad partnership among all or several players connected along the agri-food value chain is thus necessary in order to develop an effective PES scheme.

Several players could participate in a PES scheme for certified agricultural production. For example, acting as buyers for these schemes may be exporters, traders, processors and manufacturers, while small to large-scale producers are mainly sellers of services. Furthermore, regulators or standard setters, policy makers, advocates, associations and multi-bilateral development agencies may be market shapers, while acting as market service providers there may be traditional agricultural or philanthropic investors, major consultants or brokerages, land management service providers, technical service providers (legal, monitoring, verification), financial service providers, academics and information providers. Moreover, also processing, transport and other activities along the value chain could participate improving their environmental impacts on value chain. In fact, they utilize and interact with the local resources (e.g. biodiversity, soil and water resources) and affect the environment by the emissions of nutrients, toxic substances and GHG.

As stated above, a cooperative approach to coordinate the actions of involved economic agents seems to be relevant in the case of certification, in particular as it often allows a reduction in transaction costs.

It seems to be worthwhile pointing out that an increase of these PES schemes benefits at the same time both agricultural biodiversity and the agri-food value chain. A larger demand and production of biodiversity-friendly agricultural and agri-industrial products may benefit habitats, species, soils, rural landscapes and water protection with an increase in positive environmental externalities and support better performance of the agri-food value chain by integrating the principles of sustainable development and increasing incomes.

19.5 The role of public intervention

Although PES schemes could be a useful tool for integrating economic, social and environmental factors in decision making and securing financial resources for the protection of agricultural biodiversity, widening the earning opportunities for agri-food value chain players and incentivising environmental friendly behaviours, there are some reasons for hindering PES establishment.

The difficulties in obtaining the environmental valuation of ecosystem services have been considered as the main reasons for not establishing PES schemes (Troiano and Marangon, 2011). The valuation is a laborious and expensive part of PES creation that depends on monitoring or surveying data that in most cases is not readily available. Consequently, PES pricing is often based on the costs of the PES scheme creation and implementation instead of on the benefits deriving from ecosystem service provision. Moreover, sometimes the price of an ecosystem service is established on a political basis, hampering scheme efficiency.

Besides ecosystem services valuation, there could be other barriers to PES development such as, for example, the lack of an enabling policy and legal framework including property rights, or the lack of market information or high transaction costs.

Although PES was born as a market solution for the sustainable management of ecosystem services with the specific goal of creating an alternative to public management, the role of government in developing PES could be decisive (Pellegrino and Marino, 2014; Uthes and Matzdorf, 2016; Vatn, 2009), also regarding PES schemes created to support agricultural biodiversity and agri-food value chain sustainable management. In particular, its role in reducing transaction costs related to nature of traded goods could be relevant.

Public intervention in favour of PES diffusion can occur with several degrees of engagement. Firstly, we can have the traditional role of institutional decision maker, secondly the government can act as an intermediary or a promoter/financier of PES, but it could also be a seller of ecosystem services. The latter is the case in which the government has the ownership of the rural landscape and environmental resources. In this case, the creation of a PES scheme is necessary to fund the conservation activities carried out by the government (Pagiola *et al.*, 2002).

Moreover, government may be present in a PES scheme in order to remove barriers that could prevent or cause difficulties in creating a market between suppliers and users of ecosystem services. In fact, among these the presence of high transaction costs related to the implementation of a PES scheme and connected negotiating agreements could prevent its realisation. These high costs are often due to the presence of supply and demand composed by several individual economic agents. In this case the key role that government can play consists in bringing together buyers and sellers or stimulating the market mechanism by providing appropriate information, training and awareness in the community (Gutman, 2007). In particular, as regards agricultural biodiversity providing local, regional and global social benefits, the transaction costs associated with bringing together individual buyers and sellers can often be prohibitively high. In these circumstances the government could have an important role to play in facilitating PES schemes creation.

An additional institutional task is the growth of public awareness about the benefits received from the sustainable use of agricultural biodiversity resources, inviting citizens to ensure their protection, through the payment of a price for the benefit they receive. Moreover, as regards this task a fundamental role is that of environmental economic valuation, which is able to measure the benefits society derived from ecosystem services and to favour market construction, as stated above. Economic valuation supports the increase of the awareness that biodiversity conservation is essential for economic development: using measurable tangible monetary values it is possible to get the attention of citizens about benefits derived from biodiversity.

Furthermore, the government has to guarantee to the citizens the right to enjoy the essential ecosystem services, even if they have not enough financial resources to pay for their delivery. In fact, it is important they do not look at the ecosystem services as luxury goods. Taking into account ecosystems produced by biodiversity, the incentives to free-ride may preclude the establishment of direct user-financed PES programmes. In this case the government has to intervene directly by financing the creation of a PES.

19.6 Conclusions

In this paper we have illustrated PES as an innovative tool to support improved environmental management of agricultural biodiversity and agri-value chain development.

For creating more and better performance-based payments and PES schemes for conservation of ecosystem services from agricultural biodiversity, it is necessary to measure and verify the benefits and more in general the environmental impact of PES such as agricultural eco-certification schemes to improve credibility of eco-labelled agricultural products. In spite of financial crises, demand for certified agricultural products is growing, but the cost of such certification and of production of these products need to be reduced to increase the demand for such products, also in developing countries, to compete with conventional products.

Supporting farmers in overcoming these barriers is one of the main reasons why donors and governments are looking at PES schemes with high expectations.

The use of PES does not exclude the presence of other tools (Engel *et al.*, 2008). In particular, considering PES features, it is possible to state that it could not be considered the most suitable approach in every field and to achieve any goal. The choice of the best instrument depends on the characteristics of the ecosystem service, considering also the degree of linkage between the ecosystem service and other benefits provided by agricultural biodiversity. Nevertheless, the role of this innovative instrument seems to be of great importance for ecosystem services conservation and for agri-food value chain improvement. In particular, in this framework, it seems that PES developed according to Coasian approaches could play an important role in managing and conserving agricultural biodiversity and improving agri-food value chain management (Marangon and Troiano, 2013). Although these tools need to be supported by the work of specific organisations and government by implementing appropriate and flexible schemes, they seem to offer interesting opportunities for the agricultural sector.

PES schemes seem to provide excellent prospects for socio-economic as well as environmental resources, both in developing countries and in developed countries, where PES schemes have been implemented almost exclusively under public schemes. Further research in this area is therefore needed in order to enhance their implementation as these schemes require a clear definition of the rights and responsibilities of the buyers, sellers and intermediaries, the reduction of transactions costs of implementation and efficient monitoring systems to link payments to performance. Last but not least, the economic valuation of ecosystem services proves to be essential in order to point out the value of the benefits provided to society.

Joint learning and wide collaboration among all stakeholders and a multidisciplinary approach can help to overcome difficulties in creating PES schemes to favour agricultural biodiversity conservation, but also agri-food value chain enhancement, avoiding both abuse and a de facto privatisation of environmental resources.

References

Alix-Garcia, J., De Janvry, A. and Sadoulet, E., 2003. Targeting payments for environmental services: the role of risk. Agricultural and Resource Economics Update 7(4): 9-11.

Arsil, P., Li, E., Bruwer, J. and Lyons, G., 2014. Exploring consumer motivations towards buying local fresh food products: a means-end chain approach. British Food Journal 116(10): 1533-1549.

Convention on Biological Diversity (CBD), 2000. Conference of the parties. Fifth Ordinary Meeting of the Conference of the Parties to the Convention on Biological Diversity. May 15-26, 2000. Nairobi, Kenya. Available at: http://tinyurl.com/yckvkr4t.

Ecosystem Marketplace, 2013. Innovative markets and market-like instruments for ecosystem services. The matrix. Available at: http://tinyurl.com/y9yk5rgy.

Ecosystem Marketplace, 2016. Market watch. Available at: http://tinyurl.com/yb6k8wnb.

Engel, S., Pagiola, S. and Wunder, S., 2008. Designing payments for environmental services in theory and practice: an overview of the issues. Ecological Economics 65: 663-674.

European Commission, 2007. Green Paper on market-based instruments for environment and related policy purposes. COM(2007) 140. Brussels, Belgium.

Fisher, J.A. and Brown, K., 2014. Ecosystem services concepts and approaches in conservation: just a rhetorical tool? Ecological Economics 108: 257-265.

Food and Agriculture Organization (FAO), 2007. The state of food and agriculture. Paying farmers for environmental services. FAO Agriculture Series 38, Rome, Italy.

Food and Agriculture Organization FAO, 2009. Second report on the state of the world's plant genetic resources for food and agriculture. FAO, Rome, Italy.

Food and Agriculture Organization (FAO) 2013, Remuneration of Positive Externalities (RPE)/ Payments for Environmental Services (PES). FAO, Rome, Italy.

Garbach, K., Milder, J.C., Montenegro, M., Karp, D.S. and De Clerck, F.A.J., 2014. Biodiversity and ecosystem services in agroecosystems. Encyclopedia of Agriculture and Food Systems 21: 21-40.

Ghazoul, J., Garcia, C. and Kushalappa, C.G., 2009. Landscape labelling: a concept for next-generation payment for ecosystem service schemes. Forest Ecology and Management 258(9): 1889-1895.

Ghazoul, J., Garcia, C. and Kushalappa, C.G., 2011. Landscape labelling approaches to PES: bundling services, products and stewards. In: Ottaviani, D. and Hage Scialabba, N.E. (eds.) Payments for ecosystem services and food security. FAO, Rome, Italy, pp. 171-189.

Gomez-Baggethun, E., De Groot, R., Lomas, P.L. and Montes, C., 2010. The history of ecosystem services in economic theory and practice: from early notions to markets and payment schemes. Ecological Economics 69: 1209-1218.

Gutman, P., 2007. Ecosystem services: foundations for a new rural-urban compact. Ecological Economics 62: 383-387.

Gutman, P. and Davidson, S., 2007. A review of innovative international financial mechanisms for biodiversity conservation. World Wide Fund For Nature, Morges, Switzerland.

Huberman, D. and Shepherd, G., 2010. Delivering environmental services in landscapes: experiences with PES through IUNC's livelihoods and landscapes strategy. Mountain Forum Bulletin: 14-17.

Krishna, V.V., Drucker, A.G., Pascual, U., Raghu, P.T. and King, I.O.E.D., 2013. Estimating compensation payments for on-farm conservation of agricultural biodiversity in developing countries. Ecological Economics 87: 110-123.

Landell-Mills, N. and Porras, I.T., 2002. Silver bullet or fools' gold? IIED, London, UK.

Lockie, S., 2013. Market instruments, ecosystem services, and property rights: assumptions and conditions for sustained social and ecological benefits. Land Use Policy 31: 90-98.

Lombardi, G.V., Berni, R. and Rocchi, B., 2016. Environmental friendly food. Choice experiment to assess consumer's attitude toward 'climate neutral' milk: the role of communication. Journal of Cleaner Production 142(1): 257-262.

Marangon, F. and Troiano, S., 2013. New tools for the EU agricultural sector and rural areas. What role for public policy in promoting payments for ecosystem services? International Agricultural Policy 2: 7-19.

Millennium Ecosystem Assessment, 2005. Millennium Ecosystem Assessment, ecosystems and human wellbeing, synthesis. Island Press, Washington, DC, USA.

Muradian, R., Corbera, E., Pascual, U., Kosoy, N. and May, P., 2010. Reconciling theory and practice: an alternative conceptual framework for understanding payments for environmental services. Ecological Economics 69: 1202-1208.

Narloch, U., Drucker, A.G. and Pascual, U., 2011. Payments for agrobiodiversity conservation services for sustained on-farm utilization of plant and animal genetic resources. Ecological Economics 70: 1837-1845.

Organisation for Economic Co-operation and Development (OECD), 1972. Recommendation of the council on guiding principles concerning international economic aspects of environmental policies. Available at: http://tinyurl.com/ybo8xdcb.

Organisation for Economic Co-operation and Development (OECD), 2010. Paying for biodiversity: enhancing the cost-effectiveness of payments for ecosystem services. Available at: http://dx.doi.org/10.1787/9789264090279-en.

Ottaviani, D. and El-Hage Scialabba, N., 2011. Payments for ecosystem services and food security. FAO, Rome, Italy.

Page, G. and Bellotti, B., 2015. Farmers value on-farm ecosystem services as important, but what are the impediments to participation in PES schemes? Science of the Total Environment 515: 12-19.

Page, G., Ridoutt, B., Creeper, D. and Bellotti, B., 2015. A framework for assessing local PES proposals. Land Use Policy 43: 37-41.

Pagiola, S., Landell-Mills, N. and Bishop, J., 2002. Making market-based mechanisms work for forests and people. In: Pagiola, S., Bishop, J. and Landell-Mills, N. (eds.) Selling forest environmental services. Earthscan, London, UK, pp. 261-290.

Palm, C., Blanco-Canqui, H., DeClerck, F., Gatere, L. and Grace, P., 2014. Conservation agriculture and ecosystem services: an overview. Agriculture, Ecosystems and Environment 187: 87-105.

Pascual, U., Muradian, R., Rodríguez, L.C. and Duraiappah, A., 2010. Exploring the links between equity and efficiency in payments for environmental services: a conceptual approach. Ecological Economics 69: 1237-1244.

Pellegrino, D. and Marino, D., 2014. Quale governance per i pagamenti per i servizi ecosistemici nel settore agricolo? Agriregionieuropa 36 (in Italian).

Perrot-Maitre, D., 2006. The vittel payment for ecosystem services: a 'perfect' PES case? International Institute for Environment and Development, London, UK. Available at: http://tinyurl.com/ydb5qkgk.

Pettenella, D., Vidale, E., Gatto, P. and Secco, L., 2012. Paying for water-related forest services: a survey on Italian payment mechanism. iForest-Biogeosciences and Forestry 5(4): 210.

Robinson, R. and Keenan, L., 2010. The Rhoen Biosphere Reserve: developing new marketing strategies to conserve a traditional agricultural landscape. Mountain Forum Bulletin: 60-64.

Sattler, C. and Matzdorf, B., 2013. PES in a nutshell: from definitions and origins to PES in practice-approaches, design process and innovative aspects. Ecosystem Services 6: 2-11.

Scherr, S.J. and McNeely, J.A., 2008. Biodiversity conservation and agricultural sustainability: towards a new paradigm of 'ecoagriculture' landscapes. Philosophical Transactions of the Royal Society of London B: Biological Sciences 363(1491): 477-494.

Schomers, S. and Matzdorf, B., 2013. Payments for ecosystem services: a review and comparison of developing and industrialized countries. Ecosystem Services 6: 16-30.

Smith, S., Rowcroft, P., Everard, M., Couldrick, L., Reed, M., Rogers, H., Quick, T., Eves, C. and White, C., 2013. Payments for ecosystem services: a best practice guide. Defra, London, UK.

Stephenson, J., 2012. Business, biodiversity and ecosystem services. Policy priorities for engaging business to improve the health of ecosystems and conserve biodiversity. 28th round table on sustainable development. Available at: http://tinyurl.com/ydaxd6do.

The Economics of Ecosystems and Biodiversity (TEEB), 2010. Ecological and economic foundations. Earthscan, London, UK, 456 pp.

Troiano, S. and Marangon, F., 2011. Payments for ecosystem services: development opportunities from landscape and environmental resources management. Economics and Policy of Energy and the Environment LIII 3: 87-113 (in Italian).

United Nations (UN), 1992. Convention on biological diversity. Available at: https://www.cbd.int/doc/legal/cbd-en.pdf.

Uthes, S. and Matzdorf, B., 2016. Budgeting for government-financed PES: does ecosystem service demand equal ecosystem service supply? Ecosystem Services 17: 255-264.

Vatn, A., 2009. An institutional analysis of payments for environmental services. Ecological Economics 69: 1245-1252.

World Wide Fund For Nature (WWF), 2014. The Danube PES project. Available at: http://tinyurl.com/6acjtp5.

Wunder, S., 2005. Payments for environmental services: Some nuts and bolts. CIFOR Occasional Paper No. 42. Available at: http://tinyurl.com/ybed6whz.

Wunder, S., 2015. Revisiting the concept of payments for environmental services. Ecological Economics 117: 234-243.

Wunder, S., Engel, S. and Pagiola, S., 2008. Taking stock: a comparative analysis of payments for environmental services programs in developed and developing countries. Ecological Economics 65: 834-852.

20. Dissonance in the food traceability regulatory environment and food fraud

M. Bitzios[1,2], L. Jack[1], S.C. Krzyzaniak[1] and M. Xu[1]*

[1]Portsmouth Business School, Faculty of Business and Law, University of Portsmouth, Richmond Building, Portland Street, PO1 3DE, United Kingdom; [2]University of Aberdeen, Health Economics Research Unit, Foresterhill, Polwarth Building, Aberdeen, AB25 2ZD, United Kingdom; michail.bitzios@abdn.ac.uk

Abstract

Central to the inquiry is whether the issue of dissonance in traceability legislation leaves gaps that might be exploited by those engaging in food fraud for economic gain. We review the academic and professional literature to understand the implications for a hypothetical company manufacturing functional foods in the United Kingdom and trading globally, of the variances in traceability legislation across different domains in which they trade. We draw on concepts of creative dissonance from economic sociology to frame our discussion. The paper suggests changes needed to reduce the constraints under which such companies function. The paper will be of value to academics, food supply network organisations and industry regulators seeking to understand the additional risks and constraints placed on food businesses importing and exporting products where there is lack of harmony in international legislation.

Keywords: regulation, functional foods, food supply chain

20.1 Introduction

Food legislation is complex, particularly concerning high risk foods (such as foods of animal origin or foods aimed at vulnerable consumer groups like infants and young children) and 'functional foods'.[1] The global food supply chain is increasingly interconnected but the regulatory environment (e.g. good traceability regulations and standards for the food chain) remains fragmented. Isolated traceability systems (TS) and disharmonised traceability regulations and standards throughout global supply networks cause concern. Lu and Gilmore (2011), calling for harmonisation, suggest that the main threats to global food safety are pathogens, fraud and weak regulatory standards. Food fraud is a 'new' issue to be tackled by the industry and regulators – or rather a very old issue that has come back into prominence.

To address issues concerning dissonance in food regulatory frameworks we use a hypothetical company, ('Company A'), and outline the issues that exist in the current regulatory environment of such organisations. This is followed with an outline of the concepts informing our evaluation of the issues around traceability, including the concept of dissonance that we draw upon. We review the literature to look at two specific areas where the dissonance in regulations relating to traceability leads to problems for Company A, namely: (1) genetically modified organisms (GMO) regulations; and (2) trade with China; and conclude by considering the implications of food fraud driving future traceability policy for the industry and for companies such as Company A.

20.2 Hypothetical situation and challenges

Company A is a food manufacturing company based in the United Kingdom (UK), producing products fortified with vitamins and minerals and marketed with claims about nutrition and health, i.e. so called 'functional foods'. Many of the raw materials purchased by Company A are sourced from other countries, both in the European Union (EU) and from other regions such as China and the United States of America (USA). Company A now wishes to expand and sell products in other international markets. The common thread in food legislation across the world is the need to ensure that food is safe and of an acceptable quality for human consumption. However, despite these principles, Company A faces many challenges in understanding and meeting different regulatory requirements, both in terms of ensuring that imported raw materials meet UK and EU requirements, and in ensuring that finished products meet the legal requirements of export markets, both compositionally and taking

[1] There is no universally accepted definition for functional foods, with definitions varying, as stated by Arvanitoyannis and Van Houwelingen-Koukaliaroglou (2005), from country to country or even from company to company. According to Roberfroid (2000), functional foods are foods similar in appearance to conventional foods that are intended to be consumed as part of a normal diet, but have been modified to subserve physiological roles beyond the provision of simple nutrient requirements.

account of any requirements for notification or registration with the appropriate authorities before placing products on the market.

First, to successfully manufacture in the UK and sell products to the global market, Company A must take into account the requirements of key customers and consumers, and meet the responsibilities for food business operators (FBOs) set out under UK and EU law. Expansion to other markets requires understanding and meeting additional regulatory demands in those markets. Legislation pertaining to food is frequently based on scientific information (e.g. microbiological limits for foods) but there are often major differences in how legislation is framed and enforced. For example, in the UK, food supplements and foods fortified with vitamins and minerals are freely available for sale, with no requirement for registration or notification to the authorities of intention to sell products prior to placing them on the market (Department of Health, 2011). Rather the onus is on the FBOs to know of, understand and comply with the relevant legislation and for local authorities check the FBO's compliance to the legislation during enforcement inspections. However in Denmark, before any fortified product can be sold, the FBOs must submit a specific application (and pay a fee) to gain permission to sell the product (Danish Veterinary and Food Administration, 2014). This contrast between 'rules' based and 'procedures' based regulatory systems can be challenging to understand and pose a cultural barrier to trade, in this case even when these markets exist within the same overall regulatory framework.

Another challenge is the differences between enforcement regimes. For example, in the UK, enforcement of food law including on-pack claims comes under the purview of local authorities, and prosecutions for breaches of the law are brought under criminal law (Food Standards Agency, 2014b). In contrast, in the USA, many cases of alleged breaches of labelling law pertaining to product descriptions and claims are brought to court as civil actions under the Lanham Act by other actors in the food chain, for example, the case of POM Wonderful LLC v. Coca-Cola Co. (Supreme Court of the United States, 2014). In China, the enforcement of food safety law fall into the hands of multiple authorities and agencies at different levels and regions (central, provincial, county, etc.) resulting in disconnected, inconsistent or even contradictory legislations and standards that often cause problems and gaps for effective implementation and enforcement (Jia and Jukes, 2013). In all of these countries the aim of legislation is to ensure that consumers are not misled about the nature, substance or quality of a product, but differences in enforcement regimes adds complexity and uncertainty for FBOs.

Finally, a key challenge relating to food safety concerning Company A is due to different requirements for product traceability. Under EU law, product traceability along the food chain is required. Although most major food producing countries have some legal requirement for product traceability, few match the depth and breadth of the EU (Charlebois *et al.*, 2014). This can lead to challenges for Company A when importing key raw materials from outside of the EU.

20.3 Concepts of dissonance

The concept of dissonance is used in economic sociology to denote situations in which 'diverse, even antagonistic, performance principles overlap' (Stark, 2009). Dissonance can present opportunities for creativity that, exploited appropriately, can lead to innovative solutions to difficult problems (*ibid.*). Importantly though, dissonance is not a problem but part of an on-going process: the important issue is how people evaluate the worth of practices (Lamont, 2012) and understanding processes through which practices are valued (Helgesson and Muniesa, 2013; Stark, 2011). Performance principles are outcomes of evaluative principles. Working out what is valued and why is a necessary step to understanding why conflicts and confusion arise in food supply systems, and in finding pragmatic solutions that are perceived to be of greater worth.

In terms of this research, food supply chains and food safety regulations provide a site of dissonance (Stark, 2009, 2011). There appears to be a lack of harmony between the regulatory systems in place to achieve food safety, traceability and food integrity, and to counter fraud. We could explore performance principles that underpin food regulations for safety, traceability and integrity, the relative values placed on each by organisations in food supply chains and the principles of evaluation used by individuals or groups by organisations. We can reflect on the mechanisms in place to influence policy and regulation, and suggest where changes might to be made. We can also reflect on the extent, based on the literature, to which food regulations help or hinder the development of food supply systems providing safe and authentic foods for consumers.

20.4 Review of the regulatory environment and traceability requirements

20.4.1 Complex and disharmonious food safety law and traceability requirements

As outlined in Section 20.2, the drive for food safety is the common denominator in food law across the world. In the EU, to harmonise and strengthen food safety law across all member states, Regulation (EC) No 178/2002 was established (European Parliament and Council, 2002), setting out the general principles and requirements of food law, establishing the European Food Safety Authority and laying down food safety procedures. This legislation is considered as the backbone of food safety law in the EU, as it establishes that the primary responsibility for food safety rests with FBOs; all other food legislation provides the necessary detail to ensure that FBOs can meet this responsibility. The regulation also establishes the need for key tools used by FBOs and authorities, including traceability. Figure 20.1 shows these and demonstrates the inter-relationship between some of the EU laws which would impact Company A.

Even where harmonised legislation exists in the EU, gaps may still enable Member States to enforce specific national legislation, guidelines or procedures. The principle of mutual recognition should allow free sale of products, i.e. products sold in any

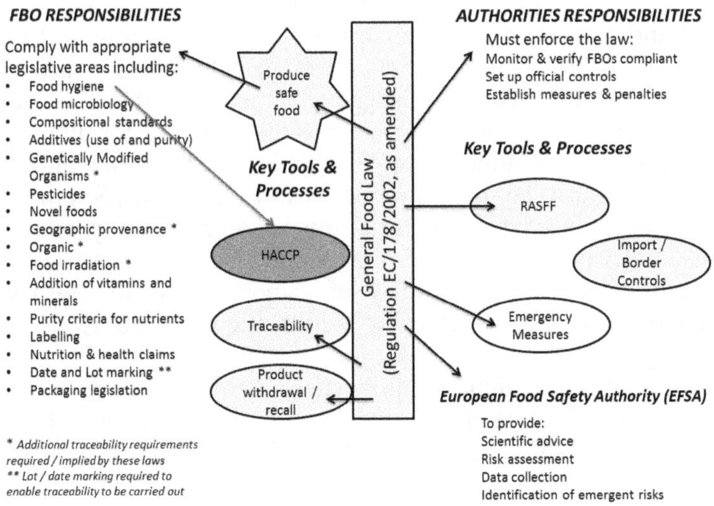

Figure 20.1. Regulation (EC) 178/2002 (as amended): roles, responsibilities, tools and relationship to other laws.

Member State should be legal for sale in any other, unless the competent authority of a new market can prove there is a risk to human health, safety or the environment (European Parliament and Council, 2008). However, practical application of the mutual recognition principle can be challenging and many FBOs choose to comply with the differing national laws as a more expedient course of practice.

For example, EU legislation on the addition of vitamins, minerals and other substances for foods (European Parliament and Council, 2006) sets out the requirements for the addition of these nutrients to foodstuffs, provides a list of specific sources of vitamins and minerals which may be added to foods and allows for the creation of a list of substances which cannot be legally added to foods in the EU. A significant gap in the legislation is the lack of maximum levels of vitamins and minerals which can be added to foods. Underlying this regulatory gap is a longstanding disagreement about the role of fortified foods versus food supplements (vitamin and mineral pills) and whether maximum levels should simply be stated as a fixed percentage of the recommended intake for each nutrient or whether levels should be based on the scientific assessment of the maximum safe upper level of intake for each nutrient. This situation remains unresolved some eight years after the publication of a Commission discussion document aimed at resolving this matter (European Commission, 2006) and it is this regulatory gap which allows for differing interpretations of requirements and the co-existence of the UK 'rule based' approach to fortified foods and the Danish 'procedures based' approach.

When considering non-EU markets, the regulatory environment becomes more complex still and to market products in the USA, Company A will need to meet stringent border control requirements and many differences in product labelling requirements. Lalor and Wall (2011) suggest that whilst both EU and USA regimes base the approval of health claims on scientific substantiation, the different regulatory regimes lead to different claims being permitted in each market. Should Company A wishes to make health claims on a product exported to the USA, they would need to significantly alter their labelling and claims language versus product labelled for UK sale. The 2011 working paper from the United Nations Industrial Development Organisation (UNIDO) reports that 58% of US border rejections for foods relate to contravention of labelling requirements (Henson and Olale, 2011).

For Company A, the traceability of imported ingredients from other countries is a significant concern. Traceability forms a cornerstone of general food law in the EU where it is defined in Regulation (EC) No 178/2002 (European Parliament and Council, 2002) as:

> the ability to trace and follow food, feed, food-producing animals or substances intended to be, or expected to be incorporated into food or feed through all stages of production, processing and distribution.

In other words, if a food safety or quality issue occurs, the traceability allows both for product recall and for identification of the source of the problem. The EU regulation requires that FBOs must be able to identify as a minimum their immediate suppliers and customers – the so called 'one step back, one step forwards' approach. In practice FBOs may need more comprehensive traceability information to ensure that they comply with other relevant EU legislation. Figure 20.2 illustrates traceability for one product made by Company A across the food supply chain.

Charlebois *et al.* (2014) review laws on product traceability in 21 OECD (Organization for Economic Co-Operation and Development) countries and conclude that only EU countries have 'superior' traceability requirements. Recent legislative changes in large markets such as the USA and China require additional TS for a range of foods that should improve the overall picture but the current mismatch of standards and expectations challenges Company A when importing key raw materials from outside of the EU. Using China as an example, which has been developing traceability regulations (Charlebois *et al.*, 2014), Food Safety Law requires food producers, processors, packers, and retailers to implement testing and record-keeping systems for all inputs and outputs and to archive the records for at least 2 years (Zhang *et al.*, 2012). Food operators are also required to establish records of the source of the products and sales (Liu, 2010). The status of traceability regulations in three major food trading areas (EU, USA, and China) is outlined in Table 20.1.

Another challenge relates to the enforcement of traceability regulations. Within the EU, enforcement of food safety legislation differs significantly between Member States,

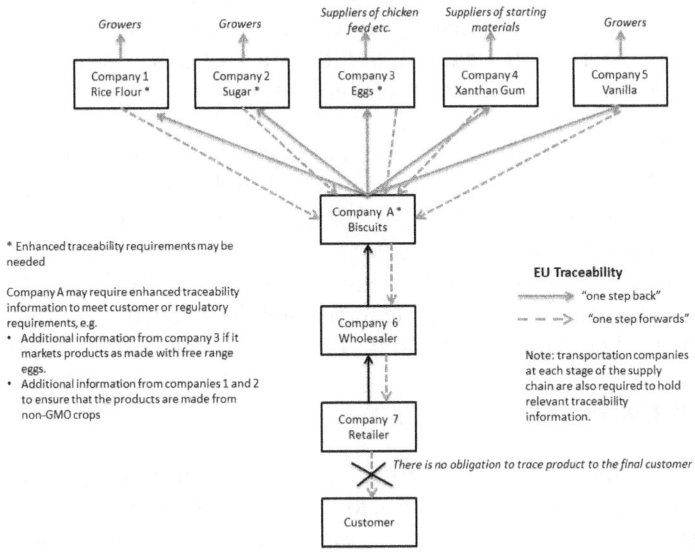

Figure 20.2. Product traceability of Company A.

Table 20.1. Evaluation of the status of traceability regulations in three major food trading zones based on several criteria (adapted from Charlebois et al., *2014).*

	European Union	USA	China
Traceability regulations for domestic products	progressive	moderate	regressive
Traceability regulations for imported products	progressive	moderate	regressive
Clarity of the system of authority responsible for traceability regulations	progressive	moderate	moderate
Voluntary traceability practices	progressive	progressive	moderate
Breadth of products regulated for traceability	progressive	regressive	regressive
Use of identifiers for tracking/registering of imports	progressive	moderate	progressive
Availability of GFSI auditing bodies	progressive	progressive	progressive
Extent of GS1 services used	progressive	progressive	moderate
Electronic livestock tracking systems	progressive	regressive	regressive
Comprehensiveness of labelling	progressive	progressive	moderate

and can be fragmented and difficult for FBOs to understand even within a single country. The UK, for example, splits responsibility for food legislation between the Department of Health, the Department of the Environment, Food and Rural Affairs, and the Food Standards Agency. Enforcement of food law is managed at a local level by trading standards officers (TSOs) and environmental health officers (EHOs), each of these reporting to different levels of local government. During any inspection, a TSO or EHO needs to assess many points of compliance with a wide range of laws and guidelines, so a full traceability audit is only one of a number of inspection points. Similar complexity exists elsewhere, including the USA where the United States Department of Agriculture, and the Food and Drug Administration regulate different foodstuffs, and where differing enforcement responsibilities can occur for products which are traded locally (within state) and those in interstate commerce. Company A, even when trading with companies in markets with comprehensive legal requirements, may lack confidence that there is adequate enforcement to assure confidence in legal compliance. For this reason, many FBOs choose to become certified with organisations such as the British Retail Consortium or the International Organization for Standardization and to trade with other certified companies.

20.4.2 Different drivers for traceability

Germain (2003) finds that traceability is used as a tool to manage food-related risks associated with animal and plant health, to improve the quality of products and ensure good practices in the production stage, as well as identifying issues of non-compliance. Canavari *et al.* (2010), found that traceability can be used as a tool for competitive strategy whilst Moe (1998) and Wang and Li (2006) argue it is used operationally to optimise production planning. TS can generate information that could, according to Resende-Filho and Hurley (2012), facilitate the synapse of contractual agreements between firms in order to endorse food safety.

Figure 20.3 expands a model by Olsen (2009) and Karlsen *et al.* (2013) to provide a picture of what drives traceability in the food supply chains. Several studies identify food safety as an important driver for the implementation of TS by companies. Additionally, the introduction of EC 178/2002 requires food companies to put into effect traceability procedures. Bio-terrorism legislation in the USA also requires documentation of food distribution. We add food fraud as a supplemental driver for traceability. Traceability will not eliminate food fraud cases but may reduce such events. It should be easier to identify the origin of fraud with the assistance of one-step-backward – one-step-forward traceability. Another traceability driver is corporate social responsibility. There is increasing demand by retailers and consumers for more information about animal products origin, including their feed, and also about the health and welfare of the animals. Certification is another driver. Consumers demand proof that the products they buy are of excellent quality and have been produced responsibly and sustainably.

There is considerable variation in the driving forces for the implementation of traceability. This relates to the different requirements of FBOs trading different products and to the position of each business in the supply chain. End users show significant dissimilarity in stating the driving forces for traceability and the information that should be made available both within and across countries (Van Rijswijk and Frewer, 2012). This asymmetry is significant because those FBOs that

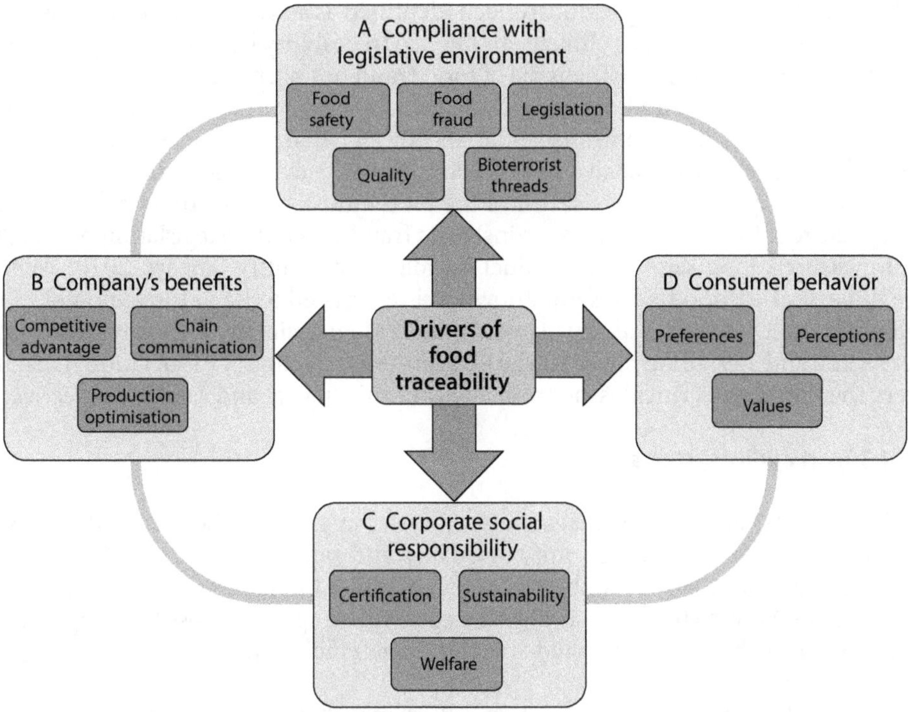

Figure 20.3. Drivers of food traceability.

adopt traceability schemes from other jurisdictions may gain a competitive advantage in export markets. The same is found with animal health documentation and animal welfare certification (Manos and Manikas, 2010; Schulz and Tonsor, 2010).

Whilst food safety, food defence (bioterrorist threats) and food quality are all identified as drivers of traceability, there appears to be a gap in identifying traceability as a tool to tackle economically motivated food fraud, which has been identified by Spink and Moyer (2011) as the fourth category to be considered when assessing food safety risk.

Food fraud is defined by the UK Food Standards Agency (2014a) as 'deliberately placing food on the market, for financial gain, with the intention of deceiving the consumer…' and includes activities such as selling food which is unfit for consumption (such as the sale of meat with an unknown origin) and the deliberate misdescription of food (such as selling farmed salmon as wild). Spink (2012) offers a broader definition, such that food fraud encompasses deliberate substitution, addition, tampering or misrepresentation of food, food ingredients, or food packaging, or false or misleading statements made about a product for economic gain.

Whilst food fraud has not previously been identified as a driver for food TS, it is likely to be seen as such in today's climate, and hence traceability legislation may not be 'fit for purpose' in this respect. The issue of food fraud has been brought to the attention of FBOs, enforcement authorities and the public over the last 18 months, primarily due to the so called 'horsemeat scandal' in Europe, arising from the substitution of cheap horsemeat for expensive beef (Elliott, 2014). Nevertheless food fraud is not new, nor is it restricted to one sector of the food industry. Many of today's food laws have their root in historical fights against food frauds, including legislation on weights and measures, food safety and product standards of identity (Shears, 2010). Whilst the full extent of food fraud is not known, it is estimated to be in the region of £11.2 billion in the UK (Gee *et al.*, 2014) whilst individual incidents are estimated to cost between 2 and 15% of turnover by the Grocery Manufacturers Association. It can be seen, therefore, as a critical issue from an economic, moral and safety perspective.

20.4.3 Traceability systems

McKean (2001) argues that traceability can assist in growing the competitiveness of FBOs as it can expedite conformity of quality and safety standards both nationally and internationally. However, the cost for traceability implementation is high and an issue that remains to be solved is how the cost will be spread across the supply chain actors for traceability to be applied effectively and efficiently.

To that end we should consider that FBOs may have different motivations for traceability information. A study by Karlsen *et al.* (2011) demonstrates that motivation (for traceability) varied between the different links of the chain firms, e.g. the sale company is motivated by new legislation which demands better documentation of the fish they offer for sale; the supermarket is motivated by the desire to be able to trace fresh fish for customers and also to help identify responsibility for bad quality; the distribution company needs efficient information for managing storage and satisfying its customer, whereas the fish landing company has less motivation to implement TS. Companies downstream perceive higher value in traceability than companies upstream, which is in line with the findings from Mai *et al.* (2010), who suggest that the costs of increased traceability tends to be comparably higher upstream than downstream.

Differences in motivation to implement TS exist between firms sizes. Large firms see less financial burdens than small ones. For small and medium enterprises (SMEs), technological investment is the major burden for implementing TS (Szewczyk *et al.*, 2008). Most food enterprises in China are small or medium, with limited financial resources for TS, although firms may be entitled to state sponsorship. Implementing TS in China is largely driven by government incentives, rather than consumer push (Chen, 2008; Li, 2006; Xu and Deng, 2008; Zhu, 2008).

Overall, therefore, it is likely that larger companies and those downstream closer to the consumer, may be more highly motivated to assume some responsibility for TS implementation costs.

The complexity of food supply chains, the participation of many and different shareholders and food safety crisis incidents during the last twenty years have resulted in the loss of consumers trust in food safety. Implicitly, food and food ingredient TS constitute a valuable means for the improvement of food safety and consequently the recovery of public confidence (Kher *et al.*, 2010). The motivation of FBOs to implement TS is also influenced by the way the general public accepts and reacts to their application.

Equally important is the public's understanding of food TS. When shopping, consumers seek information relating to food safety and security of food products on labels. The Food Standards Agency (2010) found that food labels in the UK are read by approximately half the population. However, consumers in Scotland and England may have less interest in labels when compared to people in other European countries. Given the limited conception people have about traceability, educating the public and communicating the message about what traceability can offer with respect to assuring food safety and preventing food fraud could potentially boost confidence in food supply chains. The studies on consumer perceptions of traceability conducted thus far have focused on issues related to food risk and food safety (Giraud and Amblard, 2003). They have centred around meat, a potentially risky food product (Dickinson and Bailey, 2002; Gellynck and Verbeke, 2001; Hobbs *et al.*, 2005). Only a few studies have examined the issue of the public's perceptions, attitudes and preferences for traceability. Kher *et al.* (2010) discovered that effective communication of food recalls that would occur for optimising consumer protection rather than for economic reasons would improve consumer trust about food safety. Chinese consumers seem to prefer traceable food products, while the certificate issuer has a significant effect on people's choices, with government being the most trusted authority (Bai *et al.*, 2013; Ortega *et al.*, 2011). Regarding fraud, the public's reactions are in accordance with the fact that food chain participants are perceived to be held responsible for ensuring and implementing traceability. Consumers tend to complain to the authorities and the retailer, whom they hold responsible for providing guarantees of safety and quality. However, while they also held producers responsible, they were less inclined to complain to them (Van Rijswijk and Frewer, 2012). Given the absence of quality verification, Hobbs *et al.* (2005) concluded that a simple traceability assurance is not highly valued by consumers and has little impact on their decision making as their interest in and understanding of such cues is rather debatable. The consensus, according to Verbeke and Ward (2006), seems to be that food labels have some potential to impact quality expectations, at least if well understood by consumers. This will be even more the case if they are accompanied by traceability assurance. This is in line with other studies which suggest that provision of information regarding the technical aspects of traceability does not interest consumers (Gellynck and Verbeke, 2001).

20.4.4 Lack of uniformed traceability systems

An effective TS is an enabler for compliance with food safety regulations. A TS is characterized by identification of units/batches of all ingredients and products and information on when and where the units/batches are moved or transformed, in a system that links these data (Food Standards Agency, 2002). Various data identification and tracking technologies are used for developing TS including bar code technology, 2D QR (quick response) code, Radio Frequency Identification, and DNA. For data transmission, technologies include radio frequency, GSM/GPRS, internet, satellite, and Wireless Sensors Networks, etc. yet, fully operational TS have not been implemented equally in all chain companies. It is suggested that the establishment of TS for an entire food supply chain can identify product attributes, process attributes, environmental conditions and participant node attributes effectively, and preserve the information record of food logistics through each node. This will achieve top-down tracking of raw materials, processing, packaging, storage, transportation, and marketing in the food supply chain, as well as bottom up tracing when food safety problem occurs (Lin and Zhou, 2005).

Barriers in implementing TS often refer to data issues, system performance and resource constraints. Karlsen *et al.* (2011) report that manual recording of information is time-consuming, and the chances of errors were relatively high, as the same information may need to be recorded several times causing data redundancy; as a result, system reliability and the level of accuracy are in question. In developing countries, it remains a great challenge to collect large amounts of quality data accurately because of the generally low educational level of employees working in farms and food processing centres, compounded by the lack of the sense of the responsibility of some owner managers (Chen, 2008; Li, 2006; Xu and Deng, 2008).

The mismatch in TS function and quality hampers effective traceable information flow between chain firms and thus, undermines compliance requirements set by food traceability regulations. The Food Standards Agency (2002) suggests that a better integration of electronic data exchange is needed for improving information tracking. However, not all firms in a supply chain are using compatible software for internal traceability, and the diversity of the systems used makes the integration difficult (Bechini *et al.*, 2005; Zen, 2005). Karlsen *et al.* (2011) suggests that owning to the lack of uniformity in the TS used, coordination and allocation of costs and benefits among various actors across the chain are affected.

The use of traceability technologies may not help address all food fraud concerns. Spink (2013) argues that traceability is not a single magic-bullet to stop fraud. Authentication – proving the product genuine or proving it fake – is also increasing in importance, especially when combined with traceability within the supply chain. Traceability and authentication together provide transparency, and help reduce the fraud opportunity.

20.5 Discussion

To illustrate the site of dissonance in which our hypothetical food manufacturer, Company A, operates, we consider two specific case studies of challenges that arise from traceability legislation: the use of GMO and trade with China.

20.5.1 Genetically modified organisms

Under EU legislation a GMO is defined as:

> an organism, with the exception of human beings, in which the genetic material has been altered in a way that does not occur naturally by mating and/or natural recombination; …. (Directive 2001/18/EC article 2).

The first GMO crop (tomatoes) was planted in 1994. Crops produced using GM technology are now widely available, with over 160 million hectares of land planted with GM crops in 2012. GM soya, maize, cotton and canola being the most common crops, grown predominately in North and South America and increasingly across Asia (Brookes and Barfoot, 2014). Although some GM crops have been approved for cultivation in Europe, there remain wide differences in opinion over the cultivation of GM crops, both within Europe and between Europe and other markets. For example, procedures for the approval of GM crop cultivation in the EU and USA differ substantially, with the EU requiring a detailed risk assessment for each new strain, whilst many GM crops have now been 'deregulated' in the USA on the basis of their being equivalent to non-GM variants (Dunwell, 2014). This leads to wide differences in traceability requirements, with EU legislation demanding full traceability from 'farm to fork', with records kept for a minimum of 5 years for all transactions involving GMO products, whilst no specific requirements are laid down for GM crops in the USA (Library of Congress, 2014). In the EU positive labelling of all GMO products containing or produced from GMOs is required, with the exception of an allowance up to 0.9% for adventitious (technically unavoidable) cross-contamination by approved GMO strains (European Parliament and Council, 2003a).

As a purchaser of raw materials, and producer of finished products, Company A is required to comply with the EU legislation concerning the traceability and labelling of GMO products and any products produced from these, as set out in EU Regulations (EC) 1829/2003 and 1830/2003 (European Parliament and Council, 2003a,b). It is still highly unusual to find products positively labelled as 'containing GMO' on sale in the EU, due to a perceived lack of acceptance by consumers and retailers (Davison and Bertheau, 2007). Therefore, Company A, in common with the vast majority of EU FBOs, is concerned with ensuring that the products they produce are not sourced from GMO and do not require positive labelling. At first glance, it appears that they can operate safely outside of the GM legislation. However, purchasing any plant based raw materials where there are known commercialised GMO strains available and/or from markets where GMO crops are routinely grown should trigger additional investigation to understand the materials and ensure adequate traceability is provided. In either of these situations, Company A needs to understand if a non-GMO source material has been selected for the production of the ingredient in question, and if sufficient cross-contamination controls are in place to avoid mixing GMO and non-GMO materials (so that any presence of GMO is likely to be at levels less than 0.9% and can be defended as adventitious). Of particular importance to note is that (1) any use of a GMO strain unapproved in the EU would make the ingredient illegal for use and (2) the presence or absence of GMO protein or DNA is not the distinguishing test of whether a product must be labelled; even highly refined ingredients such as natural vitamin E produced from soya oil or glucose syrup produced from maize starch should be labelled unless the starting crop can be verified as non-GMO. These nuances of EU legislation are frequently misunderstood by suppliers in markets with different legislation on GMO, hence Company A is at risk not only of deliberate fraud resulting from the sale of GMO ingredients as non-GMO, but also of fraud through ignorance of the rules on crop selection and cross-contamination.

20.5.2 Trade with China

For several years China has been a key supplier of raw materials for global food businesses and is increasingly being viewed as an export market for many agri-food businesses. Trade with China brings many challenges, not least due to the nature of the national food industry, featuring long industrial chains with numerous links and components, decentralised regulation, differing requirements and multiple enforcement agencies (Sang, 2011). Additionally, the integrity of the food industry in China has been damaged by an increasing number of food safety incidents, involving both local and international supply chain firms. As a result of this the Chinese Government promulgated a series of food safety and quality laws and regulations requiring firms to implement TS to assure food safety and public health. Nevertheless, the regulatory infrastructure, industry technologies, and industry and consumer awareness of food safety concerns are at a relatively early stage of development in China (Lu and Gilmore, 2011). According to the USTR, 'China remains among the least transparent and predictable of the world's major markets for agricultural products, largely because of selective intervention in the market by China's regulatory authorities' (Cited in Lu and Gilmore, 2011). Sang (2011) suggests that relevant supporting laws and regulations, as well as agency regulations, are not yet completed. The most notable gaps in comparison to international food safety practice are in: (1) the science of food safety; (2) food hygiene operating systems in food manufacturing; (3) traceability and verification equipment systems; (4) overlapping agency authority; and (5) weak food safety practitioner qualifications and training.

These gaps and inconsistencies in national legislation, together with the complex nature of the food industry in China mean that the opportunities for food fraud in products supplied from China are high. For example, maximum residue levels of pesticides often differ between the EU and China. An importer to the EU may declare that the product meets all EU requirements, but for crops such as table grapes, which are frequently grown by small producers and sold via cooperatives, not only is full traceability back to the farm very difficult, but controlling the exact nature and level of pesticides used is difficult. The declaration of compliance may be fraudulent, either deliberately, inadvertently because of a lack of knowledge of the EU import requirements or due to a lack of technology and resources to manage the crop and/or check pesticide residues. One of the possible risk scenarios for Company A is that when the company chooses to purchase raw materials from China, it may need to accept an increased risk, given the differences in regulatory requirements between the exporting and importing markets. For functional food and ingredients exporting to EU market, there are no specific legislations to regulate Chinese manufacturers/suppliers, as long as the importer satisfied that it meets UK and EU requirements. This puts Company A at risk if the materials are not entirely traceable and cannot be authenticated. In the scenario of Company A exporting finished functional food to China, its products must fit into one of the 22 categories of functional food as approved by the Ministry of Health which are different from what is required within EU. In addition, the product must satisfy national legislation on food safety and labelling, including, standard GB16740-1997 – the General Standard for Health (functional) Food. Any specification, which is inconsistent with EU standards and regulations, may require Company A to adjust its production and operation in order to achieve compliance. Furthermore, multiple authorities have the power to approve and certify functional food in China, which can cause inconsistency or even conflicting results that can have an adverse impact on Company A. For example, export of functional foods to China is subject to certification and approval by Ministry of Health, whereas domestically manufactured functional food is subject to certification and approval of The Administration of Quality Supervision, Inspection and Quarantine Department.

20.6 Conclusions

The dissonant environment in traceability regulations creates complexity in requirements, motivation, systems and enforcement practice. It also increases extra time and cost to market for FBO products. Dissonance also increases risks for FBOs. Lack of comparable data on compliance by companies and across jurisdictions means that they risk losing trading opportunities; incurring additional cost due to over-compliance or inadvertently breaching regulatory requirements that can bring penalties. Furthermore, they are at risk of procuring fraudulent products which in turn means that they themselves are at risk of being involved in fraud on consumers and supply chain partners.

Greater harmonisation of traceability would not eliminate fraud but would to some extent mitigate the risks faced by FBOs. Improvements in traceability for raw materials and products being traded globally could also result in other benefits beyond fraud reduction, such as reducing barriers to trade. The dissonance that we have identified in this paper does create opportunities for international dialogue about TS that work across countries. The organisation that builds relationships and sets up systems with suppliers and customers to give a functional supply network, may be able to create confidence with local enforcement authorities, reduce enforcement costs, build consumer confidence in their brands and secure a long term future.

We have examined the wider implications of dissonance in law and regulation surrounding traceability, and its effects on organisations engaged in international trade in functional foods. Our study is limited in scope and represents a general review of the literature and issues already identified. We link the issue of traceability to growing concerns about food fraud, and suggest that food fraud might be a driver that encourages harmonisation between different requirements for traceability imposed on such organisations. To inform policy in this area, further work is required to understand the extent to which a more harmonised approach to traceability can be achieved.

References

Arvanitoyannis, I. and Van Houwelingen-Koukaliaroglou, M., 2005. Functional foods: a survey of health claims, pros and cons, and current legislation. Critical Reviews in Food Science and Nutrition 45: 385-404.

Bai, J., Zhang, C. and Jiang, J., 2013. The role of certificate issuer on consumers' willingness to pay for milk traceability in China. Agricultural Economics 44(4-5): 537-544.

Bechini, A., Cimino, M.G.C.A., Lazzerini, B., Marcelloni, F. and Tomasi, A., 2005. A general framework for food traceability, applications and the internet workshops, 2005. Saint Workshops 2005 Symposium. January 31-February 4, 2005. Trento, Italy, pp. 366-369.

Brookes, G. and Barfoot, P., 2014. Economic impact of GM crops. GM Crops and Food 5: 65-75.

Canavari, M., Centonze, R., Hingley, M. and Spadoni, R., 2010. Traceability as part of competitive strategy in the fruit supply chain. British Food Journal 112: 171-186.

Charlebois, S., Sterling, B., Haratifar, S. and Naing, S.K., 2014. Comparison of global food traceability regulations and requirements. Comprehensive Reviews in Food Science and Food Safety 13: 1104-1123.

Chen, H.H., 2008. GS1 system applies in Hainan aquatic product quality and safety traceability system. World Standardization and Quality Management 8: 56-58.

Danish Veterinary and Food Administration, 2014. How to market fortified food in Denmark. Available at: http://tinyurl.com/ycw8uuvw.

Davison, J. and Bertheau, Y., 2007. EU regulations on the traceability and detection of GMOs: difficulties in interpretation, implementation and compliance. CAB Reviews: Perspectives in Agriculture, Veterinary Science, Nutrition and Natural Resources 2: 14.

Department of Health, 2011. Food supplements: summary information on legislation relating to the sale of food supplements. Available at: http://tinyurl.com/c6x5ryc.

Dickinson, D.L. and Bailey, D., 2002. Meat traceability: are US consumers willing to pay for it? Journal of Agricultural and Resource Economics 27(2): 348-364.

Dunwell, J.M., 2014. Genetically modified (GM) crops: European and transatlantic divisions. Molecular Plant Pathology 15: 119-121.

Elliott, C., 2014. Elliott review into the integrity and assurance of food supply networks – final report: a national food crime prevention framework. HM Government, London, UK.

European Commission, 2006. Discussion paper on the setting of maximum and minimum amounts for vitamins and minerals in foodstuffs, June 2006. Available at: http://tinyurl.com/y9mc58da.

European Parliament and Council, 2002. Regulation (EC) No 178/2002 of the European Parliament and of the Council of 28 January 2002 laying down the general principles and requirements of food law, establishing the European Food Safety Authority and laying down procedures in matters of food safety. Official Journal of the European Union L31: 1.

European Parliament and Council, 2003a. Regulation (EC) No 1829/2003 of the European Parliament and of the Council of 22 September 2003 on genetically modified food and feed. Official Journal of the European Union 46 L268: 1.

European Parliament and Council, 2003b. Regulation (EC) No 1830/2003 of the European Parliament and of the Council of 22 September 2003 concerning the traceability and labelling of genetically modified organisms and the traceability of food and feed products produced from genetically modified organisms and amending Directive 2001/18/EC. Official Journal of the European Union L268: 24.

European Parliament and Council, 2006. Regulation (EC) No 1925/2006 of the European Parliament and of the Council of 20 December 2006 on the addition of vitamins and minerals and certain other substances to foods. Official Journal of the European Union L404: 26.

European Parliament and Council, 2008. Regulation (EC) No 764/2008 of the European Parliament and of the Council of 9 July 2008 laying down procedures relating to the application of certain national technical rules to products lawfully marketed in another Member State and repealing Decision No 3052/95/EC. Official Journal of the European Union L218: 21.

Food Standards Agency, 2002. Traceability in the food chain – a preliminary study. Available at: http://tinyurl.com/ydglwskd.

Food Standards Agency, 2010. Public attitudes towards, and use of, general food labelling. Available at: http://tinyurl.com/ya5jabbd.

Food Standards Agency, 2014a. Food fraud. Available at: http://tinyurl.com/yd92as7a.

Food Standards Agency, 2014b. Food law code of practice. Available at: http://tinyurl.com/yd4y2qop.

Gee, J., Jack, L. and Button, M., 2014. Minimising fraud and maximising value in the UK food and drink sector. PFK, London, UK.

Gellynck, X. and Verbeke, W., 2001. Consumer perception of traceability in the meat chain. Agrarwirtschaft 50: 368-373.

Germain, C., 2003. Traceability implementation in developing countries, its possibilities and its constraints: a few case studies. Available at: http://tinyurl.com/yc5nzqo3.

Giraud, G. and Amblard, C., 2003. What does traceability mean for beef meat consumer? Food Science 23: 40-46.

Helgesson, C.-F. and Muniesa, F., 2013. For what it's worth: an introduction to valuation studies 1: 1-10.

Henson, S. and Olale, E., 2011. What do border rejections tell us about trade standards compliance of developing countries? Analysis of EU and US data 2002-2008. United Nations Industrial Development Organisation (UNIDO), Vienna, Austria.

Hobbs, J.E., Bailey, D., Dickinson, D.L. and Haghiri, M., 2005. Traceability in the Canadian red meat sector: do consumers care? Canadian Journal of Agricultural Economics/Revue Canadienne d'Agroeconomie 53: 47-65.

Jia, C. and Jukes, D., 2013. The national food safety control system of China – a systematic review. Food Control 32: 236-245.

Karlsen, K.M., Dreyer, B., Olsen, P. and Elvevoll, E.O., 2013. Literature review: does a common theoretical framework to implement food traceability exist? Food Control 32: 409-417.

Karlsen, K.M., Sørensen, C.F., Forås, F. and Olsen, P., 2011. Critical criteria when implementing electronic chain traceability in a fish supply chain. Food Control 22: 1339-1347.

Kher, S.V., Frewer, L.J., De Jonge, J., Wentholt, M., Howell Davies, O., Lucas Luijckx, N.B. and Cnossen, H.J., 2010. Experts' perspectives on the implementation of traceability in Europe. British Food Journal 112: 261-274.

Lalor, F. and Wall, P.G., 2011. Health claims regulations. British Food Journal 113: 298-313.

Lamont, M., 2012. Toward a Comparative sociology of valuation and evaluation. Annual Review of Sociology 38: 201-221.

Li, X., 2006. The urgency and feasibility for establish China aquatic products traceability system. China Fishery 9: 21-26.

Library of Congress, 2014. Restrictions on genetically modificed organisms: United States. Available at: http://tinyurl.com/ydcxxa39.

Lin, L. and Zhou, D.-Y., 2005. On the construction of food quality and safety traceability system. Commercial Research 21: 41-44.

Liu, P., 2010. Tracing and periodizing China's food safety regulation: a study on China's food safety regime change. Regulation and Governance 4: 244-260.

Lu, J. and Gilmore, R., 2011. Regimes: a patchwork of surveillance and enforcement mechanisms. In: Global Food Safety Forum (ed.) The China path to global food safety. Available at: http://tinyurl.com/y73yeetb.

Mai, N., Bogason, S.G., Arason, S., Árnason, S.V. and Matthíasson, T.G., 2010. Benefits of traceability in fish supply chains – case studies. British Food Journal 112: 976-1002.

Manos, B. and Manikas, I., 2010. Traceability in the Greek fresh produce sector: drivers and constraints. British Food Journal 112: 640-652.

McKean, J., 2001. The importance of traceability for public health and consumer protection. Revue Scientifique et Technique 20: 363-371.

Moe, T., 1998. Perspectives on traceability in food manufacture. Trends in Food Science and Technology 9: 211-214.

Olsen, P., 2009. Food traceability process mapping. Standard method for analyzing material flow, information flow and information loss in food supply chains., Workshop hosted by Nofima in association with the TRACE project. February 25-26, 2009. Tromsø, Norway.

Ortega, D.L., Wang, H.H., Wu, L. and Olynk, N.J., 2011. Modeling heterogeneity in consumer preferences for select food safety attributes in China. Food Policy 36: 318-324.

Resende-Filho, M.A. and Hurley, T.M., 2012. Information asymmetry and traceability incentives for food safety. International Journal of Production Economics 139: 596-603.

Roberfroid, M., 2000. Defining functional foods. In: Gibson, G. and Williams, C.M. (eds.) Functional Foods – concept to product. Woodhead Publishing, Cambridge, UK, pp. 9-29.

Sang, L., 2011. Food safety legislation, regulation, and practices in China. In: Global Food Safety Forum (ed.) The China path to global food safety. Available at: http://tinyurl.com/y73yeetb.

Schulz, L.L. and Tonsor, G.T., 2010. Cow-calf producer preferences for voluntary traceability systems. Journal of Agricultural Economics 61: 138-162.

Shears, P., 2010. Food fraud – a current issue but an old problem. British Food Journal 112: 198-213.

Spink, J., 2012. Defining food fraud and the chemistry of the crime. In: Ellefson, W., Zach, L. and Sullivan, D. (eds.) Improving import food safety. Institute of Food Technologists, Washington, DC, USA. pp. 195-216.

Spink, J., 2013. Does traceability really help fight fraudsters? Available at: http://tinyurl.com/y8nptzsc.

Spink, J. and Moyer, D.C., 2011. Defining the public health threat of food fraud. Journal of Food Science 76: R157-R163.

Stark, D., 2009. The sense of dissonance: accounts of worth in economic life. Princeton University Press, Princeton, NJ, USA.

Stark, D., 2011. What's valuable? In: Aspers, P. and Beckert, J. (eds.) The worth of goods: valuation and pricing in the economy. Oxford University Press, New York, NY, USA.

Supreme Court of the United States, 2014. Pom wonderful LLC v. Coca-cola Co. No. 12-761. Available at: http://www.supremecourt.gov/opinions/13pdf/12-761_6k47.pdf.

Szewczyk, R., Rzeplińska, K. and Winiarski, W., 2008. Application of the industrial automation standards and methodologies for reliable and continuous European food traceability system. Journal of Automation Mobile Robotics and Intelligent Systems 2: 55-57.

Van Rijswijk, W. and Frewer, L.J., 2012. Consumer needs and requirements for food and ingredient traceability information. International Journal of Consumer Studies 36: 282-290.

Verbeke, W. and Ward, R.W., 2006. Consumer interest in information cues denoting quality, traceability and origin: an application of ordered probit models to beef labels. Food Quality and Preference 17: 453-467.

Wang, X. and Li, D., 2006. Value added on food traceability: a supply chain management approach. IEEE International Conference on Service Operations and Logistics, and Informatics. June 21-23, 2006. SOLI '06, pp. 493-498.

Xu, J. and Deng, H., 2008. Analyze institution building of agricultural products quality and safety traceability. Journal of Hunan Agricultural University 9: 24-28.

Zen, Y., 2005. Barcode vegetables explore the road of quality and safety traceability in China. Vegetables 10: 1-4.

Zhang, C., Bai, J. and Wahl, T.I., 2012. Consumers' willingness to pay for traceable pork, milk, and cooking oil in Nanjing, China. Food Control 27: 21-28.

Zhu, X., 2008. Promote quality and safety traceability system to ensure vegetables consumption safety. Vegetables 7: 34-35.

www.ingramcontent.com/pod-product-compliance
Lightning Source LLC
LaVergne TN
LVHW080631011026
849160LV00014B/1185